PERSPECTIVES IN RUNNING WATER ECOLOGY

Headwater region of the Afon Hirnant, North Wales, site of one of the earlier comprehensive studies on running water (see Hynes, 1961).

Photograph by D.D. Williams

PERSPECTIVES IN RUNNING WATER ECOLOGY

EDITED BY

MAURICE A. LOCK
University College of North Wales
Bangor, Gwynedd, U.K.

AND

D. DUDLEY WILLIAMS
Scarborough College, University of Toronto
West Hill, Ontario, Canada

PLENUM PRESS • NEW YORK AND LONDON

Library of Congress Cataloging in Publication Data

Main entry under title:

Perspectives in running water ecology.

   Bibliography: p.
   Includes index.
   1. Stream ecology. I. Lock, M. A. II. Williams, D. D. (D. Dudley).
QH541.5.S7P47   574.5'26323   81-17838
ISBN 0-306-40898-8   AACR2

© 1981 Plenum Press, New York
A Division of Plenum Publishing Corporation
233 Spring Street, New York, N.Y. 10013

Printed in the United States of America

On behalf of all his students,
this book is warmly dedicated to

PROFESSOR H.B. NOEL HYNES,

our teacher and friend.

PREFACE

The discipline of stream ecology has grown exponentially along with most other areas of science in the last three decades. The field has changed from a fish management-dominated poor-sister of limnology to a discipline with theoretical constructs and ecological insights as rich as those in any area of ecology. A focus on energy transformations, nutrient turnover and the storage and processing of organic substrates has greatly enhanced the development of current paradigms. For example, the data base on microbial-biochemical-invertebrate interactions in streams is now very extensive.

A survey of the stream ecology literature reveals the central role played by H.B. Noel Hynes, whom I believe to be the world's premier lotic ecologist. Professor Hynes produced the major text in the field ("The Ecology of Running Waters") which has served both as an integrative review and as a bridge between the fish-water quality orientation of the fifties and sixties to the process oriented, nutritional resource-watershed perspectives of the seventies. It should also be noted that Professor Hynes' book, "The Biology of Polluted Waters", served as a basic reference for the earlier period and he has published over 150 research papers, the great majority in stream ecology, spanning four decades.

Mix portions of brilliant, intuitive scientist, seasoned natural historian, masterful linguist and writer and sparkling wit and you have H.B. Noel Hynes -- to many of us his name is synonymous with stream ecology. Of course, as this book attests, he has produced not only a significant share of the important literature of running water ecology but also he has trained a group of outstanding scientists.

> Kenneth W. Cummins
> Oregon State University

CONTENTS

SECTION III

ASPECTS OF AQUATIC ENVIRONMENT PERTURBATIONS:
MANAGEMENT AND APPLICATION

SECTION IV

REGIONAL RUNNING WATER ECOLOGY

GENERAL INTRODUCTION

The primary purpose of this book is to honour Professor
Noel Hynes on the occasion of his official retirement from
university teaching. A "festschrift", primarily on running water
ecology, and produced by some of his past graduate students and
postdoctoral fellows was felt to be a fitting tribute. All of
us have benefited greatly from his enthusiasm, knowledge and
council and we hope that he finds the contents of this volume an
acceptable way of saying thank you not only from the contributors
but from all his students.

The contents, then, reflect the areas of expertise of the
contributors and are not an attempt to cover the entire field
of running water ecology - indeed, a few reviews deal with
aspects of lentic freshwaters, especially where principles are
equally applicable to both. In the sixteen chapters presented
here, we have a very broad spectrum of prospectives in running
water ecology, ranging from groundwater studies to palaeoecology
and this in itself reflects the wide research interests of
Noel Hynes.

The review chapters are divided into four sections:

I.   Energy inputs and transformations covers the routes of
      entry and subsequent fate of light and organic energy in
      streams. One chapter deals with the structure and
      function of river epilithon as an energy trap. Two
      chapters deal with the seasonality and processing of
      coarse particulate organic matter, an important energy
      source. A fourth considers the uptake of dissolved
      organic matter in groundwater by stream beds, while
      the nature and role of certain types of organic matter
      as energy controllers rather than energy sources are
      discussed in a fifth. In the last chapter, aspects of
      nitrogen transport and transformation in headwater
      streams are considered.
II.   Benthos in time and space includes reviews of our
      rapidly growing knowledge of the movements and distribution

1

of benthic invertebrates, and the ecology of an important
organic matter processor and fish-food organism, *Gammarus*.
One chapter deals with the shallow-water benthos of large
lakes, a community very similar to that found in running
water, while another deals with a little-known, but
potentially very important area, that of lotic protozoan
communities.

III. Aspects of aquatic environment perturbations examines three
very different aspects of changes in running water habitats.
One chapter examines the changes that took place after the
damming of four large African rivers, while another presents
an account of the present status of aquatic invertebrates
in palaeoecology, one of the newest areas of river ecology.
In a quite different vein, the third chapter discusses the
problems associated with government-administered
environmental management and research.

IV.   Regional running water ecology contains reviews of three
large geographical areas. The chapters on Australia and
Africa represent the first attempts to specifically
characterize running water habitats on these continents,
while the coldwater streams of the vast, high latitude,
circumpolar areas are characterized in a third chapter.

We hope that this coverage will be of use to the research
specialist seeking a state-of-the-art update and also to the
general reader.

Production of this book has been greatly aided by the
facilities and funding provided by the Division of Life Sciences,
Scarborough College and the Department of Zoology, U.C.N.W.
Valuable technical assistance has been rendered by the editorial
staff of Plenum Publishers, Mrs. Anne Lock and Mrs. Kathy Moore.
Mrs. Betty Boonstra painstakingly typed the entire manuscript.
Professor C.J. Duncan of the University of Liverpool and Professors
J.K. Morton and J.E. Thompson of the University of Waterloo kindly
provided bibliographical information. We would like to thank also
the various chapter contributors for making the editors' job easier
and our respective families for their indulgence.

                                    Maurice A. Lock
                                    Bangor

                                    D. Dudley Williams
June, 1981   Toronto

SECTION I

ENERGY INPUTS AND TRANSFORMATIONS

RIVER EPILITHON – A LIGHT AND ORGANIC ENERGY TRANSDUCER

M.A. Lock

School of Animal Biology
University College of North Wales
Bangor, Gwynedd, U.K.

The energy pathways within running water systems have been the
focus of many river ecologists since Noel Hynes' paper on "Imported
organic matter and secondary productivity in streams" (Hynes, 1963).
This paper emphasized the role played by allochthonous organic
energy sources to the overall energy economy of a river reach and
spawned a massive research effort on this topic (see Anderson and
Sedell, 1979; Cummins and Klug, 1979; Dance, Chapter 3; Bird and
Kaushik, Chapter 2). However, lately a view of running waters has
been put forward by Vannote and his colleagues (Vannote et al.,
1980) where the relative importance of light energy over organic
energy to rivers and streams is considered to be variable in time
and space both within and between rivers. Such a perspective is an
appropriate starting point for this chapter, since the epilithon,
the heterogeneous slime community coating all submerged rock surfaces
in rivers, has the capacity (in many instances) to process both light
and organic energy simultaneously. It is important to remember from
the outset that the epilithon is a mixed community of bacteria,
fungi, algae, protozoa and micrometazoa, and although algae may at
times apparently dominate the epilithon, at such times it will still
contain a well developed heterotrophic community.

This chapter will begin with an outline of the biotic components
of the epilithon and a description of the polymer matrix (slime) in
which the biotic components are supported. This will be followed
by a consideration of the two major metabolic processes occurring
within the epilithon, carbon fixation and the heterotrophic uptake
of organic energy. The chapter will conclude with a discussion of
the implications of the polymer matrix to energy transduction and
of the anticipated biological interactions occurring within the
epilithon.

3

For ease of discussion, it will be assumed that autotrophic
organisms are always present within the epilithon although it is
acknowledged that the epilithon at the bottom of deep rivers will
be primarily heterotrophic. However, since most of our information
is based upon shallow water epilithon communities this viewpoint is
reasonable for the present.

BIOTIC COMPONENTS OF THE EPILITHON

Bacteria: it is only within the last few years that any attempt
has been made to determine the densities of bacteria in the natural
epilithon of rivers and streams. In the main this was probably due
to the absence of a reliable method, however, with the advent of
fluorescent staining techniques for epilithic bacteria (Geesey et al.,
1978), direct bacterial enumeration is now feasible. The advantage
of this technique is that it enables the investigator to easily
distinguish bacteria from the extremely heterogeneous background
material of the epilithon (detritus, slime, mineral particles etc.,)
by staining them with a U.V. fluorescent dye and viewing with an
epifluorescent microscope. The first major study using this
technique was that of Geesey et al. (1978) working on three small
mountain streams in the Canadian Rockies where they sampled the
epilithon over a twelve month period. Bacterial densities (Table 1)
ranged from 7×10^5 cells cm^{-2} to 1×10^8 cells cm^{-2}. Lowest
densities were found in all three streams just after ice break-up

Table 1. Density of bacteria in the epilithon of several rivers.

| River | Order[1] | Bacteria $cm^{-2} \times 10^7$ | | Author |
		Mean	Range	
Twin fork	3	2	0.3-6.0	Geesey et al. (1978)
Middle fork	3	4	0.07-10.0	Geesey et al. (1978)
Cabin	3	2	0.1-6.0	Geesey et al. (1978)
Muskeg	4	9	0.9-20.0	Lock & Wallace (1979)
Hartley	3	6	0.7-10.0	Lock & Wallace (1979)
Steepbank	5	3	0.5-10.0	Lock & Wallace (1979)
Ogilvie	-	2	- -	Albright et al. (1980)
Swift	-	5	2.0-8.0	Albright et al. (1980)
Canmore	2	3	- -	Ladd et al. (in prep.)

[1] Strahler (1957)

in the spring, while the highest densities were found in early
winter (October-December) in two of the streams and throughout
the summer in the third stream. The average bacterial density for
the three streams was around 10^7 cells cm^{-2}.

Lock and Wallace (1979) working on three sub-arctic brown water
rivers (Clifford, 1968) over a 12 month period found a range of
bacterial densities between 5×10^6 to 2×10^8 cells cm^{-2} (Table 1).
In the largest river, the Steepbank (order 5), bacterial densities
were generally around 10^7 cells cm^{-2} with a single peak of 10^8 cm^{-2}
occurring in early summer (July).   In contrast the other two rivers
were much more variable with peaks occurring in early summer
($1 - 2 \times 10^8$ cells cm^{-2}) and early winter (10^8 cells cm^{-2}) and the
lowest densities occurring in mid-summer (8×10^6 cells cm^{-2}).
Summer densities (Table 1) for two sub-arctic rivers were reported
by Albright et al. (1980) with values ranging from 2×10^7 to 8×10^7
cells cm^{-2} while Ladd et al. (in prep.), working on a sub-alpine
stream in Alberta, Canada, found densities of around 3×10^7 cells
cm^{-2} during a summer/autumn study.

Although the data as yet are few and any generalization needs
to be viewed with caution, it would appear that the densities of
epilithic bacteria are frequently between 10^7 and 10^8 cells cm^{-2}.
However, not all of these bacteria are likely to be metabolically
active, as several studies of suspended bacteria in freshwater and
seawater indicate that the proportion of active bacteria is
variable (Ramsay, 1974; Peroni and Lavarello, 1977; Faust and Correl,
1977; Meyer-Reil, 1978). In the study by Ladd et al. (in prep.) a
determination was made of the proportion of the bacteria which were
metabolically active using the radiorespirometric most-probable
number (MPN) technique of Lehmicke et al. (1979). They found, in
August, that 34% of the cells were active and in October, 29%.
Clearly more work needs to be done in determining both bacterial
densities and proportions of bacteria which are actively metabolizing.
It seems reasonable to anticipate that the dynamics of bacterial
populations of the epilithon will be complex as a consequence of
variable pulses of internally produced organic energy from photosynthesis, and from variable external pulses of labile dissolved
and colloidal organic energy.

Algae: in contrast with bacteria there is a considerable body of
information upon the algae of the epilithon, and this has already
been reviewed by Hynes (1970), Whitton (1975) and more recently an
annotated reference list has been produced by Johansson et al.
(1977). For the purpose of this chapter a compilation of selected
recent quantitative studies on epilithic chlorophyll a concentration
and algal cell densities will be made (Table 2) which will indicate
the size of algal populations of river epilithon. Although there
are problems in both the significance of chlorophyll a as an "algal
biomass indicator" (e.g. variable carbon:chlorophyll a ratios) and

Table 2.   Chlorophyll a concentrations and algal cell densities
           in the epilithon of several rivers.

River	Order[2]	Shade	Chlorophyll $\underline{a}$ µg cm^{-2} Mean	Range		
WS 10[1]	-	HS	0.3	0.0	-	0.5
Upper Mack[1]	3	HS	2.5	0.9	-	3.6
Lower Mack[1]	3	LS	2.9	0.8	-	4.5
Lookout[1]	5	LS	3.0	0.8	-	4.6
Bere I[1]	-	US	10.3	2.0	-	30.0
Ober water[1]	-	US	2.2	0.5	-	4.3
Avon	high	-	-	-		
Highland water	low	-	-	-		
Rybi Potok[1]	-	US	3.4	2.9	-	3.8
Middle fork	3	HS	1.2	0.3	-	3.3
Twin fork	3	HS	2.9	0.3	-	11.7
Cabin	3	HS	0.01	0.002	-	0.01
Muskeg	4	US	5.4	0.2	-	22.6
Steepbank	5	HS	1.3	0.1	-	3.0
Hartley	3	LS	1.5	0.1	-	9.6
Athabasca	8-9	US	0.3	0.25m depth, single		
			0.2	1.50m depth, single		
Ogilvie	-	-	-	-		
Swift	-	-	-	-		
Devils Club[1]	1	HS	2.0	1.2	-	2.6
Mack[1]	3	HS	5.8	3.6	-	8.4
Lookout[1]	5	LS	5.3	1.8	-	9.6
McKenzie[1]	7	US	5.8	3.1	-	8.9
Muskeg[1]	4	US	3.1	0.1	-	20.3
Steepbank[1]	5	HS	2.3	0.0	-	23.0
MacKay[1]	6	-	0.8	0.01	-	3.1
Hangingstone[1]	5	-	2.2	0.1	-	20.6
Ells[1]	4	-	4.3	0.8	-	11.2

1 Calculated directly from original value expressed as mg m^{-2}
   - new value probably an overestimate

2 Strahler (1957)

3 US - unshaded   LS - low shade   HS - high shade

Table 2. Chlorophyll a concentrations and algal cell densities in the epilithon of several rivers (continued).

Total x 10^6		Algal cell density cm^{-2} Blue-green x 10^6		Diatoms x 10^5	
Mean	Range	Mean	Range	Mean	Range
–	–	–	–	–	–
–	–	–	–	–	–
–	–	–	–	–	–
–	–	–	–	–	–
–	–	–	–	–	–
–	–	–	–	–	–
0.5	0.0 – 2.0	–	–	–	–
0.6	0.1 – 1.3	–	–	–	–
–	–	–	–	–	–
1.5	0.8 – 2.5	0.5	0.4 – 0.7	2.3	0.3 – 6.1
1.9	0.4 – 3.1	1.2	0.4 – 3.1	3.8	0.0 – 1.1
0.8	0.1 – 2.3	0.4	0.04 – 1.0	0.3	0.04 – 0.7
14.9	0.1 – 116.4	13.8	0.02 – 112.6	14.8	0.03 – 112.7
0.5	0.02 – 3.2	0.4	0.008 – 2.8	0.2	0.004 – 0.8
0.3	0.06 – 0.6	0.2	0.03 – 0.4	0.5	0.08 – 1.7
sample in autumn		–	–	–	–
sample in autumn		–	–	–	–
0.6	0.02 – 1.0	–	–	–	–
0.2	0.002 – 0.6	–	–	–	–
–	–	–	–	–	–
–	–	–	–	–	–
–	–	–	–	–	–
–	–	–	–	–	–
2.9	1.3 – 5.6	1.9	1.0 – 2.9	8.0	0.1 – 23.0
2.4	1.0 – 3.9	1.5	0.9 – 2.0	7.0	0.3 – 16.0
20.3	0.1 – 137.5	15.4	0.0 – 136.5	34.0	0.0 – 355.0
75.8	29.0 – 653.1	46.7	0.0 – 613.5	56.0	3.0 – 315.0
63.8	3.7 – 197.7	50.8	0.4 – 165.1	85.0	5.0 – 239.0

Table 2. Chlorophyll a concentrations and algal cell densities
          in the epilithon of several rivers (continued).

Green x 10^5			Author
Mean	Range		
-	-		Lyford and Gregory (1975)
-	-		Lyford and Gregory (1975)
-	-		Lyford and Gregory (1975)
-	-		Lyford and Gregory (1975)
-	-		Marker (1976a)
-	-		Marker (1976a)
-	-		Moore (1976)
-	-		Moore (1977)
-	-		Bombowna (1977)
0.6	-		Geesey et al. (1978)
2.8	-		Geesey et al. (1978)
<1%	-		Geesey et al. (1978)
1.6	0.2	- 0.5	Lock & Wallace (1978)
0.5	0.02	- 2.9	Lock & Wallace (1978)
0.6	0.2	- 1.0	Lock & Wallace (1978)
-	-		Barton & Lock (1979)
-	-		Barton & Lock (1979)
-	-		Albright et al. (1980)
-	-		Albright et al. (1980)
-	-		Naiman & Sedell (1980)
-	-		Naiman & Sedell (1980)
-	-		Naiman & Sedell (1980)
-	-		Naiman & Sedell (1980)
1.0	0.0	- 6.0	Hickman et al. (1980)
2.0	0.0	- 15.0	Hickman et al. (1980)
14.0	0.0	- 71.0	Hickman et al. (1980)
238.0	0.0	- 2997.0	Hickman et al. (1980)
46.0	0.0	- 155.0	Hickman et al. (1980)

in the varying methods for its extraction and measurement, it is
still the algal biomass indicator used in most routine surveys,
and as such, the comparisons between rivers (Table 2) must be made
with caution. Although determinations of algal numbers are perhaps
less prone to methodological problems they cannot be directly
equated to algal biomass because of variable cell volumes and again
the between river comparisons must be made with caution. Lastly,
all studies in Table 2 were of algae removed from rock surfaces and
do not include studies which have used unnatural substrata such as
glass slides, polystyrene etc. However, the algae have been removed
from the rocks in two distinct ways. One of these involves scraping
a few square centimetres of rock surface where the algal density is
usually expressed on a cm^{-2} basis and the other involves scrubbing
rocks from a unit area of stream bed and expressing the algae on a
m^{-2} of stream bed basis. In the absence of any study for arriving
at a conversion factor between these two measurements I have adopted
the simplest approach of expressing the chlorophyll a concentrations
or algal numbers m^{-2} on a cm^{-2} basis for the sake of comparability,
remembering that this will probably be an over-estimation.

In approximately half of the rivers investigated (Table 2),
mean values for chlorophyll a were around the $2.0 - 3.0 \mu g\,cm^{-2}$
level. Although the studies were confined to latitudes 44-55 N,
such a finding is somewhat surprising given the variety of river
orders, basin types, water chemistries and local climates. Some of
the more extreme mean values may be a consequence of unusual water
chemistry (Bere Stream 1), extreme shading (WS-10 and Cabin Creek)
or as a consequence of high river order (Mackay). The maxima and
minima of chlorophyll a values are much more variable but these are
probably a consequence of site specific phenomena such as the local
river bed, local light regime and local flow patterns or characteristic basin phenomena such as discharge and temperature patterns or
temporal nutrient fluctuations.

The data on the number of algal cells per unit area (Table 2)
are much more variable and once again algal numbers m^{-2} have been
expressed on a cm^{-2} basis. Average densities of total algal cells
for the rivers ranged from $0.2 - 75.8 \times 10^{6}$ algal cells cm^{-2}, a
difference of over 2 orders of magnitude. The greater numerical
proportion of these algae are blue-green algae and population
density means ranged from $0.2 - 50.8 \times 10^{6}$ cells cm^{-2}. However,
because of their small size (~ 1-10 μm in diameter) their contribution to total algal biomass is low (but this does not necessarily
imply that their contribution to total primary production would
also be low). Mean densities for diatoms were much less variable,
ranging from 0.2 to 85×10^{5} cells cm^{-2} for the rivers studies,
while green algae ranged from almost zero to 238×10^{-2}.
Examination of the ranges at each river for algal cells cm^{-2}
reveals differences of many orders of magnitude, a probable
consequence of local biotic and abiotic conditions.

Fungi, protozoa and micrometazoa: these last three groups have been placed together because they have one thing in common, a dearth of quantitative information. In the case of the fungi, there is a growing body of qualitative information (Jones, 1976; Cooke, 1979) but studies which quantified fungal mycelium or yeast cells within the epilithon apparently have not been done. Although fungal quantification will undoubtedly be difficult this information is badly needed to expand the structural model of the epilithon. The same also applies to the protozoa; while Cairns and his co-workers (Cairns and Yongue, 1977) have provided much valuable qualitative information on river and stream protozoa, the quantitative studies are seemingly lacking. Once again these will pose methodology problems, but perhaps a modification of the method of Finlay et al. (1979) for determining the ciliates in sediments could prove useful. And finally, we consider the micrometazoa (< 2 mm in length). Although these metazoans from the epilithon have been collected in the many thousands of benthic invertebrate samples taken every year, it is not possible to know which metazoans are living within the epilithon and which were just casually associated with it. Karlstrom (1978) reported that an "abundant fauna of protozoans and micro-metazoans" could be observed in the epilithon of a 5th order Swedish river and that "small chironomids may sometimes be very abundant and in such cases the organic layer must be dramatically molded". Lock and Wallace (unpublished data) followed the population dynamics of the micrometazoans of 3 rivers in northern Alberta, Canada, and found that the dominant organisms were chironomid larvae and they showed well defined seasonal changes in density. Maxima were reached in early winter (October-December) where densities of ~ 10 chironomid larvae cm^{-2} were reached. Large numbers of sessile protozoa were also noted in this study from the undersurfaces of rocks. These were *Vorticella* sp., *Vaginicola* sp. and *Carchasium* sp. Curiously though, sessile protozoa were never observed on upper rock surfaces.

Epilithon structure - the polysaccharide matrix: it is now apparent that the bacteria, fungi and algae of river epilithon are embedded within a polymeric matrix considered to be composed of polysaccharide fibrils (Geesey et al., 1977; Geesey et al., 1978; Costerton et al., 1978; Lock et al., in prep.). However, recent work by Fletcher (1980) indicated a high protein content for the extracellular polymer from a marine pseudomonad. Its general appearance when viewed with a transmission electron microscope is that of a fine fibrilla net-work (Fig. 1) in which are dispersed the microorganisms. In contrast, Allanson (1973) working on epiphyton, found the matrix to be composed of coarse strands when viewed with a scanning electron microscope. However, such differences may be due to variations in fixation and dehydration of the material where fine polysaccharide fibrils can become condensed under certain conditions to form thick strands (Roth,

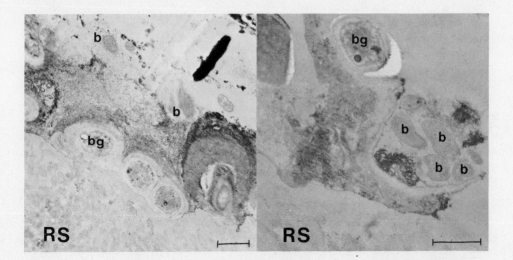

Figure 1.   Transmission electron micrograph of epilithon growing
            on epoxy resin substrata (RS) from the Muskeg River,
            Alberta, Canada, showing the extensive polysaccharide
            matrix (PSM) in which are embedded bacteria and blue-
            green algae (bar = 1 μm).

1977). Mackie et al. (1979) were able to demonstrate that the
capsular polysaccharides of group B streptococci when stabilized
with type specific antibodies were more extensive and of a finer
fibrillar nature than unstabilized capsules suggesting that fine
fibrils are closer to the natural state. Also, Roth (1977),
using freeze-etching (a technique which involves the minimum of
processing) demonstrated the presence of a fine fibrillar net-work
in a flock of *Enterobacter aerogenes*.

     Further evidence for the generality of polysaccharide
matrices binding together epilithic communities comes from work
carried out on trickling filters (Jones et al., 1969; Mack et al.,
1975). In particular, the findings of Jones et al. (1969) are
very close to those for river epilithon. For a general discussion
of the structure of prokaryote capsules (polysaccharides adherent
to the cell) and to slime (the polysaccharide matrix holding the
community together) the reader is referred to Roth (1977).
Information on the primary, secondary and tertiary chemical
structures of these extracellular polysaccharides can be found in
Powell (1979). Lastly, the presence of a polysaccharide matrix
binding together the various biotic components of river epilithon
has a number of potentially significant ecological implications but
these will be dealt with later.

SOURCES OF ENERGY TO THE EPILITHON

Energy available to the epilithon is in the form of allochthonous organic energy and light energy of which the former may be
dissolved, colloidal or as fine particles. For reasons of convenience, the dissolved and colloidal organic components are usually
measured together as "dissolved" organic matter ("DOM") with this
being operationally defined as material passing a ~ 0.5 µm filter.
However, it is apparent even on a $priori$ grounds, that colloidal
organic matter warrants distinction from dissolved organic matter
in terms of availability to the epilithon heterotrophs (Allan, 1976).
Opinion does vary as to the size limits for colloidal organic
matter but 0.4 µm - 0.01 µm seems to be a convenient and acceptable
range, with dissolved organic matter being defined as material
< 0.01 µm (Lock et al., 1977).

The amount of work on organic energy trapping by epilithon is
small and this section will use information from a number of very
different studies. These will include re-circulating laboratory
streams, intact stream cores, in $situ$ experiments and most recently
in $vitro$ studies using detached and undisturbed epilithon.

$Uptake$ of $high$ $concentration$ $undefined$ $organic$ $matter$: possibly
the first studies on the uptake of high concentration undefined
(unknown chemical composition) organic matter by a stream community
were those of Cummins et al. (1972) and Wetzel and Manny (1972)
using a 12m recirculating artificial stream designed to simulate a
natural hard-water woodland stream. In these studies they followed
the disappearance of water soluble extracts of autumn shed leaves
from the circulating water. The leaf extracts were monitored as
"dissolved" organic carbon and nitrogen of which the organic
nitrogen was further fractionated into a labile and refractory form
on the basis of ultraviolet combustion. They observed a rapid
disappearance of "DOC" over the first few days of the experiment
(Table 3) and then a much slower rate from approximately day 10 to
the termination at day 69. Half-lives were computed for these two
decay rates (Wetzel and Manny, 1972) with the initial rapid rate
having a $t\frac{1}{2}$ of 2.0 days and the second slower rate having a $t\frac{1}{2}$ of
80 days. The "dissolved" organic nitrogen (DON) analyses revealed
a rapid increase in refractory organic nitrogen to 3.2 mg 1^{-1} over
the first 24 hours and then a gradual rise to 3.8 mg 1^{-1} by day 24.
However, the labile "dissolved" organic nitrogen (LDON) increased
rapidly to a peak of 0.9 mg 1^{-1} at 30 hours and reached a minimum
(0.2 mg 1^{-1}) at day 10. Since the control stream ranged from
0.05 - 0.1 mg 1^{-1} of "DON", this indicated that a significant
amount of "LDON" had been removed from the water. Because of the
design of the experiment, an instantaneous addition of leaf extract
could not be made and the water soluble material leached into the
stream over a period of 30 hours, therefore the computed rate of
removal (Table 3) of 150 mg C m^{-2} h^{-1} over the first 48 hours is

Table 3. Uptake of high concentration defined and undefined dissolved organic matter by rivers.

River	Substrate	Starting concentration (mg l^{-1})	Uptake rate (mg m^{-2} h^{-1})	Author
High concentration undefined DOM [1] (Water soluble leaf leachates)				
Lab stream	*Carya glabra, Acer saccharum*	>61	150	Wetzel & Manny (1972)
Speed	*Populus tremuloides*	72.6	390(max)	Lock & Hynes (1975)
	Tsuga occidentalis	22.8	74(min)	Lock & Hynes (1975)
S. Kashe	*Populus tremuloides*	78.0	503(max)	Lock & Hynes (1975)
	Pinus restinosa	17.8	39(min)	Lock & Hynes (1975)
Roaring	*Fagus grandifolia, Betula lenta, Acer saccharum, Tsuga canadensis*	?	9(max)	McDowell & Fisher (1976)
			3(min)	McDowell & Fisher (1976)
	Thuja occidentalis	<20.7	1100(max)	Lush & Hynes (1978)
	Acer saccharum	<24.7	~500(min)	Lush & Hynes (1978)
Oak	*Alnus rubra*	19.6	28	Dahm (1980)
Middle trib	Groundwater DOM	4.9	207	Wallis et al. (1979)
Laurel	Groundwater DOM	45.1	16	Wallis et al. (1979)
High concentration defined DOM				
Outdoor channel	Glucose	0.2	2.8	Wuhrmann et al. (1975)
	Sucrose	1.8	21.4	Wuhrmann et al. (1975)
	Fructose	0.2	1.8	Wuhrmann et al. (1975)
	Glutamic acid	1.4	10.6	Wuhrmann et al. (1975)
Middle trib.	Glucose	7.5	363	Wallis et al. (1979)
	Glutamic acid	1.5	540	Wallis et al. (1979)

[1] May also contain colloidal organic matter

almost certainly an under-estimation. Additionally, this disappearance was attributed to a large increase in suspended bacteria.
However, subsequent work by Lock and Hynes (1976) and Ladd et al.
(1979) would suggest that a considerable proportion of the removal
would probably have been due to the microbial benthic community and
microbial growth on the channel walls.

Lock and Hynes (1976) examined the capacity of natural stream
bed cores to remove maple leaf leachates from the water. They over-
came the problem of determining the amount of leachate added by
making the additions in a freeze dried form. Using this approach
they demonstrated that over 24 hours, the "DOC" levels of the leachate in stream water alone decreased by only 20% while over the same
period with the addition of an intact stream bed core the leachate
was reduced to 80-90% of the initial level. Thus in that stream,
the dominant removal mechanism of leaf leachates resided with the
stream bed. When these leachate carbon concentrations were converted
to organic matter ("DOC" concentration x 2) and expressed on a m^{-2}
basis, the uptake was around 220 mg m^{-2} h^{-1}. Finally, in this study
they also examined the nature of the removal mechanisms using
sterilized and un-sterilized stream bed cores and they concluded
that at least 40% and up to 100% of the uptake was due to biotic
processes, presumably heterotrophic microorganisms.

Lock and Hynes (1975) also carried out a comparative study of
the uptake of leachates of maple, aspen, cedar and pine in a hard
(Ca ~ 90 mg l^{-1}) and soft (Ca ~ 2 mg l^{-1}) water river. Once again
the principal mechanism for leachate removal appeared to be the
stream bed but differences were found between leachate uptake rates
on the basis of leaf species and river type (Table 3). In the
hard water stream the disappearance rates of the leachate were in
the order pine > maple > aspen > cedar and in the soft water stream,
aspen > cedar > maple > pine. However, this difference may be a
seasonal one with the uptake mechanism of the soft water stream
beds being more variable throughout the year. Interestingly, in
neither of the rivers was the commonest riparian leaf the one
which disappeared most rapidly and it is indeed strange that the
leaf leachate with the highest removal rate in the hard water stream
should exhibit the lowest uptake rate in the soft water stream.
This would suggest that the microbial communities were rather
specifically adapted to the existing spectrum of organic compounds
present in the streams at that time and that the community was
slow to adapt (in days as opposed to hours) to new organic compounds.

Confirmation by field study of a massive disappearance of leaf-
-derived "dissolved" organic matter during autumn was provided by
McDowell and Fisher (1976) using an organic matter budget approach
(Table 3). During the autumn they calculated that 77% of the
leachate from autumn-shed leaves entering the study reach was

retained within it. This represented an average rate of "DOM"
uptake of 0.28 mg l^{-1} h^{-1} and a peak uptake rate of 0.88 mg l^{-1} h^{-1}.
If it is assumed that much of this uptake was due to the stream bed
(Lock and Hynes, 1975, 1976) then this represents ~ 3 mg m^{-2} h^{-1}
and ~9 mg m^{-2} h^{-1} respectively. Lush and Hynes (1978) also conducted
a field experiment where leaf leachates were added to a small stream
(Table 3). They observed a maximum uptake rate of 1100 mg m^{-1} h^{-1},
the highest of all studies to date using natural undefined organic
matter (Table 3). Most recently, Dahm (1981) in a critical
evaluation of the loci of leaf leachate uptake within streams found
that adsorption accounted for approximately 23% and microbial
utilization approximately 77% of the removal of ^{14}C labelled alder
leaf leachate over 48 hours. The rate of microbial uptake was 28 mg
OM m^{-2} h^{-1}. Finally, Wallis et al. 1979 (see also Chapter 4)
found a substantial removal of groundwater "DOM" by a small sub-
alpine stream where an average rate of 207 mg m^{-2} h^{-1} was recorded
(Table 3) using an organic matter budget approach, while in an
Ontario stream using a laboratory experiment where groundwater
was passed upwards through stream bed material he reported an average
uptake rate of 16.4 mg m^{-2} h^{-1}. The reason for the difference
between these two rates is unclear but this is clearly an important
area of study which requires further investigation.

Uptake of high concentration defined organic matter: a second
approach in the examination of the uptake of organic matter by
streams has been through the use of defined organic compounds.
Such studies have been pioneered by Wuhrmann and his Swiss co-
workers using long (100 - 500 m) groundwater fed outdoor channels
(Wuhrmann et al., 1975). Although their work has been primarily
directed toward the understanding of the self-purification process
of rivers subjected to organic pollutants, their controls using non-
pollution stressed communities are comparable to other stream models.
In a study examining influence of water velocity upon substrate
uptake they found when using a 19 day old epilithic algae-dominated
community, at the highest velocity tested (18 cm s^{-1}), that the rate
of uptake per m^2 varied between 1.8 mg fructose m^{-2} h^{-1} and 21.4 mg
sucrose m^{-2} h^{-1}. The respective initial starting concentrations
were 0.2 mg l^{-1} of fructose and 1.8 mg l^{-1} of sucrose. The uptake
of glutamic acid under the same conditions was 10.6 m^{-2} h^{-1} and
glucose 2.8 mg m^{-2} h^{-1}. The highest rate of uptake recorded in
that study was for a filamentous *Sphaerotilus natans*/algal
community where the uptake of sucrose was 36.0 mg m^{-2} h^{-1} (starting
concentration = 2.4 mg l^{-1}. Wallis et al. (1979; Chapter 4) also
reports an *in situ* experiment on the uptake of glucose and glutamic
acid in the same stream in which he examined the uptake of ground
water DOM. He found removal rates of 363 mg m^{-2} h^{-1} for glutamic
acid, initial starting levels were 7.5 and 1.5 mg l^{-1} respectively.
Given the comparatively low inputs, these are very high uptake
rates (Table 3) considering that it is a pristine mountain stream.

In the studies discussed so far, the uptake, in all instances, of organic matter has been related to a unit area of stream bed and although it seems reasonable to conclude that a large proportion of this uptake will be due to the epilithon of the rocks, pebbles and gravel, some of the uptake may have been due to the interstitial organic flocs (FPOM) composed of detritus and micro-organisms or to microbially coated fine inorganic particles. However, with this provision in mind, these studies stand as a reasonable first approximation for river epilithon uptake, particularly since data on this subject are so few.

Uptake of low concentration defined organic matter: the above studies have all been concerned with the uptake rates of high concentration organic matter (in the mg l^{-1} range). With the exception of possibly the groundwater "DOM" inputs and extra-cellular release from dense macrophyte stands (Wetzel, 1975a), concentrations or supplies of this magnitude are only likely to occur intermittently during autumn leaf fall, during and/or after algal blooms and during macrophyte die-back, etc. Yet in many streams and rivers the concentrations of identifiable microbially labile organic compounds (monosaccharides, disaccharides, amino acids, fatty acids) are likely to be quite low and measured in $\mu g\ l^{-1}$ as opposed to mg l^{-1} (Telang et al., 1976, Larson, 1978). As a consequence of this, measurement of their uptake by microbial communities is more difficult. The principal approach to this problem has been to use a modification of Wright and Hobbie's (1966) kinetic approach. With this technique, confirmation is required that the relationship between microbial uptake and substrate concentration can be described by Michaelis-Menten kinetics, and if that is so, then the theoretical maximum uptake velocity (V_{max}), and the uptake velocity at the normal substrate concentration (if this is known) can be generated (see Write and Burnison, 1979, for the most recent discussion of this technique). This type of measurement on river epilithon has been pioneered in Costerton's laboratory at the University of Calgary, Alberta (Ladd et al., 1979; Wallis et al., 1979; Ladd et al., in prep.). The information to date is limited (Table 4) and restricted to a single compound, glutamic acid, but they serve to give an impression of the theoretical maximum uptake rates associated with a labile organic substance present in river water at a concentration of only a few $\mu g\ l^{-1}$. V_{max} specific activity indices (Wright, 1978) ranged from 2 - 615 $\mu g\ cm^{-2}\ litre^{-1}\ cell^{-1}$ 10^{-12}. The actual uptake velocity for a particular epilithon community depends upon the number of actively metabolizing bacteria present and the concentration of the substrate (uptake velocity increases with substrate concentration). The major limitation of this method is that it provides information for a single substrate only and it would be a daunting if not impossible task to determine the total labile DOM uptake by carrying out a 'complete' analysis

Table 4. V_{max} and specific activity indices of epilithon from several rivers for a low concentration defined organic matter input, glutamic acid (~ 0.2 μg l^{-1})

River	Bacteria (cm^{-2} x 10^3)	V_{max} specific activity index (μg cm^{-2} $cell^{-1}$ $litre^{-1}$ 10^{-12})	V_{max} (μg cm^{-2} h^{-1})	Author
Middle trib.	0.3	-	0.8	Wallis et al. 1979
Middle	4.8	-	11.3	Wallis et al. 1979
Marmot	0.7	615	-	Ladd et al. 1979
Luscar	1.0	331	-	
Canmore	-	64(2-144)	-	Ladd et al. (in prep.)
-	-	55	-	McFeters and Dockins 1980

Table 5. Input/output study of low concentration defined organic matter in a segment of Marmot Creek over a year (After Telang and Costerton et al., 1976).

Substrate	Average input concentration (μg l^{-1})	Main input	Net local input	Total input	Output	Uptake	Uptake as a % of input
		(in normalized units)					
Carbohydrate	65	100	15	115	60	55	48
Fatty acids	2	100	15	115	55	60	52
"Combined" amino acids	4.4	100	15	115	95	20	17
"Free" amino acids	2.3	100	20	120	80	40	33

of the water followed by kinetic uptake studies on each of the
compounds isolated. At present there are apparently no grounds for
supposing that the uptake of any one substrate, or even suite of
substrates, would be a reflection of the actual uptake of the entire
labile DOM pool.

One other approach to measuring the uptake of low concentration
labile DOM has been through budget studies for a defined length of
river. Prerequisites for this are the analytical capability to
quantify substrates at the $\mu g\ l^{-1}$ concentration and also to
determine the local input of water within the river section under
investigation. This approach has been adopted by Telang et al.
(1977) and the inputs and outputs of the substrate under question
were determined on a mass flow basis. The difference in discharge
between the input and output points of the river section enabled
the local input or output of water to be determined. Throughout
the study, water entered the experimental section and the
assumption was made that the DOM content of this water was the
same as that of the upstream input. This assumption may or may not
be valid (see Wallis, Chapter 4). They reported budgets for tannins
and lignins, "hydrocarbons", carbohydrates, fatty acids and
"combined" and free amino acids; the results of the last three are
given in Table 5.

Carbohydrates showed massive depletions in the river section
and they also noted the presence of seasonal effects, not so much
in the overall loss but in the pattern of losses and gains
between a pool of seven monosaccharides and the "total" carbohydrate
pool. In the winter, 35 units of "total" carbohydrate were
generated while almost twice as much mass of individual sugars were
lost within the river section. In the summer the opposite pattern
was observed with a net loss of "total" carbohydrate (45 units)
and a net generation of individual sugars (1000 units). Such shifts
strongly suggest that seasonal activity of epilithic microorganisms
could be responsible. Fatty acids were also taken up within the
study reach by as much as 50% of the inital input. Amino acid
uptake was dependent on whether they were free or in the combined
state (proteins, polypeptides etc.,) and although a much higher
concentration of combined amino acids (Telang et al., 1976) were
recorded (average of $4.4\ g\ l^{-1}$) the proportionate amount of uptake
was lower (17% as opposed to 33% of input) than that of the free
amino acids (average of $2.3\ g\ l^{-1}$). The extent of uptake of amino
acids was in the order serine > glycine > isoleucine > aspartic
acid > phenyl alanine > glytamic acid > alanine > ornithine
> leucine > valine > threonine > proline. However, a major problem
does exist with this approach, namely the estimation of the local
input of organic matter. An overestimation of local input would
mean an overestimation in the amount of uptake in the study reach,
but since the estimated input amounted to only 15% of the total

input the difference would not be great. Conversely, an under-
estimation of the local input would mean that a greater amount of
labile organic matter was being retained within the study reach.

DOM/COM uptake by epilithon - an epilogue: the inputs of organic
energy to rivers can be divided into pulsed high concentration
microbially labile inputs and the ambient organic matter content
of river water comprising only of a small proportion of microbially
labile compounds. Of the former, only the water soluble components
of autumn shed leaves have been investigated but one would
anticipate that other pulsed inputs would occur e.g. during and/or
after algal blooms, extracellular release from macrophyte beds,
leaching and decomposition of macrophytes, as a consequence of
overland flow during storm events and probably others. And, it
seems reasonable to expect that these inputs would be trapped
with the same apparent efficiency as the leaf leachates. However,
pulsed organic matter inputs apart, there is the continuous background of dissolved and colloidal organic matter flowing over the
epilithon. Investigations on the uptake of part of this pool, the
low level labile compounds (amino acids, monosaccharides etc.,)
have already begun but since these form only a very small fraction
of the total DOM/COM pool it is tempting to dismiss these as being
of only minor importance in the functioning of a river. The study
of Telang et al. (1976) in the Marmot Creek basin showed that
carbohydrates, total amino acids and fatty acids constituted only
4% of the DOM/COM pool, while Larson (1974) working on White Clay
Creek showed that carbohydrates, protein, amino acids, peptides
and lipids constituted on average 12% of the DOM/COM pool. Clearly
in proportionate terms, the concentrations of known labile
compounds in river water are indeed low. But, since we do not
know the magnitude of the (small?) enzymatically released flux of
labile organic substrates which supports the ecologically important
'conditioning' microflora of submerged autumn-shed leaves (Kaushik
and Hynes, 1971; Suberkrop and Klug, 1976; Bird and Kaushik,
Chapter 2) it is premature to assign the low concentration labile
DOM of riverwater a minor ecological role.

     Since only 5 - 10% of the DOM/COM pool has been accounted for
so far, what of the rest? Telang et al. (1976) found that 36% of
the DOM/COM pool was humic and fulvic acids, tannins and lignins
and total phenolics while Larson (1978) found very little humic
material present (a maximum of $150\,\mu g\ l^{-1}$ or 1% of the DOM/COM pool)
while phenolic compounds constituted 8% of the DOM/COM pool. Yet a
considerable proportion of the DOM/COM pool remains unclassified;
60% in the case of Telang et al. (1976) and 80% in the case of
Larson (1978). The tendency at the moment is to regard everything
other than the known microbially labile compounds as being
refractory or "relatively" refractory. However, de Haan (1974)
has suggested that humic acids can be co-metabolized and Lock and

Hynes (1976) observed on one occasion that 75% of river DOM was
taken up by a stream bed core over a period of 4 days. Thus, it is
perhaps also premature to assume that much of river DOM is microbially refractory. Work is now underway in my laboratory to
directly determine the proportion of the DOM/COM pool which is
utilizable by the heterotrophic microorganism of the epilithon.

AUTOTROPHIC ENERGY INPUTS

Minshall (1978) has recently written a cogent and most timely
paper on the extent of autotrophic production in rivers and it is
apparent that the generalization that all (or even many!) rivers
are primarily heterotrophic in nature is in considerable doubt.
As embodied in the River Continuum Hypothesis, it is also likely
that even in the same river, the light energy: organic energy
input ratios will vary in time and space. And the same is probably
no less true for the energy transduction components within a
section of river, in this case the balance between autotrophic
and heterotrophic microorganisms of the epilithon.

Energy inputs from coarse particulate organic matter apart,
the greatest amount of information on energy inputs to rivers is
for primary production. Since this subject has been recently
reviewed (Wetzel, 1975b; Johansson et al., 1977; Minshall, 1978)
reference will only be made to selected recent studies.

In Table 6 I have tabulated the results from 24 studies,
indicating whether the production estimate is closer to gross or
net primary production. It is apparent that the mean gross
fixation of carbon for these temperate rivers varies approximately
between 2 and 10 μg C cm^{-2} h^{-1} (assuming net production to be
approximately 50% of gross production). The range however is much
wider, spanning 5 orders of magnitude from 0.06 - 47 μg C cm^{-2} h^{-1}.
It must again be stressed that these figures are only indicative
of the rates of primary production because of the wide variation in
experimental methodologies and basic assumptions. However, they
serve to arrive at some generalizations for one of the metabolic
aspects of the epilithon. Additionally, it is also important to
stress at this point, that the large pulses of primary production
indicated by the maxima in Table 6 may be of great significance
in the overall functioning of the river ecosystem. It is likely
that several mechanisms operate in the river to retain this energy
within a reach and these will be discussed in the synthesis.

ECOLOGICAL IMPLICATIONS OF THE POLYSACCHARIDE MATRIX

It is now apparent that the bacteria, fungi and algae of
river epilithon are embedded within a polymer (slime) matrix
probably composed of polysaccharide (Geesey et al., 1977;
Costerton et al., 1978; Lock et al., in prep.).

Table 6. Primary production of epilithic algae of several rivers[1].

River	Order[2]	Shade[3]	Primary production (μg C cm^{-2} h^{-1}) Mean	Range	Note	Author
Ottawa	9+	US	0.8	-	Uptake of ^{14}C by epilithon coated glass slides. Annual mean value (~net)	Rosemarin 1975
Ottawa	9+	US	1.5	-	Summer mean value	Rosemarin 1975
14 rivers	-	-	11.3	0.02 - 46.9	14 rivers cited in Minshall (1978), oxygen methods (~gross)	Minshall 1978
Devils Club	1	HS	0.5	0.2 - 0.8	Oxygen method (~gross)	Naiman & Sedell 1980
Mack	3	HS	1.4	0.8 - 2.5		
Lookout	5	LS	2.5	1.3 - 3.4		
McKenzie	7	US	2.9	1.7 - 3.8		
Muskeg	4	LS	2.5	0.1 - 10.8	^{14}C method (~net)	Hickman et al. 1980
Steepbank	5	HS	1.6	0.06 - 13.1		
Mackay	6	-	0.8	0.05 - 2.6		
Hanging-stone	5	-	1.2	0.006- 4.2		
Ells	4	-	2.3	0.1 - 5.3		

[1] Calculated directly from original value expressed on a m^{-2} basis; new value probably an overestimate but no conversion factor available

[2] Strahler (1957)

[3] US - Unshaded   LS - Low shade   HS - High shade

The presence of this polysaccharide matrix binding together
the various components of the epilithon (bacteria, fungi, algae,
protozoa and micrometazoa) has a number of potentially ecologically
significant implications. These have been outlined in Lock et al.
(in prep.) in an expansion of the Madsen (1972) model of the
organic layer and it is proposed to expand further on these here.
Basically the model assumes that the epilithon receives light
energy, organic energy and inorganic matter in varying proportions
according to the type of stream or river section under consideration
(Vannote et al., 1978; Edelmann and Wuhrmann, 1978) and that the
fluxes of organic energy and inorganic nutrients in particular
are influenced by the structure of the epilithon as determined by
the polysaccharide matrix. The following aspects will be discussed:
resistance to diffusion, a site for attached enzymes and capacity
for adsorption.

Diffusion resistance of the polysaccharide matrix: it has been
apparent to chemical engineers for some time, that the uptake of
organic compounds by attached slime layers or suspended flocs from
sewage treatment processes are influenced by slime diffusional
resistance and it can be a substantial rate-limiting factor. Unfortunately, for river ecologists, all of this type of work has
been carried out at high substrate concentrations (2 - several
hundred mg l^{-1}) and since diffusion rates are partially dependent
upon the substrate concentration gradient then the role of
diffusion resistance in natural river epilithon as an uptake-rate
limiter will be exacerbated by the generally low ambient concentrations of labile substrates (Telang et al., 1976; Larson, 1978).

La Motta (1976) analysing the problem of substrate removal by
attached sewage microorganisms, simplified the overall process
into three major steps:
   (i)   Diffusion of the substrate from the bulk liquid
          to the interface between the liquid and the
          biological film.
   (ii)   Diffusion of the substrate within the biological
          slime.
   (iii) Biochemical reactions (substrate consumption) within
          the biological film.
He also assumed that the diffusion of the reaction products from
the internal pores of the slime had no effect upon the rate of
substrate utilization. This latter assumption is reasonable in
the short term since the reaction is irreversible, with biomass,
a small amount of extracellular release and CO_2 being the
principle reaction products.

Step (i) has been widely recognized as a potential rate-
limiting step in river epilithon metabolism and has been referred

Table 7. Diffusion coefficients of glucose through several types of slime.

Slime	External glucose concentration mgl^{-1}	Media C.N.	Diffusion coefficient $cm^{-2}S^{-1}x10^{6}$	Diffusion coefficient % of rate in water	Author
Zooglea ramigera	-	-	0.5	8	Baillod (1969)[1]
Mixed culture	-	-	0.6-6.0	10-100	Pipes (1974)[1]
Aerobic sludge – 20 days old	7100.0	5	1.7	28	Matson & Characklis (1976)
Aerobic sludge – 20 days old	7100.0	20	0.6	10	Matson & Characklis (1976)
Aerobic sludge – 20 days old	7100.0	50	2.1	35	Matson & Characklis (1976)
Attached laboratory film, media velocity >0.8 m s^{-1}	2.2	-	4.3	67	Lamotta (1976)
Attached laboratory film, media velocity >0.8 m s^{-1}	5.0	-	1.3	20	Lamotta (1976)
Attached laboratory film, media velocity >0.8 m s^{-1}	200.0	-	2.5	39	Lamotta (1976)

[1] Cited in Matson and Characklis (1976).

to as the 'current effect', where elevated water velocities or
levels of turbulence enhanced the biological rate under consideration,
(Whitford and Schumacher, 1961; Schumacher and Whitford, 1965;
Sperling and Grunewald, 1969; Rodgers and Harvey, 1976; Lock and
John, 1979). Step (iii) has already been considered in the section
on organic energy trapping and it is step (ii), the lower diffusion
rate of substrates within the microbial film as compared with water
that is dictated by the presence of the polysaccharide matrix. For
example, the diffusion coefficients of glucose for microbial films
have been determined by several workers at varying external
substrate concentrations and these results are presented in Table 7.
Diffusion coefficients for the microbial film expressed as a % of
the diffusion coefficients for water varied from 10% - 100%. It
is therefore clear that reduced diffusion rates of organic substrates
within these 'sewage' slimes will potentially limit their uptake
by the slime microorganisms. It also seems reasonable to
anticipate that similar events will also take place during the
diffusion of naturally occurring organic substrates through the
epilithon of running waters. Of course this phenomenon is not
limited to organic substrates but has also been demonstrated for
oxygen. Diffusion coefficients of oxygen for several microbial
slime systems are presented in Table 8 and these range from 2 - 100%
of the equivalent coefficient for pure water. Empirical evidence
was also provided for a diffusion resistance to carbon dioxide in
the slime sheath of the blue-green bacterium, *Oscillatoria rubescens*.
Chang (1980) showed that the removal of a 3 nm sheath resulted in
a 30% increase in the carbon dioxide uptake rate. It also seems
likely that the metabolic activity of slime communities may be
similarly affected by the diffusion characteristics of the inorganic
nutrients. Williamson and McCartey (1976) showed that the diffusion
coefficients of ammonium, nitrite and nitrate were 80 - 90% of the
value for water.

As yet there is no direct evidence that the polysaccharide
matrix of running water epilithon does retard the diffusion of
organic substrates, oxygen, carbon dioxide, inorganic nutrients
etc., but on *a priori* grounds the case would appear to be strong.
One piece of supportive evidence comes from a study by Ladd et al.
(1979). They compared the uptake of glutamic acid by an intact
river epilithic community with a corresponding disrupted one and
found a 2.7 fold increase in glutamic acid uptake by the disrupted
community suggesting that diffusional resistance to glutamic acid
was reduced by physical disruption. Certainly, other explanations
for this finding are possible and direct investigations of
diffusion within river epilithon are needed.

Another related aspect to diffusion resistance, which also
requires study, is the capacity of the polysaccharide matrix to
exclude macromolecules or colloidal aggregates. Preliminary studies

Table 8. Diffusion coefficient of oxygen through several types of slime.

Slime	Diffusion coefficient		Author
	$cm^{-2} s^{-1} \times 10^{-5}$	% of rate in water	
Bacterial slime	1.5	70	Tomlinson & Snaddon (1966)[1]
Sewage sludge (laboratory)	0.4 - 2.0	20 - 100	Matson (1975)[1]
Sewage sludge (domestic)	1.2	60	
Zooglea ramigera	0.2	8	Mueller (1966)[1]
Microbial slime	0.04	2	Bungay et al. (1969)

[1] Cited in Matson and Characklis 1976

on a 2 month old epilithon film from a soft water mountain stream
(Lock unpublished data) indicated that a molecule of $>$ 700,000
daltons would be excluded from the matrix, and work is underway to
determine the lower molecular exclusion point of river epilithon.
This information is important in determining how colloidal and high
molecular weight organic matter might become available to epilithic
microorganisms.

The polysaccharide matrix - a site for attached-enzymes: Zobell
(1943) first suggested that attached bacteria might have a
strategic advantage over suspended bacteria in terms of conserving
or concentrating exoenzymes. He postulated that exoenzymes and
hydrolysates would accumulate at the tangent of the bacterial cell
and the attachment surface. This idea has since been extended to
include the polysaccharide matrix where it has been suggested by
Costerton et al. (1978) that it might also have a conserving and
concentrating function for exoenzymes, while Lock et al. (in prep.)
suggest it may also act as a site for the attachment of exoenzymes.
Support for this latter idea comes from two sources, firstly from
studies on enzyme secretion by *Bacillus cereus* and *Micrococcus
sodonensis* cited in Dudman (1977) where the cosecretion of polysaccharide enhanced the activity of the enzyme by as much as 50%
and also served to protect the enzyme from proteolytic attack.
The second source of support comes from the soil literature where
Burns and his co-workers have suggested that enzymes trapped inside
porous humus polymers are protected from proteolysis and yet are
accessible to substrates (Burns, 1978). Credence to this model
comes from Ladd and Butler (1975) who produced enzyme-humic
complexes which retained enzymic properties. Thus it seems feasible
that a similar relationship could occur between exoenzymes and
enzymes released by cell lysis and the polysaccharide matrix. A
number of strategic advantages would accrue to epilithic microorganisms which facilitated the development of an attached enzyme
system and these have been outlined by Burns (1978) in the context
of attached soil enzymes:

(i)   "microorganisms would be 'on the spot' when a
       suitable low molecular product of enzymic activity
       appeared".
(ii)   "the existence of a battery of enzymes from previous
       generations would mean that less energy would be
       required for de-novo enzyme synthesis by microorganisms
       in that area".
(iii) "the levels of product released by the attached
       enzymes may be instrumental in initiating enzyme
       induction in adjacent microorganisms since macro-
       molecules incapable of passing the cell wall cannot
       possibly induce enzyme synthesis".

Such advantages would be equally true for a polysaccharide matrix-bound enzyme system, and it also seems a distinct possibility that humic material known to be present in river water, could also be incorporated into the epilithon matrix, perhaps as a consequence of adsorption processes. Thus both humic and polysaccharide enzyme complexes could be present in the epilithon of rivers and streams, but evidence for either of these has yet to be obtained.

The polysaccharide matrix - a site for adsorption: the polyanionic nature of the polysaccharide matrix led Costerton et al. (1978) to suggest that it might behave as an ion exchange compound, retaining nutrient ions and charged organic molecules within itself. Consequently, the binding kinetics for such interactions would be pH dependent, which in turn would be partially governed by microbial metabolism, in particular by the CO_2 transactions of respiration and photosynthesis. Clearly there is considerable potential for an extremely dynamic ion exchange system with perhaps marked diurnal differences in binding capacity occurring, governed by localized shifts in the P:R ratio of the epilithon.

Adsorption of dissolved and colloidal organic matter by stream sediments was observed by Lock and Hynes (1976) and although not measured, it seems certain that these sediments would have been coated with epilithon. They found that approximately 20% of the uptake of a maple leaf leachate by river sediment could be accounted for by adsorption. Lush and Hynes (1979) speculated that a very high uptake rate of leaf leachate by a small spring stream sediment was also due to adsorption by an "organically rich substratum". Dugan et al. (1971) in a study of microbial polymer flocs in lakes and their relation to eutrophication reported that the extracellular polymer produced by the bacterium *Zoogloea ramigera* could adsorb dissolved organic matter such as amino acids. This was regarded as an "amplification effect" of the already known high capacity of *Z. ramigera* to take up organic substances, and acted as an explanation for its exceptionally high rates of uptake for organic matter reported by Butterfield 1935, cited in Dugan et al. (1971).

Finally, Karlstrom (1978) provided evidence that the epilithon can also trap fine particulate organic matter. Rocks which had been allowed to develop an epilithon film in the dark, and which were apparently devoid of living algae, accumulated a large amount of algal debris. Similar accumulations of debris were also noted in an equivalent study by Lock et al. (in prep.) and it seems likely that the epilithon of both the upper and lower surfaces of rocks can trap fine particulate organic matter. Such accumulations are then potentially available for microbial invasion and are a further source of energy and nutrients to the epilithon system.

SOME INTERNAL ENERGY FLOWS

Energy transfer from the algae to the rest of the epilithon: it
has been accepted for sometime that planktonic algae release some
of their photosynthetically fixed carbon. Traditionally this is
measured as percentage extracellular release (PER) and is usually
calculated from radioactive dissolved carbon $DO^{14}C$ found in
filtrates from algal populations incubated with ^{14}C-bicarbonate
(Fogg, 1966; Watt, 1966). PER values from a number of selected
recent studies are presented in Table 9 and these fall into the
range of 1 - 32%. However, such studies have recently been
challenged by Sharp (1977) who suggests that 'average' extracellular releases are probably in the 5% region with higher values
being methodological artifacts. This viewpoint was subsequently
refuted by Fogg (1977) in connection with an early paper (Fogg et
al., 1965). In any event, strict comparisons are not possible
because of the very different growth conditions found in the various
studies and recently Smith and Wiebe (1976) suggested that there
may in fact be a constant release rate of extracellular products
and differences between studies may be solely due to changes in
the rate of primary production. To overcome this problem they
suggested that extracellular release should be reported as an
absolute value.

    Of course the values quoted in the above papers are closer to
net rates rather than gross extracellular release rates. This is
because the algal populations under study were natural ones which
would also contain heterotrophic microorganisms. Nalewajko et al.
(1976) tackled this problem by examining the kinetics of extracellular release of axenic algae along with mixed algae/bacteria
cultures. They found that the extracellular, release rate of
axenic *Chlorella pyrenoidsa* was linear with time (PER ~ 10%)
"plateau-type" curves were obtained when bacteria were present.
Similar "plateau-type" curves were obtained using a mixed lake
algal culture and they suggested that an initial lag in bacterial
uptake of the algal extracellular products was responsible and
cited supporting evidence. They went on to suggest that since
both axenic and algal/bacterial mixtures had similar initial
release rates (over the first 10 minutes), then the initial
release rates (10%) for mixed populations might be close to the
gross extracellular release rate.

    If similar phenomena occur within the epilithon, and there
seems no reason to expect otherwise, then the heterotrophic
microorganisms are particularly well placed to trap the organic
energy resulting from algal extracellular release as has been
demonstrated for other pelagic microheterotrophs (Smith et al.,
1977; Pearl, 1978). Many of the compounds released (Fogg, 1966;
Hellebust, 1974) are readily assimilated by bacteria. Thus what

Table 9. Percentage extracellular release (PER) of algae.

Location		PER	Notes	Author
Lake 227 (eutrophic)		7-15	Low biomass	Nalewajko & Schindler (1976)
		1-4	High biomass	Nalewajko & Schindler (1976)
Lake 239 (oligotrophic)		~20		Nalewajko & Schindler (1976)
Lake Kinneret	mean	3.7	Inverse relationship between PER and	Berman (1976)
	range	1.9-31.7	chlorophyll a concentration	
Coastal Subtropic Sea		<1-23	Highest release rates associated with intermediate levels of primary production	Williams & Yentsch 1976
Off the Coast of NW Africa		7-8	A direct correlation of PER with primary	Smith et al. (1977)
Lake		~2	Maximum rates of PER observed in first hour of the experiment	Steinberg (1978)

appears to be a potential loss of high quality energy may in the
short term never leave the confines of the epilithon, being directly
incorporated into bacterial biomass or being held within the polysaccharide matrix of river epilithon by adsorption processes or
diffusional resistance. Of course what has been said of algal
extracellular release is probably also true for bacterial and
possibly fungal extracellular release. Dunstall and Nalewajko
(1975) demonstrated that *Pseudomonas fluorescens*, a bacterium
common to freshwater, released both high and intermediate
molecular weight compounds. Release rates over the first $9\frac{1}{4}$ hours
amounted to 7.3% of the net carbon uptake increasing to nearly 30%
over 48 hours. Thus we can also anticipate energy transfers
between the heterotrophic microorganisms of the epilithon with the
concomitant conservation of organic energy sources.

*The polysaccharide matrix as a reserve energy source for the
heterotrophic microorganisms:* Lange (1976) in a discussion of the
function of the polysaccharide sheath (capsule) of blue-green algae
presented evidence which suggested that the sheath material may
serve as an energy source to the sheath associated bacteria under
conditions of low external organic matter concentrations. He found
when incubating blue-green algae in the presence of compounds
readily assimilated by the sheath bacterial (glucose and sucrose)
that the sheath was "voluminous and fluffy" in appearance.
Conversely, if the algae were grown in an organic matter free
medium then the sheaths were "compact, thin and smooth", suggesting
that the blue-green algal sheath was being consumed by the sheath
bacteria.

     If the polysaccharide matrix of river epilithon is also an
energy source to heterotrophic microorganisms during low organic
nutrient conditions then such a process would facilitate the
disintegration of the epilithon at such times. One would
anticipate that the structural integrity of the epilithon would
be eroded from within and the film would be expected to fall apart.
Such a phenomenon was observed (Lock, unpublished data) on the
Muskeg River in Alberta, Canada, where a thick epilithon appeared
to fall apart during a period of stable discharge. This suggests
that some agent other than elevated shear forces was responsible,
possibly matrix degradation by bacteria.

Internal grazing: Karlstrom (1978) and Lock et al. (in prep.)
have observed large numbers of micrometazoans (< 2 mm in length)
within river epilithon. Chironomid larvae were frequently found
to be the dominant micrometazoan in both of these studies, with
densities up to ~ 10 cm^{-2} occurring at times. Nematodes,
oligochaetes and other insect larvae were also noted but these
were always are very much lower densities (0.1 - 1.0 cm^{-2}).
At the times of high chironomid larval densities dramatic

sculpturing of the epilithon was observed presumably due to
burrowing activities of the larvae. An example of this can be
seen in Figure 2 where numerous ~ 500 μm and ~ 200 μm holes were
observed, corresponding to the sizes of chironomid larvae present
in the river at that time. It seems reasonable to assume that
these micrometazoans are grazing within the epilithon and at times
of high animal density and high grazing activity that the
structural integrity of the epilithon would be ultimately
destroyed (Fig. 2).

However, it would be misleading to suggest that the
micrometazoans and protozoans would have only a negative effect
upon the epilithon. It is becoming increasingly clear in detrital
systems that animal feeding processes have a positive effect upon
the microorganisms of that community (Fenchel, 1977; Gerlach,
1977) and overall rates of community metabolism have been
increased, through increased nutrient regeneration and enhancement
of primary production and decomposition processes. Such phenomena
are probably also taking place within the epilithon communities of
rivers and streams.

SYNTHESIS

The epilithon of rivers and streams is both a site of organic
energy generation via the algae and a trap for allochthonous and
autochthonous organic energy, and these processes are summarized
in Figure 3. During algal photosynthesis, extracellular products
are released and these diffuse into the polysaccharide matrix to
be either intercepted directly by microheterotrophs or to first
undergo enzymic breakdown by enzymes attached to or held within
the matrix. Upon the death of the algal cell, this will itself
be utilized by the bacteria and fungi of the epilithon. Trapping
of allochthonous organic energy is by two processes, firstly
adsorption of low molecular weight organic matter to the surface
polysaccharide fibrils or the trapping at the surface of molecules
or particles too big to penetrate the polysaccharide matrix. These
will build-up in successive layers on the surface of the epilithon,
to be continually invaded and metabolized by bacteria and fungi
from the epilithon beneath. The second process is the diffusion
of molecules and colloidal particles capable of penetrating the
polysaccharide matrix into the interior of the epilithon. As it
passes through the matrix it may be subjected to attack from
enzymes attached to the polysaccharide matrix or in concentration
gradients from bacterial and fungal cells. The products of this
enzymic activity may be suitable for direct metabolism by the
bacteria and fungi or may undergo further breakdown by surface
exoenzymes or by different suites of enzymes held within the
polysaccharide matrix. The very low molecular weight organic
compounds such as simple sugars and amino acids could of course
simply diffuse through the matrix and be available for direct

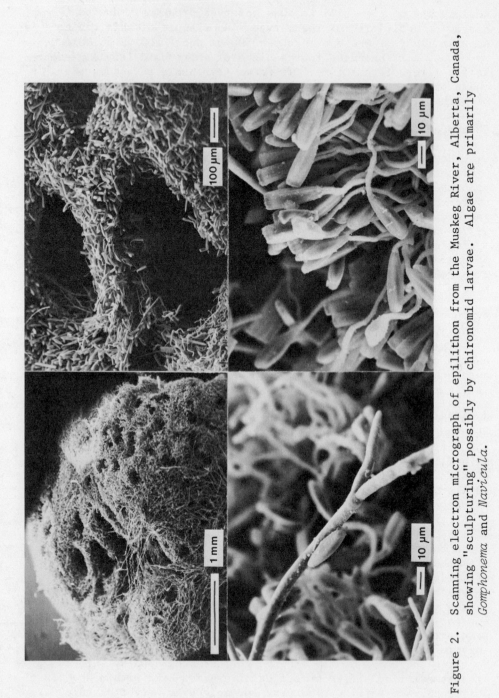

Figure 2. Scanning electron micrograph of epilithon from the Muskeg River, Alberta, Canada, showing "sculpturing" possibly by chironomid larvae. Algae are primarily *Gomphonema* and *Navicula*.

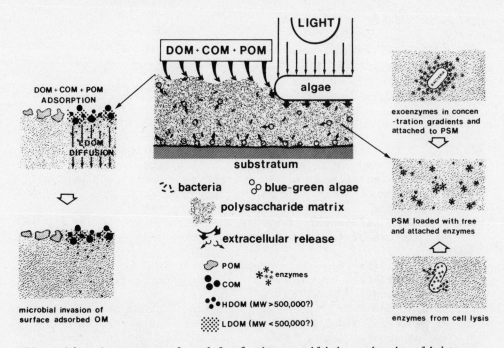

Figure 3.   A structural model of river epilithon showing light
            and organic energy pathways and some proposed aspects
            of the polysaccharide matrix (PSM). POM = particulate
            organic matter, COM = colloidal organic matter, HDOW =
            high molecular weight dissolved organic matter, LDOM =
            low molecular weight dissolved organic matter.

uptake by the microheterotrophs. The enzymes held within the
matrix may be the result of active secretion by the bacteria and
fungi or may be released as a consequence of cell lysis. The
polysaccharide matrix is visualized as a 3 dimensional lattice
work of fibrils which hold and protect a vast suite of enzymes
capable of degrading organic matter into the low molecular
weight compounds which can then be transported into the bacteria
and fungi. Thus a polysaccharide matrix/enzyme system would serve
as an energy cost effective digestive complex for the microheterotrophs of the epilithon.

    Subsequent transfers of energy from epilithon to the higher
trophic levels within rivers will take place through the activity
of the "scrapers" and "grazers" (Cummins and Klug, 1979) but
perhaps, more importantly, the epilithon may be a continuous major
source of fine and very fine particulate organic matter through
progressive sloughing and thus be a supplier of high quality
energy to the large 'collector' community (Cummins and Klug, 1979).

ACKNOWLEDGEMENTS

I would like to thank Miss L.D. Owen and Mrs. N.G.W. Jones for typing the initial manuscript.

REFERENCES

Albright, L.J. 1980. Microbial dynamics of two sub-arctic Canadian Rivers. *Wat. Res.* 14: 1353-1362.

Allanson, B.R. 1973. The fine structure of the periphyton of *Chara* sp. and *Potamogeton natans* from Wytham Pond, Oxford, and its significance to the macrophyte - periphyton metabolic model of R.G. Wetzel and H.L. Allen. *Freshwat. Biol.* 3: 535-542.

Anderson, N.H. and Sedell, J.R. 1979. Detritus processing by macroinvertebrates in stream ecosystems. *Ann. Rev. Entomol.* 24: 351-377.

Allan, H.L. 1976. Dissolved organic matter in lake water: characteristics of molecular size fractions and ecological implications. *Oikos* 27: 64-70.

Barton, D.R. and Lock, M.A. 1979. Numerical abundance and biomass of bacteria, algae and macrobenthos of a large northern river, the Athabasca. *Int. Revue ges. Hydrobiol.* 64(3): 345-359.

Berman, T. 1976. Release of DOM by photosynthesizing algae in Lake Kinneret, Israel. *Freshwat. Biol.* 6: 13-18.

Bombowna, M. 1977. Biocoenosis of a high mountain stream under the influence of tourism I Chemistry of the Rybi Potok waters and the chlorophyll content in attached algae and seston in relation to pollution. *Acta Hydrobiol.* 19(3): 243-255.

Burns, R.G. 1979. Interactions of microorganisms, their substrates and their products with soil surfaces. *In* "Adhesion of Microorganisms to surfaces", Ellwood, D.C., Melling, J. and Rutter, P. eds. Society for General Microbiology, Academic Press, London.

Burns, R.G. (ed) 1978. Soil Enzymes. Academic Press, London.

Bungay, H.R., Whalen, W.J. and Sanders, W.M. 1969. Microprobe techniques for determining diffusivities and respiration rates in microbial slime systems. *Biotechnol. Bioeng.* 11: 765-772.

Cairns, J. Jr. and Yongue, W.H. Jr. 1977. Factors affecting the number of species in freshwater protozoan community. *In* "Aquatic Microbial Communities". Cairns, J. Jr. ed. Garland Publishing Inc., N.Y.

Chang, T.P. 1980. Mucilage sheath as a barrier to carbon uptake in a cyanophyte *Oscillatoria rubescens*. *Arch. Hydrobiol.* 88: 128-133.

Clifford, H.F. 1969. Limnological features of a northern brown-water stream with special reference to life histories of aquatic insects. *Am. midl. Nat.* 82: 578-597.

Cooke, W.B. 1979. "The ecology of fungi". CRC Press Inc.,
    Florida.
Costerton, J.W., Geesey, G.G. and Cheng, K.J. 1978. How bacteria
    stick. *Sci. Am.* 238(1): 86-95.
Cummins, K.W., Klug, J.J., Wetzel, R.G., Petersen, R.C.,
    Subertropp, K.F., Manny, B.A., Wuychechk, J.C. and Howard, F.O.
    1972. Organic enrichment with leaf leachate in experimental
    lotic ecosystems. *BioScience* 22(12): 719-722.
Cummins, K.W. and Klug, M.J. 1979. Feeding ecology of stream
    invertebrates. *Ann. Rev. Ecol. Syst.* 10: 147-172.
Dahm, C.N. 1981. Pathways and mechanisms for removal of
    dissolved organic carbon from leaf leachate in streams.
    Can. J. Fish. Aquat. Sci. 38: 68-76.
De Haan, H. 1974. Effect of a fulvic acid fraction on the
    growth of a Pseudomonas from Tjeukemeer (The Netherlands).
    Freshwat. Biol. 4: 301-310.
Dudman, W.F. 1977. The role of surface polysaccharides in natural
    environments. *In* "Surface Carbohydrates of the Prokaryotic
    Cell". Sutherland, I. ed. Academic Press, London.
Dugan, P.R., Pfister, R.M. and Frea, J.I. 1971. Implications of
    microbial polymer synthesis in waste treatment and lake
    eutrophication. *In* "Advances in Water Pollution Research,
    Vol. 2". Jenkins, S.H. ed. III-20-III-20/10.
Dunstall, T.G. and Nalewajko, C. 1975. Extracellular release in
    planktonic bacteria. *Verh. Internat. Verein. Limnol.*
    19: 2643-2649.
Edelmann, W. and Wuhrmann, K. 1978. Energy of running water
    systems. *Verh. Internat. Verein. Limnol.* 20: 1800-1805.
Faust, M.A. and Correll, D.L. 1977. Autoradiographic study to
    detect metabolically active phytoplankton and bacteria in
    the Rhode River Estuary. *Mar. Biol.* 41: 293-305.
Fenchel, T. 1977. The significance of bactiverous protozoa in
    the microbial community of detrital particles. *In* "Aquatic
    Microbial Communities", Cairns, J. ed. Garland Publishing Inc.,
    N.Y.
Finlay, B.J., Laybourn, J. and Strachan, I. 1979. A technique
    for the enumeration of benthic ciliated protozoa. *Oecologia*
    39: 375-377.
Fisher, S.G. 1977. Organic matter processing by a stream
    segment ecosystem, Fort River, Massachusetts, U.S.A.
    Int. Revue ges. Hydrobiol. 62(6): 701-727.
Fisher, S.G. and Likens, G.E. 1973. Energy flow in Bear Brook,
    New Hampshire: an integrative approach to stream ecosystem
    metabolism. *Ecol. Monogr.* 43: 421-439.
Fletcher, M. 1980. Adherence of marine micro-organisms to
    smooth surfaces. *In* "Bacterial Adherence" Beachey E.H. ed.
    Chapman and Hall, London.
Fogg, G.E. 1966. The extracellular products of algae.
    Oceanogr. Mar. Biol. Annv. Rev. 4: 195-212.

Fogg, G.E. 1977. Excretion of organic matter by phytoplankton.
    Limnol. Oceanogr. 22(3): 576–577.
Fogg, G.E., Nalewajko, C. and Watt, W.D. 1965. Extracellular
    products of phytoplankton photosynthesis. *Proc. R. Soc.
    Lond. Ser.* B 162: 517–534.
Geesey, G.G., Richardson, W.T., Yeomans, H.G., Irvin, R.T. and
    Costerton, J.W. 1977. Microscopic examination of natural
    sessile bacterial populations from an alpine stream.
    Can. J. Microbiol. 23: 1733–1736.
Geesey, G.G., Mutch, R., Costerton, J.W. and Green, R.B. 1978.
    Sessile bacteria: an important component of the microbial
    population in small mountain streams. *Limnol. Oceanogr.*
    23(6): 1214–1223.
Gerlach, S.A. 1978. Food chain relationships in subtidal silty
    sand marine sediments and the role of meiofauna in stimulating
    bacterial productivity. *Oecologia* 33: 55–69.
Hellebust, J.A. 1974. Extracellular products. *In* "Algal
    physiology and Biochemistry". Stewart, W.D. ed. Univ.
    of California Press.
Hickman, M., Charlton, S.E.D., and Jenkerson, C.G. 1980. A
    comparative study of benthic algal primary productivity
    in the AOSERP study area. Alberta Oilsands Environmental
    Research Program, Project WS 1.3.4. Edmonton 133 pp.
Hynes, H.B.N. 1963. Imported organic matter and secondary
    productivity in streams. *Int. Congr. Zool.* 16, 4: 324–329.
Hynes, H.B.N. 1970. The ecology of running waters. Liverpool
    University Press, Liverpool.
Johansson, C., Kronborg, L. and Thomasson, K. 1977. Attached
    algal vegetation in running waters. *Excerpta botanica Section*
    B 16: 126–178.
Jones, E.B.G. (ed.) 1976. "Recent advances in aquatic
    mycology". Elek Science. London.
Jones, H.C., Roth, I.L. and Sanders, W.M. 1969. Electron
    microscopic study of a slime layer. *J. Bact.* 99: 316–325.
Karlstrom, U. 1978. Role of the organic layer on stones in
    detrital metabolism in streams. *Verh. Internat. Verein.
    Limnol.* 20: 1463–1470.
Kaushik, N.K. and Hynes, H.B.N. 1971. The fate of dead leaves
    that fall into streams. *Arch. Hydrobiol.* 68: 465–515.
Ladd, J.N. and Butler, J.H.S. 1975. Humus-enzyme systems and
    synthetic organic polymer analogs. *In* "Social Biochemistry,
    Vol. 4". Paul, E.A. and McLaren, A.D. eds. Marcel Dekker,
    New York.
Ladd, T.I., Costerton, J.W. and Geesey, G.G. 1979. Determination
    of the heterotrophic activity of epilithic microbial populations.
    In "Native aquatic bacteria: Enumeration, activity, and
    ecology". Costerton, J.W. and Colwell, R.R. eds. ASTM STP
    695, American Society for Testing and Materials. Philadelphia.
    180–195.

Ladd, T.I., Read, R., Ventullo, R.M., Wallis, P.M., Telang, S.A. and Costerton, J.W. (In prep.). The microbiology of a mountain stream flowing through a coal mine reclamation project.

LaMotta, E.J. 1976. Internal diffusion and reaction in biological films. *Environ. Sci. Technol.* 10 (8): 765–769.

Lange, W. 1976. Speculation on a possible essential function of the gelatinous sheath of blue-green algae. *Can. J. Microbiol.* 22: 1181–1185.

Larson, R.A. 1978. Dissolved organic matter of a low coloured stream. *Freshwat. Biol.* 8: 91–104.

Lehmicke, L.G., Williams, R.T. and Crawford, R.L. 1979. ^{14}C-most probable-number method for estimation of active heterotrophic microorganisms in natural waters. *Appl. Environ. Microbiol.* 38: 644–649.

Lock, M.A. and Hynes, H.B.N. 1975. The disappearance of four leaf leachates in a hard and soft water stream in South Western, Ontario, Canada. *Int. Revue ges. Hydrobiol.* 60(6): 847–855.

Lock, M.A., Hynes, H.B.N. 1976. The fate of "dissolved" organic carbon derived from autumn-shed maple leaves (*Acer saccharum*) in a temperate hard-water stream. *Limnol. Oceanogr.* 21(3): 436–443.

Lock, M.A. and John, P.H. 1979. The effect of flow patterns on uptake of phosphorus by river periphyton. *Limnol. Oceanogr.* 24(2): 376–383.

Lock, M.A. and Wallace, R.R. 1979. Interim report on ecological studies on the lower trophic levels of Muskeg rivers within the Alberta Oil Sands Environmental Research Program (A.O.S.E.R.P.) study area. A.O.S.E.R.P. Report 58: 105 pp. (Paper on this work in preparation).

Lock, M.A., Wallace, R.R., Costerton, W., Ventulla, R. and Charlton, S.E. (In prep.) River epilithon – its composition and structure on the upper and lower surfaces of rocks.

Lock, M.A., Wallis, P.M. and Hynes, H.B.N. 1977. Colloidal organic matter in running waters. *Oikos* 29: 1–4.

Lush, D.L. and Hynes, H.B.N. 1978. The uptake of dissolved organic matter by a small spring stream. *Hydrobiologia* 60(3): 271–275.

Lyford, J.H. and Gregory, S.V. 1975. The dynamics and structure of periphyton communites in three Cascade Mountain streams. *Verh. Internat. Verein Limnol.* 19: 1610–1616.

Mack, W.N., Mack, J.P. and Ackerson, A.O. 1975. Microbial film development in a trickling filter. *Microb. Ecol.* 2: 215–226.

Mackie, E.B., Brown, K.N., Lam, J. and Costerton, J.W. 1979. Morphological stabilization of capsules of group B streptococci, types Ia, Ib, II and III, with specific antibody. *J. Bact.* 138(2): 609–617.

Madsen, B.L. 1972. Detritus on stones in small streams. *Mem. Ist. Ital. Idrobiol. Suppl.* 29: 385–403.

Marker, A.F.H. 1975a. The benthic algae of some streams in
    southern England. I. Biomass of the epilithon in some small
    streams. *J. Ecol.* 64: 343-358.
Marker, A.F.H. 1976b. The benthic algae of some streams in
    southern England. II. The primary production of the
    epilithon in a small chalk stream. *J. Ecol.* 64: 359-373.
Matson, J.V. and Charaklis, W.G. 1976. Diffusion into microbial
    aggregates. *Wat. Res.* 10: 877-885.
McDowell, W.H. and Fisher, S.G. 1976. Autumnal processing of
    dissolved organic matter in a small woodland stream ecosystem.
    Ecology   57(3): 561-569.
McFeters, G.A. and Dockins, W.S. 1980. Ecology of a benthic
    microbial community in a high alpine stream. Abstract,
    Second International Symposium on Microbial Ecology, University
    of Warwick 7-12 September, 1980.
Meyer-Reil, L.A. 1978. Autoradiography and epifluorescent
    microscopy combined for the determination of number and
    spectrum of actively metabolizing bacteria in natural
    waters. *Appl. Environ. Microbiol.* 36(3): 506-512.
Minshall, G.W. 1978. Autotrophy in stream ecosystems. *BioScience*
    28(12): 767-771.
Moore, J.W. 1976. Seasonal succession of algae in rivers. I
    Examples from the Avon a large slow flowing river. *J. Phycol.*
    12: 342-349.
Moore, J.W. 1977. Seasonal succession of algae in rivers II
    Examples from Highland Water, a small woodland stream.
    Arch. Hydrobiol. 80(2): 160-171.
Nalewajko, C., Dunstall, T.G. and Shear, H. 1976. Kinetics of
    extracellular release in axenic algae and in mixed algal-
    bacterial cultures: significance in estimation of total
    (gross) phytoplankton excretion rates. *J. Phycol.* 12: 1-5.
Nalewajko, C. and Schindler, D.W. 1976. Primary production,
    extracellular release and heterotrophy in two lakes in the
    E.L.A., Northwestern Ontario. *J. Fish. Res. Bd. Can.* 33:
    219-226.
Paerl, H.W. 1978. Microbial organic carbon recovery in aquatic
    ecosystems. *Limnol. Oceanogr.* 23(5): 927-935.
Peroni, C. and Lavarello, O. 1975. Microbial activities as a
    function of water depth in the Ligurian Sea: an autoradiograph
    study. *Mar. Biol.* 30: 37-50.
Petersen, R.C. and Cummins, K.W. 1974. Leaf processing in a
    woodland stream. *Freshwat. Biol.* 4: 343-368.
Powell, D.A. 1979. Structure, solution properties and some
    biological interactions of some microbial extracellular
    polysaccharides. *In* "Microbial polysaccharides and
    polysaccharases". Berkeley, R.C.W., Gooday, G.W. and
    Ellwood, D.C. eds. Academic Press, London.
Ramsay, A.J. 1974. The use of autoradiography to determine the
    proportion of bacteria metabolizing in an aquatic environment.
    J. Gen. Microbiol. 80: 363-373.

Rodgers, J.H. and Harvey, R.S. 1976. The effect of current on periphytic productivity as determined using ^{14}C. *Water Resources Bulletin* 12(6): 1109-1118.

Rosemarin, A.S. 1975. Comparison of primary productivity (^{14}C) per unit biomass between phytoplankton and periphyton in the Ottawa River near Ottawa, Canada. *Verh. Internat. Verein. Limnol.* 19: 1584-1592.

Roth, I.L. 1977. Physical structure of surface carbohydrates. *In* "Surface Carbohydrates of the Prokaryotes", Sutherland, I. ed., Academic Press, New York.

Schumacker, G.J. and Whitford, L.A. 1961. Repiration and ^{32}P uptake in various species of freshwater algae as affected by current. *J. Phycol.* 1: 78-80.

Sharp, J.H. 1977. Excretion of organic matter by marine phytoplankton: Do healthy cells do it? *Limnol. Oceanogr.* 22: 381-399.

Smith, D.F. and Wiebe, W.J. 1976. Constant release of photo-synthate from marine phytoplankton. *Appl. Environ. Microbiol.* 32: 75-79.

Smith, W.D., Barber, W.T. and Huntsman, S.A. 1977. Primary production off the coast of N.W. Africa: excretion of dissolved organic matter and its heterotrophic uptake. *Deep Sea Research* 24: 35-47.

Sperling, J.A. and Grunewald, R. 1969. Batch culturing of thermophilic benthic algae and phosphorus uptake in a laboratory stream model. *Limnol. Oceanogr.* 14: 944-949.

Steinberg, C. 1978. Release of dissolved organic carbon of various molecular sizes, in plankton populations. *Arch. Hydrobiol.* 82: 155-165.

Strahler, A.N. 1957. Quantitative analysis of watershed geomorphology. *Trans. Am. Geophys. Union* 38: 913-920.

Suberkropp, K. and Klug, M.J. 1976. Fungi and bacteria associated with leaves during processing in a woodland stream. *Ecology* 57: 707-719.

Telang, S.A. and Costerton, J.W. et al. 1976. Water quality and forest management: Chemical and biological processes in a forest-stream ecosystem of the Marmot Creek Drainage Basin. Report to Environment Canada, Inland Waters Directorate.

Telang, S.A., Geesey, G.G., Ladd, T., Mutch, R.M., Wallis, P.M., Costerton, W.J. and Hodgson, G.W. 1977. Water quality in the Marmot Streams. Report to Inland Waters Directorate, Environment Canada, Ottawa.

Vannote, R.L., Minshall, G.W., Cummins, K.W., Sedell, J.R. and Cushing, C.E. 1980. The River Continuum Concept. *Can. J. Fish. Aquat. Sci.* 37: 130-137.

Wallis, P.M., Hynes, H.B.N. and Fritz, P. 1979. Sources, transportation and utilization of dissolved organic matter in groundwater and streams. Scientific Series No. 100 Inland Waters Directorate, Water Quality Branch, Ottawa.

Watt, W.D.   1966.   Release of dissolved organic material from the
    cells of phytoplankton populations. *Proc. R. Soc. Lond. Ser.*
    B.   164: 521-551.

Wetzel, R.G.   1975.   Limnology W.B. Saunders Company,
    Philadelphia.

Wetzel, R.G.   1975.   Primary production. *In* "River Ecology",
    Whitton, B.A. ed. Blackwell Scientific Publications, Oxford.

Wetzel, R.G. and Manny, B.A.   1972.   Decomposition of dissolved
    organic carbon and nitrogen compounds from leaves in an
    experimental hard water stream. *Limnol. Oceanogr.*   17:
    927-931.

Williams, P.J. Le B. and Yentsch, C.S.   1976.   An examination of
    photosynthetic production excretion and photosynthetic
    products and heterotrophic utilization of dissolved organic
    compounds with reference to results from a coastal subtropical
    sea. *Mar. Biol.*   35: 31-40.

Williamson, J. and McCartey, P.L.   1976.   A model of substrate
    utilization by bacterial films. *Jour. Wat. Pollut. Control*
    Fed.   489-524.

Whitford, L.A. and Schumacher, G.J.   1961.   Effect of current on
    mineral uptake and respiration by a freshwater alga.
    Limnol. Oceanogr.   6: 423-425.

Whitton, B.A.   1975. Algae. *In* "River Ecology", Whitton, B.A.
    ed. Blackwell Scientific Publications, Oxford.

Wright, R.T.   1978.   Measurement and significance of specific
    activity in the heterotrophic bacteria of natural waters.
    Appl. Environ. Microbiol.   36(2): 297-305.

Wright, R.T. and Burnison, B.K.   1979.   Heterotrophic activity
    measured with radiolabelled organic substrates. *In*
    "Native aquatic bacteria: enumeration, activity and ecology"
    Costerton, J.W. and Colwell, R.R. eds. ASTM STP 695,
    American Society for Testing and Materials, Philadelphia.
    140-155.

Wright, R.T. and Hobbie, J.E.   1966.   Use of glucose and acetate by
    bacteria and algae in aquatic ecosystems. *Ecology.*   47:
    447-464.

Wuhrmann, Von K., Eichemberger, E., Leidner, H.A. and Wuest, D.
    1975.   Uber den Einfluss der Stromungsgeschwindigkeit auf
    die Selbstreinigung in Fliessgewassern. *Schweiz, Z. Hydrol.*
    37(2): 253-272.

Zobell, C.F.   1943.   The effect of solid surfaces upon bacterial
    activity. *J. Bact.*   46: 39-56.

COARSE PARTICULATE ORGANIC MATTER IN STREAMS

Glen A. Bird and Narinder K. Kaushik

Department of Environmental Biology
University of Guelph
Guelph, Ontario, Canada

INTRODUCTION

Coarse particulate organic matter (CPOM) has been defined as particulate organic matter >1 mm in size. Since its transport and movement in streams is the topic of another chapter in this book (Dance, Chapter 3), our discussion will mainly deal with such aspects as sources of CPOM, its processing and decomposition and importance as a food source to stream invertebrates. Also, in view of the excellent reviews by Anderson and Sedell (1979), and Cummins and Klug (1979) that include discussions on CPOM and earlier pertinent literature, our emphasis will be more on recent publications.

SOURCES OF CPOM

CPOM may be of allochthonous origin, e.g. leaves, wood, bark, flowers, bud scales, insect frass, etc., or of autochthonous origin, e.g. macrophytes and filamentous algae. The importance of allochthonous organic matter (OM) in streams was pointed out as early as 1912 by Thieneman (1912) and subsequently by Hynes (1963). Its importance is illustrated by the fact that many headwater streams are basically heterotrophic (e.g. Fisher and Likens, 1973; Hall, 1972; Hynes, 1970, 1975). In such streams more OM is respired than is produced within the stream, and the ratio between the rate of gross primary production (P) and the rate of community respiration (R) is less than one (Fisher and Likens, 1973). However, the comparative importance of allochthonous versus autochthonous inputs varies from stream to stream and, in general, as a river increases in size towards its mouth, the importance of riparian vegetation as a source of energy diminishes (Vannote et al., 1980). In

41

shaded streams, because of restricted primary production, CPOM is
primarily of allochthonous origin whereas that in streams with
limited shading, e.g. those passing through xeric regions, meadows
or agricultural areas, may be mostly of autochthonous origin. For
example, Minshall (1978) found that autotrophy dominated in a small
desert stream and Marxsen (1980) noted that the Breitenbach, which
flows through a meadow, had a primary production of 80-200 g
C m^{-2} yr^{-1}, but received only 46.8 g C m^{-2} yr^{-1} from allochthonous
sources. In contrast, primary production (22.2 g C m^{-2} yr^{-1}) in
the nearby Rohrwiesenbach was only of minor importance as a food
source because of large allochthonous inputs (634 g C m^{-2} yr^{-1}) of
CPOM. An idea of the magnitude of input of allochthonous CPOM to
a variety of streams can be gained from the values presented in
Table 1. Although a maximum total input of 1719 g m^{-2} yr^{-1} was
recorded in Rohrwiesenbach (Marxsen, 1980), a value of about 400-
600 g m^{-2} yr^{-1} can be regarded as more common with autumn-shed
leaves constituting about 350 g m^{-2} yr^{-1} (based on Table 1). Input
of CPOM to streams is continuous (Hynes, 1975) but maximum inputs
usually occur during or shortly after peak litterfall, e.g. as
leaves during the autumn in North America or as bark during the
summer in Australia (Blackburn and Petr, 1979). Input of wood, on
the other hand, is mainly by wind and precipitation events (Bagnall,
1972; Gosz et al., 1972; McColl, 1966). Because autumn-shed leaves
may constitute a substantial component of CPOM and because they have
been the subject of a number of research investigations during the
last decade, we can start our discussions with leaves.

AUTUMN-SHED LEAVES

    Autumn-shed leaves are low in N and P as these elements are
reabsorbed by the plant from dying leaves, but are a rich source
of carbohydrates and such nutrients as Ca, Mg, K and Na (Whittaker
et al., 1979). The nutrient composition of some terrestrial litter
is given in Table 2. Petersen and Cummins (1974) divided leaf
processing in streams into: 1) leaching, 2) initial microbial
processing or conditioning, and 3) animal-microbial conversion. To
this may be added physical abrasion as another factor. The
relative importance of each of these is illustrated in Figure 1.
Naturally the extent to which any of the above subprocesses
contributes to the overall breakdown will vary with the nature of
leaves and such factors as water quality, temperature, activity of
benthos, size of leaf packs and the nature of stream substratum.

    Upon entering a stream, leaves, if not already leached, undergo
a rapid weight loss, because of leaching of soluble compounds,
within about 24 hours. This weight loss, depending upon
temperature and initial pH of water, may account for about 20-30%
of the initial weight for deciduous leaf species and 3-13% for
evergreen species (Lush and Hynes, 1973); Petersen and Cummins
(1974) reported a mean value of 15% for various deciduous leaf

Table 1. Inputs of allochthonous CPOM to streams.

Stream Location	Width (m)	Vegetation	Total Input ($gm^{-2}yr^{-1}$)	Input of Leaves	Comments	Authority
Cement Creek, Australia	1-4.9	mainly mountain ash, (Eucalyptus regnans) & myrtle beech (Nothofagus cunninghami)	604	151	Branches, and bark made 66% of the total input	Blackburn & Petr, 1979
Berestream, England (a) Roke Farm	3-4	Wooded area dominated by willow (Salix)	315	–	Leaf input was about 90% of the total input. Branches and bark made 66%	Dawson, 1976
(b) Bere Heath	8-10	dominated by alder (Alnus)	466	430	Woods on both sides	
			240	216	Woods on only 1 side	
Bear Brook, New Hampshire, U.S.A.	2.2-4.0	hardwood forest – beech, birch, maple	655	339	Leaf input was 66% of total input. On basis of total stream area of 5,877 m²	Fisher & Likens, 1973
Doe Run, Kentucky	–	mixed hardwood	–	355		*Krumholz, 1972 & Linton, 1972
Rohrwiesenbach, West Germany	0.06-1.75	mostly beech (Fagus)	1,719	–	Rohrwiesenbach is located in a gully and is fully canopied.	Marxsen, 1980
Breitenbach, West Germany	0.11-1.64	mostly beech in the upper part	643	–		
Carnation Creek,	–	60% conifer	707	598	Alder and conifer comprised about 69% of leaf input	Neaves, 1978
Stampen stream, Sweden	0.69	pasture-alder (Alnus glutinosa)	780	680		Otto, 1975
Catahoule Creek, Mississippi	2nd order	forested with pine & deciduous	386	–	83% twig in April, 75% deciduous leaves in November	Post & Cruz, de la, 1977
Coweeta, North Carolina, U.S.A.	–	white pine / old field / mixed hardwood	319 / 286 / 352	– / – / –		*Webster, 1975
Middle Bush, N.Z.	< 2	mountain beech (Nothofagus)	525	350	Value does not include fine material (< 2mm)	Winterbourn, 1976

– No information given
* From Anderson and Sedell, 1979

Table 2.   Percent nutrient composition of some leaf and macrophyte species.

	K	Ca	Mg	Mn	P	N	Ash	Comments	Authority
Branches, wood and bark									
Acer saccharum	0.24	0.82	0.06	0.11	0.07	0.88	2.70	dead, fresh-fallen	Whittaker et al., 1979
Betula lutea	0.11	0.90	0.06	0.07	0.08	0.94	2.00	dead, fresh-fallen	Whittaker et al., 1979
Fagus grandifolia	0.11	1.09	0.10	0.12	0.08	0.86	2.50	dead, fresh-fallen	Whittaker et al., 1979
Leaves									
A. saccharum	1.01	0.60	0.12	0.17	0.18	2.19	5.65	summer	Whittaker et al., 1979
A. saccharum	0.36	0.76	0.08	0.30	0.03	0.68	4.65	autumn, fresh-fallen	Whittaker et al., 1979
B. lutea	0.44	1.27	0.23	0.39	0.06	1.09	5.40	autumn, fresh-fallen	Whittaker et al., 1979
F. grandifolia	0.38	0.78	0.12	0.28	0.04	0.84	4.87	autumn, fresh-fallen	Whittaker et al., 1979
Picea rubens	0.49	0.30	0.06	0.21	0.10	1.26	2.83	summer	Whittaker et al., 1979
Pinus nigra	0.43	0.95	0.11	0.04	0.09	0.87	2.82	late summer	Ovington, 1956
Emergent macrophytes									
Typha angustifolia	1.45	0.93	0.13	–	0.10	1.45	–	dead, standing shoots	Mason and Bryant, 1975
Phragmites communis	0.52	0.37	0.15	–	0.15	1.87	–	dead, standing shoots	Mason and Bryant, 1975
Juncus effusus	0.16	0.17	0.08	–	0.22	1.58	–	dead portions of living shoots	Boyd, 1971
Submersed macrophyte									
Potamogeton spp.	1.10	7.4	0.32	0.03	0.17	1.63	29.4*	September, lake	Muztar et al., 1978
Macroalgae									
Cladophora glomerata	1.40	7.0	0.45	0.03	0.17	2.64	43.9*	September, lake	Muztar et al., 1978

* extremely high ash was the result of surface calcification with $CaCO_3$

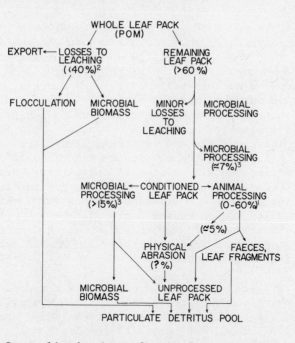

Figure 1.   Generalized scheme for leaf processing in streams.
(1) Cummins et al. (1973), (2) Hynes et al. (1974),
and (3) Petersen and Cummins (1974).

species. Leachates are important sources of dissolved organic
matter and are rapidly removed from the water by both biotic
(microbial) and abiotic (freezing, flocculation) factors (Wetzel
and Manny, 1972; Lush and Hynes, 1973; Lock and Hynes, 1975, 1976;
Lock, Chapter 1). Leachates are mainly reducing sugars,
polyphenols and amino acids (Suberkropp et al., 1976; Willoughby,
1974).

     At least in the early stages, fungi are more important than
bacteria in conditioning leaves, perhaps because fungi are superior
at digesting cellulose and lignin and because their mycelium can
penetrate the substrate. In streams the aquatic hyphomycetes
are numerically the most important fungal group (Willoughby, 1974).
They display a seasonal succession similar to the pattern of
allochthonous input and are the main agents of leaf decay at the
low water temperatures of autumn and winter. The aquatic
hyphomycetes exhibit substrate preferences, showing differences in
abundance and successional patterns on different types of leaves.
For example, Suberkropp and Klug (1976) found that *Flagellospora*

curvula and *Lemonniera aquatica* were early dominants (4 to 6 weeks)
on oak and hickory leaves, whereas *Alataspora acuminata* became
dominant at 6 to 8 weeks, especially on oak, but was co-dominant
with *Tetracladium marchalianus* on hickory. *Anguillospora* was of
major importance on hickory half-way through succession but was of
minor importance on oak. More recently Bärlocher et al. (1978a)
demonstrated that pine needles are colonized by aquatic hyphomycetes,
but only very slowly for the first 4-5 months. This is partly
because their thick cuticle and epidermal layers act as a physical
barrier to colonization (Michaelides and Kendrick, 1978) but mainly
because the needles contain antifungal inhibitors (Bärlocher and
Oertli, 1978) whose removal by decomposition or leaching is
prevented by the cuticle and epidermal layers (Bärlocher et al.,
1979). The needles are processed faster if the inhibiting factor
is removed by treatment with hot HCl, H_2O, alcohol or NaOH and
such treated needles are more palatable to *Gammarus pseudolimnaeus*
as compared with untreated ones.

The role of bacteria in decomposition is regarded as smaller
than that of aquatic hyphomycetes (Kaushik and Hynes, 1968, 1971;
Suberkropp and Klug, 1976), although Iversen (1973) and Iversen
and Madsen (1977) found that bacterial respiration accounted for a
major portion of the respiration on leaves. Suberkropp and Klug
(1976) for example attributed higher bacterial populations during
the latter stages of processing of oak and hickory leaves to an
increase in exposed surface area of the leaves because of fungi
and invertebrate activity. In their study, major bacterial genera
associated with stream water and oats and hickory leaves were
Flavobacteria, Flexibacter, Pseudomonas, Acinetobacter, Achromobacter
and *Chromobacterium*.

Based on laboratory and field observations Kaushik and Hynes
(1971) concluded that the absolute quantity of nitrogen and protein
contents of leaves increases during decomposition mainly because
of fungal colonization. This was especially true if the water
was enriched with nitrogen and phosphorus sources. Similar
observations have been made by Howarth and Fisher (1976) and
Suberkropp et al. (1976). Suberkropp et al. (1976) found that in
both hickory and oak leaves the losses in hemicellulose and
cellulose nearly paralleled loss in weight and neither type of
leaf showed an increase in the absolute quantity of these
constituents. In contrast the content of lignin increased in
absolute quantity during the first 6-12 weeks of processing.
Changes in leaf phosphorus content have been reported only recently
by Meyer (1980). She noted that phosphorus content of leaves
increased during decomposition and the increase was more rapid in
the leaves placed in depositional sections of the stream. She
also regarded the assimilation of P by microorganisms decomposing
leaves as an important sink for dissolved P.

Invertebrate processing of CPOM and its utilization as a food source are largely mediated through microbes. Before discussing this interaction in more detail, we shall focus on the importance of invertebrates and other factors in leaf processing. The role of stream invertebrates in leaf litter processing has been the subject of two recent reviews (Andersen and Sedell, 1979; Cummins and Klug, 1979). Stream invertebrates have been classified into functional groups (shredders, collectors, scrapers, piercers, and predators) based on their mode of feeding as opposed to taxonomic units or food eaten (Cummins, 1974; Cummins and Klug, 1979). In the processing of litter it is only the shredders, collectors and scrapers that concern us.

Shredders tear up leaf material, while collectors scrape softer tissue leaving the veins intact and grazers may crop only the microflora colonizing the leaves. The act of feeding (and moving about over the leaf surface) results in a reduction in particle size, which is important in the degradation of CPOM. Therefore, invertebrates function in processing leaf material (large CPOM) into animal biomass, smaller CPOM, fine particulate organic matter (FPOM), and faecal material.

Leaf processing may also be affected by experimental techniques. In the field, leaf processing has been studied by using leaf packs, either in a mesh bag or without a bag, tied to the upstream side of a brick. Leaf packs without bags are considered more similar to natural conditions though fragments of leaves can be lost without having undergone decomposition. Bags, in contrast, can reduce processing rates by impeding water-flow through the leaf pack reducing physical abrasion and by blocking the entry of shredders and other invertebrates if the mesh-size is not large enough. In studies investigating the role of stream invertebrates in leaf processing, bags also have the disadvantage that it is not possible to distinguish between animals associated with the leaves from those associated and the bags (Iversen, 1975; Winterbourn, 1978). However, Winterbourn found 2-3 times more shredders and grazers in bags with leaves than empty bags or bags with plastic squares. In a study of confined and naturally entrained leaf litter in a woodland stream, Cummins et al. (1980) found that basswood leaves naturally entrained and in leaf packs had similar processing rates in riffle areas. However, leaves in mesh bags (1 mm mesh size) were processed much more slowly at rates typical of depositional areas, pools and alcoves. Loose leaves also shift from erosional areas of rapid processing to depositional areas of slow processing with changes in discharge (Cummins et al., 1980; Karlström, 1978), something that anchored leaf packs have less tendency to do. Thus, confined leaf packs with or without mesh bags, estimate the processing capacity of a stream but do not truly reflect the processing rates of loose leaves. Processing rates are also

dependent on the size of the leaf pack, larger leaf packs
generally having lower processing rates (Reice, 1974; Benfield
et al., 1979).

Although Benfield et al. (1977), Kaushik and Hynes (1971),
Mathews and Kowalczewski (1969), Meyer (1980), and Reice (1977,
1978, 1980) found no significant difference in leaf processing
rates in the presence of invertebrates, faster rates have been
associated with greater numbers and biomass of invertebrates per
gram leaf pack (Anderson and Sedell, 1979; Hart and Howmiller,
1975; Iversen, 1975; Sedell et al., 1975). Collectors are the
most numerous group of invertebrates in leaf packs, but it is the
shredders that generally account for most of the leaf weight loss,
about 21% (Petersen and Cummins, 1974). Hence, the processing rate
is dependent on rate of conditioning (governed by the leaf species)
and the intensity of attack by invertebrates. Naturally, this will
vary in different streams because of differences in faunal
composition; the faunal composition being in tune with its energy
source. Shredders also enhance the availability of nutrient to
collectors. This has been demonstrated by Short and Maslin (1977),
who found that the collectors *Hydropsyche californica* and
Simulium arcticum accrue significantly greater amounts of ^{32}P in
the presence of a shredder, *Pteronarcys californica*.

Invertebrate processing of leaf packs in eastern woodland
streams of North America is mainly by a few large shredders
(Tipulidae and Limnephilidae) (Petersen and Cummins, 1974), whereas
that in a western mountain river is by a large number of small-
sized shredders (the stoneflies *Capnia* and *Zapada*) (Short et al.,
1980). These small-size shredders, unlike the larger ones which
bite off chunks of leaf material, apparently skeletonize the leaves
by scraping much the same as collectors and grazers (Anderson and
Sedell, 1979). Since collectors are much more numerous than other
functional groups in leaf packs their role in the degradation of
CPOM should not be overlooked. Their abundance in leaf packs is
directly related to the amount of FPOM (75-100 μm) trapped within
the leaf pack (Short et al., 1980). Since the growth rate of at
least one collector, *Paratendipes albimanus* (Meigen) is greater on
a diet of ground conditioned leaves, than on tipulid faeces, or on
natural stream FPOM (Ward and Cummins, 1979), collectors may stand
to gain a great deal nutritionally by the colonization of leaf packs.
Short and Ward (1980) investigated leaf pack processing below an
impoundment and in an unregulated river in order to test the
hypothesis that a reduction in shredders in the regulated river
would decrease the processing rate. To their surprise, processing
was faster in the regulated river (k = 0.0462) than in the
unregulated river (k = 0.0235). This they attributed to the higher
winter temperatures below the impoundment enhancing microbial
processing. In their study, similar numbers of invertebrates were
found per leaf pack in the two rivers but *Ephemerella infrequens,*

a collector, was much more abundant in the regulated river, whereas
the shredder biomass was ten times greater in the unregulated river
samples. Thus more palatable leaf material in the regulated river
due to enhanced microbial growth may have allowed greater consumption
of leaf material by *E. infrequens* than by the small-sized shredders
in the unregulated river. Anderson (unpublished in Anderson and
Sedell, 1979) attributed rapid processing of alder leaves in mid-
summer to both increased microbial activity at higher temperatures
and the feeding activity of collectors and scrapers. Faster
processing has been reported at warmer temperatures (Paul et al.,
1978; Reice, 1974; Short and Ward, 1980); however, rapid processing
rates have also been reported at colder temperatures (Reice, 1977;
Short et al., 1980; Triska and Sedell, 1976).

During incubation of leaves in streams the numbers of
invertebrates per gram of leaf material increases with time
(Petersen and Cummins, 1974; Sedell et al., 1975; Davis and
Winterbourn, 1977). This has been interpreted as a positive
response to a continual increase in microbial conditioning. However,
after initial colonization, the population of invertebrates per
leaf pack may remain the same or may actually decrease (Davis and
Winterbourn, 1977; Winterbourn, 1978). Thus the increase in
numbers of invertebrates per unit weight is caused by the decrease
in leaf weight, not by an actual increase in colonization. For
this reason they suggest that the population density of invertebrates
should be expressed per leaf pack or as a function of the surface
area available for colonization. Their findings are supported by
Reice (1980) who found that the number of animals in leaf packs,
compared with those on the substratum, decreased as the leaf packs
decomposed thus decreasing in weight and increasing in inedible
material (stems and veins).

Other important factors influencing the species-specific
processing rates of litter in streams are hydrological factors
(discharge and current velocity), geomorphology and temperature.
These are closely interrelated. Discharge and current velocity
to a large extent determine the morphology of the stream bed
(erosional, depositional), which in turn, together with temperature,
strongly influences the faunal composition; hence leaf processing
rates. Hydrological and geomorphological factors may also
function independently of biological factors in increasing the
processing rates by physical abrasion. Spates are very important
in redistributing OM (import and export of OM) (e.g. Cummins et
al., 1980; Dance, Chapter 3, Kärlstrom, 1978) although water
velocity is reported to have no direct effect on leaf processing
rates (Reice, 1977). Reice (1974) found that processing rates
were lowest on silt substrates, slightly higher on sand, and
highest on gravel and rock. He interpreted this as an increase in
processing rates with an increase in habitat heterogenity, the
community diversity showing the same trend. Anaerobic conditions

may also have played a role in inhibiting leaf processing on the
silt substrate (Reice, 1974). Similarly, Meyer (1980) and Cummins
et al., (1980) found that decay rates were slowest in areas of
sediment deposition and most rapid in erosional areas. Within
erosional areas, Reice (1980) found size made no difference in
leaf processing rates on stones of 1.0, 2.5 and 8.5 cm diameter,
although it has been demonstrated that the size of the substrate
is important in retaining various sizes of detritus particles,
which in turn affects the microdistribution of benthic insects
(Rabeni and Minshall, 1977) e.g. 1.0-2.0 cm substratum collected
more small-size detritus particles (<3.95 mm) and was colonized by
more insects than larger-size substratum.

So far we have discussed the factors that have a more direct
impact on leaf processing. We shall now return to the influence
of microbes and their interaction with invertebrates in the
processing of leaves – an aspect alluded to earlier and reviewed
recently by Bärlocher and Kendrick (1976, 1981) and Anderson and
Sedell (1979). It is known that different leaf species decay at
different rates (Table 3) and this has been attributed to
differences in chemical composition of leaves (Suberkropp et al.,
1976; Triska et al., 1975). Thus some types of leaf (e.g. elm or
maple) are colonized by microbes much faster than are others (e.g.
oak and beech). Given a choice, invertebrates preferentially
feed on leaves in the same order as their rates of decay, e.g. elm
> maple > alder = oak > beech (Kaushik and Hynes, 1971). Also it
has been shown (Fig. 2) that invertebrates prefer leaves with
microbial growth (conditioned) over those that are without microbes

Table 3.   Comparison of some decay coefficients (k) of leaf
           species in streams (from Petersen and Cummins, 1974)
           with that of macrophytes in the laboratory at about
           18°C (Jewell, 1971).

Leaf species	K	Macrophytes*	K
Fagus grandifolia	0.0025	*Callitriche* sp.	0.052
Quercus alba	0.0038	*Potamogeton* sp.	0.067
Populus tremuloides	0.0046	*Elodea canadensis*	0.087
Alnus glutinosa	0.0075	*Rorippa* sp.	0.18
Carya glabra	0.0089		
Acer saccharum	0.0107		
Betula lutea	0.0116		

* even faster rates of decay would be expected in streams at
  comparable temperatures

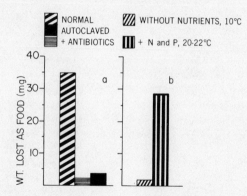

Figure 2. Quantity of elm leaves consumed by *Gammarus* when
provided with (a) elm leaves with or without microbial
growth - microbial growth inhibited either by
autoclaving or by using antibiotics or (b) elm
leaves with different degrees of microbial growth
(from Kaushik and Hynes, 1971).

(Anderson and Graficus, 1975; Bärlocher and Kendrick, 1973a, b;
Cummins et al., 1973; Kaushik and Hynes, 1968, 1971; Lautenschlager,
1976; Mackay and Kalff, 1973; Sedell et al., 1975) and that an
increase in the duration of conditioning time increases ingestion
of leaf material (Anderson and Grafius, 1975; Lautenschlager et al.,
1979). However, if a choice is not available invertebrates do feed
on leaves that are normally not preferred or that have not been
fully conditioned. For example, *Gammarus*, when given a choice
between only oak and beech, did consume oak leaves (Kaushik and
Hynes, 1971). Similarly, Haeckel et al. (1971) noticed that in
stream regions receiving alder, poplar and beech, *Gammarus*
skeletonized alder first while beech leaves remained untouched, but
in a stream reach receiving mainly beech leaves, their breakdown,
after the same duration of time, was much further advanced.

The importance of microbes, especially fungi, lies not only in
the obvious acceleration of processing rates of leaves by
invertebrates but also in making the bulk of leaf tissue,
cellulose and lignin, available to the invertebrates by converting
leaf tissue into microbial biomass and by partially decomposing it
into subunits digestible by detritus feeders (Willoughby, 1974).
This probably applies not only to those organisms that cannot
digest cellulose and lignin because they lack gut microbial
populations (Willoughby, 1974), but also to *Gammarus* which has the
digestive enzyme cellulase (Monk, 1976) and those insects, mainly

shredders and collectors, which have attached microbial populations
in their guts (Cummins and Klug, 1979).

Bärlocher and Kendrick (1973a, b) found that *Gammarus*
consumed ten times more all-leaf diet (by weight) than all-fungal
diet, but exhibited the greatest weight gain on the latter. Also
Gammarus has a preference for certain fungal species which
influence its choice of food e.g. *Gammarus* preferred oak, a species
not readily accepted, when bearing *Anguillospora* to other leaf
species bearing *Tetracladium*, the least preferred fungal species.
These authors also found that the assimilation efficiency of
Gammarus was much higher (42.6-75.6%) when fed fungal mycelium as
opposed to leaves (10%). Similar findings are reported for *Hyalella*
azteca (Hargrave, 1970) and *Asellus aquaticus* (Marcus and Willoughby,
1978; Rossi and Vitagliano Tadini, 1978). However, Willoughby and
Marcus (1979) reported that growth of *A. aquaticus* was poorer on
the fungus, *Streptomyces* (2.74% day^{-1}) than on faecal matter
(3.4% day^{-1}), *Elodea* (6.06% day^{-1}) or decaying oak leaves (6.81%
day^{-1}). More recent work by Rossi and Fano (1979) demonstrated that
not only do the isopods *Asellus aquaticus* and *A. coxalis* do better
on a fungal diet than on leaves but these two isopods utilize the
fungi differently. For example, *A. aquaticus* had greater survival
on 5 of 8 fungi than on leaves, whereas *A. coxalis* had greater
survival on only 2 of 8 fungi. Survival and growth of *Gammarus pulex*,
however, was much poorer on the fungi *Clavariopsis* and *Tricladium*,
than on decaying elm and oak leaves (Willoughby and Sutcliffe, 1976).
From the above it seems that fungi not only improve the food quality
of OM but may be a complete food source for certain invertebrates
and the different invertebrates vary in their ability to utilize
various fungal species.

Bacterial growth like that of fungal growth, increases the
food quality of decaying leaves but their importance in the diet
of stream detritivores has not received the attention it deserves.
Although Gram-positive bacteria are more resistant to proteolytic
enzymes and the action of alkali than are Gram-negative bacteria
(Burrows, 1959), Calow (1974) found that the gastropod *Planorbis*
contorus L. preferred Gram-positive cocci which were the dominant
bacteria on detritus in his study. Other work has demonstrated
that *Simulium* and *Chironomus* digest at least half of the bacteria
that they ingest *in situ* and that they do not appear to be
selective with regard to bacterial type in their digestion (Baker
and Bradman, 1976). These authors found no evidence for digestion
of bacteria by *Baetis* or by *Ephemerella*. In the laboratory,
Simulium can be reared on a culture of bacteria (Fredeen, 1964).

Based on a few investigations that lend themselves to the
calculation of microbial biomass on decomposed leaves, Bärlocher
and Kendrick (1981) suggested that on maple leaves the average
density is about 2% the dry weight while the maximum density, in

small heavily colonized areas is about 4%. On the basis of ATP
data reported by Suberkropp et al. (1976) Bärlocher and Kendrick
(1981) calculated that living microbial cells would vary between
5.9% and 11.8% of the dry weight of hickory leaves and between
1.6% and 3.3% on oak leaves.

FAECAL MATERIAL

Most of the organic fraction of river sediments consists of
fragments of plant material and faeces, which together represent
a large reservoir of food for many aquatic animals (Hynes, 1970).
Copious amounts of faecal material are continuously produced by
stream invertebrates (Table 4) because of their low assimilation
efficiencies of 5-30% and rapid gastric evacuation rate of 30
minutes to 2 hours (Bärlocher and Kendrick, 1975; Hargrave, 1970;
Ladle et al., 1972; Lautenschlager, 1976; Prus, 1971; Sedell, 1972).
Vertebrates also contribute a substantial amount of faeces to the
sediments. Winberg (1956) suggested that about 15% of the daily
energy intake of fish is lost as faeces; whereas Elliott (1976)
reported a value of 11-31% for brown trout and 2-28.2% for other
carnivorous fish. In fish the percentage loss of the daily energy
intake in the faeces increases as the ration level increases and
the temperature decreases (Elliott, 1976). In aquatic invertebrates
the ingestion rate varies inversely with the organic content of
the food and as a function of body size (Cammen, 1980), faecal
production probably following a similar trend. Both the quality
of FPOM and faeces produced by *Gammarus* increases when it is fed
leaf material in a more advanced state of decomposition
(Lautenschlager et al., 1979).

Coprophagy has been demonstrated in young *Gammarus*, in *Asellus*,
and in mayflies *Habroleptoides modesta* and *Habrophlebia lauta* (Hynes,
1970) and more recently in blackfly larvae, *Simulium*, which
aggregate in large numbers to facilitate coprophagy (Wotton, 1980).
Coprophagy may increase the utilization of refractory material and,
in the case of blackfly larvae, also increase the economy of
filtering. That is, each whole faecal pellet captured and ingested
was, in terms of number of bacteria consumed, equivalent to 23-100
ml of river water filtered.

Based on a study of a variety of invertebrates, Ladle and
Griffiths (1980) noted that faecal pellets are shaped differently
and that pellets may be compact and cohesive (from snails and
dipteran larvae) or friable and liable to disintegrate easily
(from caddisflies). They noted that *Asellus aquaticus* has faecal
pellets with a longitudinal groove, a feature also noted for the
faeces of *G. lacustris limnaeus* (Lautenschlager et al., 1978).
Faecal pellets of *Gammarus* also possess a thin, tight-fitting
peritrophic membrane which is lost about seven hours after
egestion but the cylindrical pellets, up to 0.3 mm in diameter by

Table 4. Faecal production of some stream invertebrates.

Species	Faecal production (mg mg^{-1}day^{-1})	Conditions	Authority
Heteroplectron californicum	0.15-2.15	Laboratory, 15.6°C, variety of food--wood to leaves.	Anderson et al., 1978
Lara avara	0.41	Stream, on wood.	Anderson et al., 1978
Oxytrema silicula	0.20	Stream, on wood.	Anderson et al., 1978
Clistoronia magnifica	0.27-0.46	Laboratory, 15°C, on *Alnus* leaves, and leaves + wheat grains.	Anderson & Cummins, 1979
Pteronarcys sp.	0.18	Laboratory, 5,10,15°C.	McDiffett, 1970
Hexagenia limbata	1.35	Laboratory, 20°C.	Zimmermann & Wissing, 1978
H. limbata	0.69	Laboratory, 5°C.	
Potamopyrgus jenkins	1.20	Laboratory, 20°C.	Heywood & Edwards, 1962
Asellus aquaticus	0.05	Males, laboratory, 10°C.	Prus, 1971
A. aquaticus	0.04	Females, laboratory, 10°C.	Prus, 1971
A. aquaticus	0.02	Ovigerous females, laboratory 10°C.	Prus, 1971

1.0-1.5 mm in length and weighing 19-28 µg, retain their compact
shape after the loss of peritrophic membrane (Lautenschläger et al.,
1978). Faecal material produced by shredders is in the size range
of 0.075 to 1.00 mm (Cummins, 1974) whereas the diameter of faecal
pellets from a variety of invertebrates ranged from 0.10 to 0.48 mm
(Ladle and Griffiths, 1980). These latter authors considered the
production of compact spherical faecal pellets by *Simulium*, which
sink quickly, a means by which the faeces are removed from the
water that the larvae are filtering.

Faecal pellets, especially after microbial growth, represent
a nutritious food source. Protein content of the faeces of *Gammarus*
was much higher than that of decomposed maple leaves used as food
presumably because of selective feeding on the softer leaf tissue
(Lautenschlager, 1976). A higher protein content in faeces may
also result from the more efficient utilization of carbohydrates
and lipids than protein (Prus, 1976) or the addition of microorganisms from the hindgut to the faeces (Johannes and Satomi, 1966).
Cummins and Klug (1979) recorded the presence of bacteria in the gut
of a number of invertebrates, so it is possible then that faeces are
colonized by bacteria before egestion. The subsequent rate of
colonization depends in part on the size of the pellet and the time
it remains intact; friable and particulate faeces are presumably
more readily colonized than firm cohesive pellets (Hargrave, 1972).
During microbial colonization, the nitrogen content of the faecal
pellets increases (Lautenschlager, 1976).

Despite the fact that faecal pellets represent high quality
packages of food and several animals have been shown to be
coprophagous, feeding studies in the laboratory have found that
many animals have faster growth rates on a diet of conditioned
leaves or fungi than faeces (Rossi and Fano, 1979; Ward and Cummins,
1979; Willoughby and Marcus, 1979). Even so, faecal material is
no doubt an important food source for stream invertebrates both
quantitatively and qualitatively. A turnover of fine sediments
(< 63 µm) in 71-102 days by defecation by Simuliidae and *Gammarus*
pulex and reuse by Tubificidae reported by Ladle and Griffiths
(1980) is suggestive of the importance of faecal matter in
streams.

PROCESSING OF WOOD

Large POM such as bark, branches and logs are also important
contributors of litter energy and biomass to streams. For example,
Blackburn and Petr (1980) reported that bark and branches
contributed 62% (1,759 kcal m^{-2} y^{-1}) of the litter energy and 62%
of the relative biomass to an Australian stream. In comparison,
leaves contributed slightly more energy (29%) than their relative
biomass (25%). In Bear Brook, New Hampshire, the input of branches
accounted for only 9% of the annual energy budget (Fisher and

Likens, 1973) but 46% of the standing crop of particulate detritus
in summer. This is because of the refractory nature of bark,
branches and logs, the time for their decomposition being
measured in years (e.g. 1 year for leaves and twigs, 4.2 years for
branches (Fisher and Likens, 1973), 5.5 years for 1 cm diameter
twigs (Buckley, in Anderson et al., 1978), 10-20 years for wood
about 2 cm in diameter (Anderson et al., 1978), about 114 years
for 7.5 cm diameter logs (Hodkinson, 1975) and decades or
centuries for larger logs (Swanson et al., 1976).

     Initially submerged wood is a poor quality food source. It
exhibits a very high C/N ratio (235-1343, Buckley and Triska, 1975),
however, with microbial conditioning, especially by fungi
(Willoughby and Archer, 1973; Lamore and Goos, 1978) and nitrogen-
fixing bacteria (Buckley and Triska, 1978) the outer surface of
the wood becomes more palatable and the C/N ratio decreases. Some
of the invertebrates that colonize submerged wood seem to occur
only on that substratum, e.g. *Macronychus* (Hynes, 1970), whereas
others are common members of the stream fauna (e.g. collectors and
grazers). The latter may use the wood as a substrate for attachment (e.g. filter feeders such as Simuliidae and net-spinning
caddis larvae), or graze upon the surface flora. Hynes (1970)
reported that a number of animals, for example, caddisflies
(Limnephilidae), some Elminthidae, the stonefly *Pteronarcys*, and
species of *Polypedilum* and *Brillia* (Chironomidae) feed on water-
soaked wood. The fauna associated with wood may be diverse but
its biomass (mg kg^{-1}) is low and about two orders of magnitude less
than that on leaves. Anderson et al. (1978) found that the biomass
of xylophagus species (e.g. the caddisfly *Heteroplectron californicum*,
the elmid beetle *Lara avara*, and the snail *Oxytrema silicula*) was
2.5-5.5 times less than that of other species which colonized wood.
Although only a very small portion of the submerged wood is
processed by invertebrates per year, Anderson et al. (1978)
suggested that their impact on the processing of wood over the full
decomposition cycle may be similar to that for leaf litter (10-20%
of the total decomposition).

     Adaptations of invertebrates for exploitation of wood as a
food source are: 1) long life cycles with low metabolic activity
(e.g. the elmid *Lara avara*), 2) high ingestion rates to provide
sufficient assimilative gut contact with the microflora to meet
the animal's nutrient requirements (e.g. the caddisfly *H.
californicum*), 3) supplementary feeding on higher quality food
(e.g. grazers and collectors) or 4) some combination of the above
(Anderson and Cummins, 1979). Although a large portion of the
energy and biomass of POM in many streams may be in the form of
wood, very few species have the adaptations needed to utilize it
directly and its refractory nature renders wood a much less
important source of food than leaf litter and perhaps autochthonous
production. The major importance of wood in streams is in providing

cover for fish, a substratum for stream invertebrates and in shaping
the stream channel.

MACROPHYTES AND OTHER VEGETATION

Emergent macrophytes in streams that drain marshes as well as
submerged macrophytes in streams rich in nutrients but devoid of
riparian vegetation and its shading effect (e.g. those draining
agricultural areas) may contribute substantially to CPOM after
their senescence (see Minshall, 1978). Emergent macrophytes,
because of their erect growth and lack of water support for the
aerial part of the plant, require more structural constituents
(e.g. hemicellulose, cellulose, and lignin) and are more resistant
to decomposition than are submerged plants. Faster decomposing
plants generally have less total fibre content, higher initial N
concentrations and lower C/N ratios. Godshalk and Wetzel (1978)
and Jewell (1971) reported that submerged macrophytes, because of
their lower content of refractory material, decay faster and more
completely than do algae. However, some emergent macrophytes such
as *Rorippa nasturtium-aquaticum* may decompose faster (< 2 weeks)
than submerged species such as *Ranunculus caleareus* (2-3 months)
which in turn decayed faster than *Salix viminalis* and *Fraxinus
excelsior* leaves (4-6 months) (Dawson, 1980) (see Table 3).

Mason and Bryant (1975) demonstrated that neither *Typha* nor
Phragmites conserve their nutrients at senescene as there was no
increase in the nutrient content of the root-rhizomes. Dead
standing shoots of these plants, while still out of water, showed
a loss of about 83% of their initial weight by springtime, most of
which occurred soon after senescence. When placed in water they
lost a further 10-20% of their remaining weight in the first month,
primarily because of leaching, but thereafter weight loss was at a
fairly constant but slower rate. Mason and Bryant (1975) also
noted that processing rates were 10% faster in the presence of
animal activity, mainly of chironomids and *Asellus*, and that
Phragmites required 200 days for 50% weight loss whereas the
corresponding time for *Typha* was one year. Puriveth (1980) found
that the initial losses to leaching were greater in the root-
rhizome than in the shoots and processing rates were accelerated
at summer temperatures. Over a period of 348 days the shoot litter
of *Sparganium, Typha,* and *Scirpus* showed weight losses of 73, 53,
and 49%, respectively, whereas the corresponding values for root-
rhizomes were 58, 41, and 72%.

Mason and Bryant (1975) demonstrated that the feeding
activity of the snail, *Lymnaea peregra*, significantly increased
the respiration rate of *Typha* litter, whereas that of *A. aquaticus*
and *G. pulex* did not significantly increase the respiration rate
of *Phragmites* litter. They attributed this to a difference in
their mode of feeding: *L. peregra* fed by rasping across the whole

surface of the tissue, thus increasing its surface area for
microbial colonization. In contrast, *Gammarus* and *Asellus* fed by
shredding the broken ends of the litter. These animals, when fed
sterile CPOM (> 0.5 mm), exerted little effect (1-4% more than
controls) on the comminution of the CPOM to particle sizes of
less than 0.5 mm over a period of 120 days. Recently, Smock and
Stoneburner (1980) found that the macrophyte *Nelumbo lutea*, after
its senescence, is an important source of food for a chironomid,
Polypedilum nymphaeorum, and for naidid oligochaetes.

The contribution of other riparian vegetation such as
grasses, forbs and ferns to CPOM in streams, like that of
macrophytes, has received little attention. Mundie et al. (1973)
demonstrated that hay is a poorer quality source of food than
alder leaves or cereal grain and Otto (1975) found that the
caddisfly *Potamophylax cingulatus* does not feed upon the herbs
Epilobium hirsutum and *Mentha aquatica*. However, Bärlocher et al.
(1978b) reported that grasses, ferns, forbs and deciduous leaves,
once conditioned, are fed upon by shredders in a temporary vernal
pool and that at least one caddisfly had a preference for young
green blades of grass.

PATTERNS OF DECAY

The decay pattern of most macrophytes and leaf species is
best described by the exponential decay model (e.g. Howard-
Williams and Davies, 1979; Petersen and Cummins, 1974):

$$Yt = Yoe^{-kt}$$

where Yt is the mass remaining at time t, Yo is the initial mass
and k is the rate constant. This assumes that for any amount of
material at any time there is a constant fractional loss. For
example, the macrophyte *Potamogeton pectinatus* follows an
exponential pattern of decay with k = 0.0205 per day. Decomposition
was complete after 158 days (Howard-Williams and Davies, 1979).
Although the exponential model may often give a good fit, it is no
more than a convenient form of describing the overall pattern of
decay as k may vary a great deal throughout the decomposition
cycle, e.g. Howard-Williams and Davies (1979) found that k was
three times greater at $25^{o}C$ than at $15^{o}C$. Suberkropp et al. (1976)
reported that pignut hickory and white oak follow a linear rate of
weight loss. Other plants exhibit a diphasic pattern of decay
(Boyd, 1971; Howard-Williams and Howard-Williams, 1978; Howard-
Williams and Junk, 1976). For example, the initial phase of rapid
decomposition of the rush *J. effusus* was attributed to both
leaching and losses of plant fragments, especially the pith of the
shoots. The remaining fibrous cortex and vascular bundles
decomposed at a much slower rate (Boyd, 1971).

In general, the higher the nitrogen content or the lower the

carbon-nitrogen (C/N) ratio, of litter the faster its rate of
decomposition (Hynes and Kaushik, 1969; Godshalk and Wetzel,
1978), although other factors such as the structural composition
of the litter must also be considered. For example, the rate of
decay is in the order macrophytes (C/N of 12-25 (Godshalk and
Wetzel, 1978)) > leaves (14-85 (Triska et al., 1975; Ward and
Cummins, 1979)) > Typha (57 (Puriveth, 1989)) > wood (235-1343
(Buckley and Triska, 1978)). Carbon-nitrogen ratios have also
commonly been used as an index of palatability, a C/N ratio of 17,
or lower, apparently is required by most animals in order to
obtain an adequate source of N (Russell-Hunter, 1970). C/N ratios
tend to decrease during the decomposition of plant material. This
is partially because of a loss of soluble carbon during leaching,
which may result in an increase in the precent N in the remaining
litter and also to an increase in N with microbial growth as well
as an increase in non-protein nitrogen containing compounds (Odum
et al., 1979; Puriveth, 1980; Suberkropp et al., 1976). Although
the latter may help lower the C/N ratio, such compounds do little
to increase the palatibility of the latter to invertebrates
because of their refractory nature. This has led Odum et al. (1979)
to question the use of C/N ratios as an indicator of food quality.
In fact, ATP content and substrate respiration have been found to
be better indicators of food quality than C or N content (Ward
and Cummins, 1979).

CONCLUSION

     From the foregoing as well as from an earlier review
(Anderson and Sedell, 1979), it is evident that during the last
decade or so tremendous progress has been made in the understanding
of principles and factors that dictate processing and utilization
of CPOM, especially of allochthonous origin, in streams. In fact,
as pointed out by Anderson and Sedell (1979) it seems that
biologists did follow the guidance of Hynes, who had the
foresight to point out (Hynes 1963, 1970) the significance and the
pivotal role of allochthonous organic matter in streams.
Although many fundamental questions relating to CPOM have been
elucidated or touched upon, many still remain unanswered. Such
areas as factors affecting food quality of detritus and its impact
on invertebrate populations and dynamics of microbial populations
and processes have been suggested (Anderson and Sedell, 1979) for
future research. Bärlocher and Kendrick (1981) indicate lack of
research on the role of individual fungal species in leaf
decomposition and the effect of invertebrate feeding on dynamics
of fungal populations. Our unpublished results indicate that leaf
degradation and their preference by *Gammarus* is greatly reduced in
streams affected by acid rain. Such an impact, can drastically
affect the food and thus eliminate even those detritivores that
can tolerate low pH values. This aspect needs further investigation.
That CPOM, especially of allochthonous origin, is an important food

source for invertebrates is now well recognized. Intuitively it
should be obvious that this energy source governs the activities
of heterotrophic microbes that may be involved in nutrient
transformations in streams. We (Kaushik et al. 1979) have shown
that the rate of denitrification in stream sediment is enhanced
with the addition of leaf material as an energy source. Such
areas of research still need to be investigated. Similarly, compared to the studies on leaves, much less is known about the
nature, decomposition and utilization of faecal pellets, wood,
macrophytes and also riparian herbaceous plants that may enter
streams after senescence.

It is also about time that we start documenting the role of
CPOM from the point of view of stream management. Not only
should the impact of the addition or removal (e.g. by
deforestation) of CPOM be studied but we should also study the
effect of quality of input (e.g. CPOM from forests under
monoculture of conifers as against that from mixed forest) on
stream productivity. Preliminary studies (Mundie et al., 1973)
have shown the magnitude of impact that the quality of detritus
can have on the production of benthos. Can such studies be fully
investigated and then made use of to enhance fish production?

ACKNOWLEDGEMENTS

We would like to express our special thanks to D. Barton and
J.B. Robinson for their constructive comments, and to H. Daniecki
and W. McGavin for typing the initial manuscript. We are also
grateful to the Natural Sciences and Engineering Research Council
of Canada for providing funds for research on carbon cycling in
streams.

REFERENCES

Anderson, N.H. and Cummins, K.W. 1979. Influences of diet on the
    life histories of aquatic insects. *J. Fish. Res. Bd. Can.*
    36: 335-342.
Anderson, N.H. and Grafius, E. 1975. Utilization and processing
    of allochthonous material by stream Trichoptera. *Verh. Int.
    Ver. Limnol.* 19: 3083-3088.
Anderson, N.H. and Sedell, J.R. 1979. Detritus processing by
    macroinvertebrates in stream ecosystems. *Ann. Rev. Entomol.*
    24: 351-377.
Anderson, N.H., Sedell, J.R., Roberts, L.M. and Triska, F.J. 1978.
    The role of aquatic invertebrates in processing of wood
    debris in a coniferous forest stream. *Amer. Midl. Nat.*
    100: 64-82.
Bagnall, R.G. 1972. The dry weight and calorific value of litter
    fall in a New Zealand *Nothofagus* forest. *New Zealand J. Bot.*
    10: 27-36.

Baker, J.H. and Bradnam, L.A. 1976. The role of bacteria in the nutrition of aquatic detritivores. *Oecologia* 24: 95-104.

Bärlocher, F. and Kendrick, B. 1973a. Fungi in the diet of *Gammarus pseudolimnaeus*. *Oikos* 24: 295-300.

Bärlocher, F. and Kendrick, B. 1973b. Fungi and food preference of *Gammarus pseudolimnaeus*. *Arch. Hydrobiol.* 72: 501-516.

Bärlocher, F. and Kendrick, B. 1975. Assimilation efficiency of *Gammarus pseudolimnaeus* (Amphipoda) feeding on fungal mycelium or autumn-shed leaves. *Oikos* 26: 55-59.

Bärlocher, F. and Kendrick, B. 1976. Hyphomycetes as intermediaries of energy flow in streams. *In* Recent advances in aquatic mycology. E.B. Gareth Jones, ed. Paul Elek, London.

Bärlocher, F. and Kendrick, B. 1981. The role of aquatic hyphomycetes in the trophic structure of streams. *In* The fungal community, its organization and role in the ecosystem· D.T. Wicklow and G.C. Caroll. eds. Marcel Dekkes Inc.. New York.

Bärlocher, F., Kendrick, B. and Michaelides, J. 1978a. Colonization and conditioning of *Pinus resinosa* needles by aquatic hyphomycetes. *Arch. Hydrobiol.* 81: 462-474.

Bärlocher, F., Mackay, R.J. and Wiggins, G.B. 1978b. Detritus processing in a temporary vernal pool in Southern Ontario. *Arch. Hydrobiol.* 81: 269-295.

Bärlocher, F. and Oertli, J.J. 1978. Inhibitors of aquatic hyphomycetes in dead conifer needles. *Mycologia* 70: 964-974.

Bärlocher, F., Oertli, J.J. and Guggenheim, R. 1979. Accelerated loss of antifungal inhibitors from *Pinus leucodermis* needles. *Trans. Br. mycol. Soc.* 72: 277-281.

Benfield, E.F., Jones, D.S. and Patterson, M.F. 1977. Leaf pack processing in a pastureland stream. *Oikos* 29: 99-103.

Benfield, E.F., Paul, R.W. Jr. and Webster, J.R. 1979. Influence of exposure technique on leaf breakdown rates in streams. *Oikos* 33: 386-391.

Blackburn, W.M. and Petr, T. 1979. Forest litter decomposition and benthos in a mountain stream in Victoria, Australia. *Arch. Hydrobiol.* 86: 453-498.

Boyd, C.E. 1971. The dynamics of dry matter and chemical substances in a *Juncus effusus* population. *Am. midl. Nat.* 86: 28-45.

Buckley, B.M. and Triska, F.J. 1978. Presence and ecological role of nitrogen-fixing bacteria associated with wood decay in streams. *Verh. Internat. Verein. Limnol.* 20: 1333-1339.

Burrows, W. 1959. Microbiology. W.B. Saunders Co., Philadelphia.

Calow, P. 1974. Evidence for bacterial feeding in *Planorbis contortus* Linn. (Gastropoda:Pulmonata). *Proc. malac. Soc. Lond.* 40: 145-156.

Cammen, L.M. 1980. Ingestion rate: an empirical model for aquatic deposit feeders and detritivores. *Oecologia.* 44: 303-310.

Cummins, K.W. 1974. Structure and function of stream ecosystems. *BioScience* 24: 631-641.

Cummins, K.W. and Klug, M.J. 1979. Feeding ecology of stream
    invertebrates. *Ann. Rev. Ecol. Syst.* 10: 147-172.
Cummins, K.W., Petersen, R.C., Howard, F.O., Wuycheck, J.C. and
    Holt, V.I. 1973. The utilization of leaf litter by stream
    detritivores. *Ecology* 54: 336-345.
Cummins, K.W., Spengler, G.L., Ward, G.M., Speaker, R.M., Ovink,
    R.W., Mahan, D.C. and Mattingly, R.L. 1980. Processing of
    confined and naturally entrained leaf litter in a woodland
    stream ecosystem. *Limnol. Oceanogr.* 25: 952-957.
Davis, S.F. and Winterbourn, M.J. 1977. Breakdown and
    colonization of *Nothofagus* leaves in a New Zealand stream.
    Oikos 28: 250-255.
Dawson, F.H. 1976. Organic contribution of stream edge forest
    litter fall to the chalk stream ecosystem. *Oikos* 27: 13-18.
Dawson, F.H. 1980. The origin, composition and downstream
    transport of plant material in a small chalk stream.
    Freshwat. Biol. 10: 419-435.
Elliott, J.M. 1976. Energy losses in the waste products of
    brown trout (*Salmo trutta* L.). *J. Anim. Ecol.* 45: 561-580.
Fisher, S.G. and Likens, G.W. 1973. Energy flow in Bear Brook,
    New Hampshire: An integrative approach to stream ecosystem
    metabolism. *Ecol. Mongr.* 43: 421-439.
Freedeen, F.J.H. 1964. Bacteria as a source of food for black-fly
    larvae in laboratory cultures and in natural streams. *Can.
    J. Zool.* 42: 527-548.
Godshalk, G.L. and Wetzel, R.G. 1978. Decomposition in the
    littoral zone of lakes. *In* Freshwater wetlands, ecological
    processes and management potential. R.E. Good, Whigham, D.F.
    and Simpson, R.L. eds. Academic Press Inc. Ltd., New York.
    131-144.
Gosz, J.R., Likens, G.E. and Bormann, F.H. 1972. Nutrient
    content of litter fall on the Hubbard Brook experimental
    forest, New Hampshire. *Ecology* 53: 769-784.
Haeckel, J.W., Meijering, M.P.D. and Rusetzki, H. 1973.
    Gammarus Koch als Fallaubzersitzer in Waldbächen.
    Freshwat. Biol. 3: 241-249.
Hall, C.A. 1972. Migration and metabolism in a temperate stream
    ecosystem. *Ecology* 53: 585-604.
Hargrave, B.T. 1970. The utilization of benthic microflora by
    Hyallela azteca. *J. Anim. Ecol.* 39: 427-437.
Hargrave, B.T. 1972. Prediction of egestion by the deposit-
    feeding amphipod *Hyallela azteca*. *Oikos*. 23: 116-124.
Hart, S.D. and Howmiller, R.P. 1975. Studies on the decomposition
    of allochthonous detritus in two Southern California streams.
    Verh. Internat. Verein. Limnol. 19: 1665-1674.
Heywood, J. and Edwards, R.W. 1962. Some aspects of the ecology
    of *Potamopyrgus jenkinsi* Smith. *J. Anim. Ecol.* 31: 239-250.
Hodkinson, I.D. 1975. Dry weight loss and chemical changes in
    vascular plant litter of terrestrial origin, occurring in a
    beaver pond ecosystem. *J. Ecol.* 63: 131-142.

Howard-Williams, C. and Davies, B.R. 1979. The rates of dry matter and nutrient loss from decomposing *Potamogeton pectinatus* in a brackish south-temperate coastal lake. *Freshwat. Biol.* 9: 13-21.

Howard-Williams, C. and Howard-Williams, W.A. 1978. Leaching of nutrients from swamp vegetation in an African lake. *Aquatic Bot.* 4: 257-267.

Howard - Williams, C. and Junk, W.V. 1976. The decomposition of aquatic macrophytes in the floating meadows of a Central Amazonian várzea lake. *Biogeographica* 7: 115-123.

Howarth, R.W. and Fisher, S.G. 1976. Carbon, nitrogen, and phosphorous dynamics during leaf decay in nutrient-enriched stream microecosystem. *Freshwat. Biol.* 6: 221-228.

Hynes, H.B.N. 1963. Imported organic matter and secondary productivity in streams. *Proc. XVI. Int. Congr. Zool. Washington.* 4: 324-329.

Hynes, H.B.N. 1970. The Ecology of Running Waters. Univ. Toronto Press. Toronto.

Hynes. H.B.N. 1975. The stream and its valley. *Verh. Internat. Verein. Limnol.* 19: 1687-1692.

Hynes, H.B.N. and Kaushik, N.K. 1969. The relationship between dissolved nutrient salts and protein production in submerged autumnal leaves. *Verh. Internat. Verein. Limnol.* 17: 95-103.

Hynes, H.B.N., Kaushik, N.K., Lock, M.A., Lush, D.L., Stocker, Z.S.J., Wallace, R.R. and Williams, D.D. 1974. Benthos and allochthonous organic matter in streams. *J. Fish. Res. Bd. Can.* 31: 545-553.

Iversen, T.M. 1973. Decomposition of autumn-shed beech leaves in a spring brook and its significance for the fauna. *Arch. Hydrobiol.* 72: 305-312.

Iversen, T.M. 1975. Disappearance of autumn-shed leaves placed in bags in small streams. *Verh. Internat. Verein. Limnol.* 19: 1687-1692.

Iversen, T.M. and Madsen, B.L. 1977. Allochthonous organic matter in streams. *Folia. Limnol. Scand.* 17: 17-20.

Johannes, R.E. and Satomi, M. 1966. Composition and nutritive value of faecal pellets of a marine crustacean. *Limnol. Oceanogr.* 11: 191-197.

Jewell, W.J. 1971. Aquatic weed decay: dissolved oxygen utilisation and nitrogen and phosphorus regeneration. *J. Water Poll. Control. Fed.* 43: 1457-1467.

Karlström, U. 1978. Environmental factors, detritus and bottom fauna in the Ricklean - a north Swedish forest river. *Report of the Institute of Limnology,* Univ. of Uppsala, Sweden.

Kaushik, N.K. and Hynes, H.B.N. 1968. Experimental study on the role of autumn-shed leaves in aquatic environments. *J. Ecol.* 56: 229-243.

Kaushik, N.K. and Hynes, H.B.N. 1971. The role of dead leaves
    that fall into streams. *Arch. Hydrobiol.* 68: 465-515.
Kaushik, N.K., Robinson, J.B. and Chatarpaul, L. 1979. Role of
    certain benthic microorganisms and invertebrates in nitrogen
    transformations in stream sediments. *Can. Spec. Publ. Fish.,*
    Aquat. Sci. 43: 21-30.
Ladle, M., Bass, J.A.B. and Jenkins, W.R. 1972. Studies on
    production and food consumption by the larval Simuliidae of
    a chalk stream. *Hydrobiologia* 39: 429-448.
Ladle, M. and Griffiths, B.S. 1980. A study on the faeces of some
    chalk stream invertebrates. *Hydrobiologia* 74: 161-171.
Lamore, B.J. and Goos, R.D. 1978. Wood-inhabiting fungi of a
    freshwater stream in Rhode Island. *Mycologia* 70: 1025-1034.
Lautenschlager, K.P. 1976. Consumption of autumn-shed leaves by
    Gammarus and decomposition of leaf-derived fecal pellets.
    M.Sc. Thesis, Univ. of Guelph, Guelph, Ontario.
Lautenschlager, K.P., Kaushik, N.K. and Robinson, J.B. 1978.
    The peritrophic membrane and fecal pellets of *Gammarus*
    lacustris limnaeus Smith. *Freshwat. Biol.* 8: 207-211.
Lautenschlager, K.P., Kaushik, N.K. and Robinson, J.B. 1979.
    A simple system for the collection of leaf-derived faecal
    pellets of aquatic invertebrates. *Hydrobiologia* 64: 5-8.
Lock, M.A. and Hynes, H.B.N. 1975. The disappearance of four leaf
    leachates in a hard and a soft water stream in south western
    Ontario, Canada. *Int. Revue ges. Hydrobiol.* 60: 847-855.
Lock, M.A. and Hynes, H.B.N. 1976. The fate of dissolved
    organic carbon derived from autumn-shed maple leaves (*Acer*
    saccharum) in a temperate hard-water stream. *Limnol.*
    Oceanogr. 21: 436-443.
Lush, D.L. and Hynes, H.B.N. 1973. The formation of particles
    in freshwater leachates of dead leaves. *Limnol. Oceanogr.*
    18: 968-977.
Mackay, R.J. and Kalff, J. 1973. Ecology of two related species
    of caddisfly larvae in the organic substrates of a woodland
    stream. *Ecology* 54: 499-511.
Marcus, J.H. and Willoughby, L.G. 1978. Fungi as food for the
    aquatic invertebrate *Asellus aquaticus*. *Trans. Br. mycol. Soc.*
    70: 140-146.
Marxsen, J. 1980. Untersuchungen zur Ökologie der Bakterien in
    der flieBenden Welle von Bächen. I Chemismus,
    Primärproduktion, CO_2 - Dunkelfixierung und Eintrag von
    partikulärem organischen Material. Schlitzer
    Produktionsbiologische Studien (23-1). *Arch. Hydrobiol.*
    Suppl. 58: 26-55.
Mason, C.F. and Bryant, R.J. 1975. Production, nutrient content
    and decomposition of *Phragmites communis* Trin. and *Typha*
    angustifolia L. *J. Ecol.* 63: 71-95.
Mathews, C.P. and Kowalczewski, A. 1969. The disappearance of leaf
    litter and its contribution to production in the River
    Thames. *J. Ecol.* 57: 543-552.

McColl, J.G.   1966.   Accession and decomposition in spotted gum
    forests.   *Aust. For.*   30: 191–198.
McDiffett, W.F.   1970.   The transformation of energy by a stream
    detritivore, *Pteronarcys scotti*.   *Ecology*   51: 975–988.
Meyer, J.L.   1980.   Dynamics of phosphorus and organic matter
    during leaf decomposition in a forest stream.   *Oikos*   34:
    44–53.
Michaelides, J. and Kendrick, B.   1978.   An investigation of
    factors retarding colonization of conifer needles by
    amphibious hyphomycetes in streams.   *Mycologia*   70: 419–430.
Minshall, G.W.   1978.   Autotrophy in stream ecosystems.
    BioScience   28: 767–771.
Monk, D.C.   1976.   The distribution of cellulase in freshwater
    invertebrates of different feeding habits.   *Freshwat. Biol.*
    6: 471–475.
Mundie, J.H., Mounce, D.E. and Smith, L.E.   1973.   Observations on
    the response of zoobenthos to additions of hay, willow
    leaves, alder leaves and cereal grain to stream substrates.
    Fish. Res. Bd. Can. Technical Report No. 387, 123 pp.
Muztar, A.J., Slinger, S.J. and Burton, J.H.   1978a.   Chemical
    composition of aquatic macrophytes.   I. Investigation of
    organic constituents and nutritional potential.   *Can. J. Plant
    Sci.*   58: 829–841.
Muztar, A.J., Slinger, S.J. and Burton, J.H.   1978b.   Chemical
    composition of aquatic macrophytes. III. Mineral composition
    of freshwater macrophytes and their potential for mineral
    nutrient removal from lake water.   *Can. J. Plant Sci.*
    58: 851–862.
Neaves, P.I.   1978.   Litter fall, export, decomposition, and
    retention in Carnation Creek, Vancouver Island.   *Fisheries
    and Marine Service Technical Report* No. 809.
Odum, W.E., Kirk, P.W. and Zieman, J.C.   1979.   Non-protein
    nitrogen compounds associated with particles of vascular
    plant detritus.   *Oikos*   32: 363–367.
Otto, C.   1975.   Energetic relationships of the larval populations
    of *Potamophylax cingulatus* (Trichoptera) in a South Swedish
    Stream.   *Oikos*   26: 159–169.
Ovington, J.D.   1956.   The composition of tree leaves.   *Forestry*
    29: 22–28.
Paul, R.W. Jr., Benfield, E.F. and Cairns, J. Jr.   1978.   Effects
    of thermal discharge on leaf decomposition in a river
    ecosystem.   *Verh. Int. Verein. Limnol.*   20: 1759–1766.
Petersen, R.C. and Cummins, K.W.   1974.   Leaf processing in a
    woodland stream.   *Freshwat. Biol.*   4: 343–368.
Post, H.A. and Cruz, A.A. de la   1977.   Litterfall, litter
    decomposition, and flux of particulate organic material in
    a coastal plain stream.   *Hydrobiologia*   55: 201–207.
Prus, T.   1971.   The assimilation efficiency of *Asellus aquaticus*
    L.   *Freshwat. Biol.*   1: 287–305.

Prus, T. 1976. Experimental and field studies on ecological
    energetics of *Asellus aquaticus* L. (Isopoda). I.
    Assimilability of lipids, proteins and carbohydrates.
    Ekol. pol. 24: 461-472.
Puriveth, P. 1980. Decomposition of emergent macrophytes in a
    Wisconsin marsh. *Hydrobiologia* 72: 231-242.
Rabeni, C.F. and Minshall, G.W. 1977. Factors affecting micro-
    distribution of stream benthic insects. *Oikos* 29: 33-43.
Reice, S.R. 1974. Environmental patchiness and the breakdown of
    leaf litter in a woodland stream. *Ecology* 55: 1271-1282.
Reice, S.R. 1977. The role of animal associations and current
    velocity in sediment-specific leaf litter decomposition.
    Oikos 29: 357-365.
Reice, S.R. 1978. The role of detritivore selectivity in
    species-specific litter decomposition in a woodland stream.
    Verh. Internat. Verein. Limnol. 20: 1396-1400.
Reice, S.R. 1980. The role of substratum in benthic macro-
    invertebrate microdistribution and litter decomposition in a
    woodland stream. *Ecology* 61: 580-590.
Rossi, L. and Fano, A.E. 1979. Role of fungi in the trophic
    niche of the congeneric detritivorous *Asellus aquaticus* and
    A. coxalis (Isopoda). *Oikos* 32: 380-385.
Rossi, L. and Vitagliano Tadino, G. 1978. Role of adult faeces
    in the nutrition of larvae of *Asellus aquaticus* Isopoda.
    Oikos 30: 109-113.
Russell-Hunter, W.D. 1970. Aquatic productivity: an introduction
    to some basic aspects of biological oceanography and
    limnology. MacMillan, London.
Sedell, J.R. 1972. Feeding rates and food utilization of stream
    caddisfly larvae of the genus *Neophylax* (Trichoptera:
    Limnephilidae) using ^{60}Co and ^{14}C. *In* Symposium on
    Radioecology, D.J. Nelson, ed. 486-491. Proc. Third Nat.
    Symp., Oak Ridge, Tenn., May 8-10, 1971. Conf. 710501-Pl.
Sedell, J.R., Triska, F.J. and Triska, N.S. 1975. The
    processing of conifer and hardwood leaves in two coniferous
    forest streams. I. Weight loss and associated invertebrates.
    Verh. Internat. Verein. Limnol. 19: 1617-1627.
Short, R.A. and Maslin, P.E. 1977. Processing of leaf litter by
    a stream detritivore: Effect on nutrient availability to
    collectors. *Ecology* 58: 935-938.
Short, R.A. and Ward, J.V. 1980. Leaf litter processing in a
    regulated Rocky Mountain stream. *Can. J. Fish. Aquat. Sci.*
    37: 123-127.
Short, R.A., Canton, S.P. and Ward, J.V. 1980. Detrital
    processing and associated macroinvertebrates in a
    Colorado mountain stream. *Ecology* 61: 727-732.
Smock, L.A. and Stoneburner, D.L. 1980. The response of macro-
    invertebrates to aquatic macrophyte decomposition. *Oikos*
    35: 397-403.

Suberkropp, K. and Klug, M.J.   1976.   Fungi and bacteria
    associated with leaves during processing in a woodland
    stream.   *Ecology*   57: 707–719.

Suberkropp, K., Godshalk, G.L. and Klug, M.J.   1976.   Changes in
    the chemical composition of leaves during processing in a
    woodland stream.   *Ecology*   57: 720–727.

Swanson, F.J., Lienkaemper, G.W. and Sedell, J.R.   1976.   History,
    physical effects and management implications of large
    organic debris in western Oregon streams.   *USDA For. Serv.
    Gen. Tech. Rep. PNW-56*.   15 pp.

Thienemann, A.   1912.   Der Bergbach des Sauerlandes.   *Internat.
    Revue ges. Hydrobiol. Suppl.*   4(2): 1–125.

Triska, F.J., Sedell, J.R. and Buckley, B.   1975.   The processing
    of conifer and hardwood leaves in two coniferous forest
    streams.   II. Biochemical and nutrient changes.   *Verh.
    Internat. Verein. Limnol.*   19: 1628–1640.

Triska, F.J. and Sedell, J.R.   1976.   Decomposition of four species
    of leaf litter in response to nitrate manipulation.   *Ecology*
    57: 783–792.

Vannote, R.L., Minshall, G.W., Cummins, K.W., Sedell, J.R. and
    Cushing, C.E.   1980.   The river continuum concept.   *Can. J.
    Fish. Aquat. Sci.*   37: 130–137.

Ward, G.M. and Cummins, K.W.   1979.   Effects of food quality on
    growth of a stream detritivore, *Paratendipes albimanus*
    (Meigen) (Diptera:Chironomidae).   *Ecology*   60: 57–64.

Wetzel, R.G. and Manny, B.A.   1972.   Decomposition of dissolved
    organic carbon and nitrogen compounds from leaves in an
    experimental hardwater stream.   *Limnol. Oceanogr.*   17:
    927–930.

Whittaker, R.H., Likens, G.E., Bormann, F.H., Eaton, J.S. and
    Siccama, T.G.   1979.   The Hubbard Brook ecosystem study:
    forest nutrient cycling and element behavior.   *Ecology*
    60: 203–220.

Willoughby, L.G.   1974.   Decomposition of litter in freshwater.
    In Biology of plant litter decomposition.   C.H. Dickinson and
    Pugh, G.J.F. eds.   658–681.   Academic Press, London.

Willoughby, L.G. and Archer, J.F.   1973.   The fungal spora of a
    freshwater stream and its colonization pattern on wood.
    Freshwat. Biol.   3: 219–239.

Willoughby, L.G. and Marcus, J.H.   1979.   Feeding and growth of
    the isopod *Asellus aquaticus* on actinomycetes, considered
    as model filamentous bacteria.   *Freshwat. Biol.*   9: 441–449.

Willoughby, L.G. and Sutcliffe, D.W.   1976.   Experiments on
    feeding and growth of the amphipod *Gammarus pulex* (L.)
    related to its distribution in the River Duddon.   *Freshwat.
    Biol.*   6: 577–586.

Winberg, G.G.   1956.   Rate of Metabolism and Food Requirements of
    Fishes.   Belorussian State University, Minsk.   *Fish. Res. Bd.
    Can. Transl.*   Ser. No. 194, 1960.

Winterbourn, M.J. 1976. Fluxes of litter falling into a small
    beech forest stream. *N.Z. J. Mar. Freshwater Res.* 10:
    399-416.

Winterbourn, M.J. 1978. An evaluation of the meshbag method
    for studying leaf colonization by stream invertebrates.
    Verh. Internat. Verein Limnol. 20: 1557-1561.

Wotton, R.S. 1980. Coprophagy as an economic feeding tactic in
    blackfly larvae. *Oikos* 34: 282-286.

Zimmerman, M.C. and Wissing, T.E. 1978. Effects of temperature
    on gut-loading and gut-clearing times of the burrowing
    mayfly, *Hexagenia limbata*. *Freshwat. Biol.* 8: 259-277.

SEASONAL ASPECTS OF TRANSPORT OF ORGANIC AND INORGANIC MATTER IN STREAMS

Kenneth W. Dance

Ecologistics Limited
309 Lancaster Street West
Kitchener, Ontario, Canada

INTRODUCTION

There are three basic components to the transport of matter in streams: that which is being moved, the act and agent of transport, and timing or seasonality of movement. The materials which are to be moved can be divided into two major groups, organic and inorganic with emphasis being placed on the former. The various origins of organic materials will be discussed first. The act and agents of transport will be explored using findings obtained during organic matter studies. The seasonality of transport will be discussed using principally organic matter examples, since inorganic matter transport is similar in many respects.

The implications of seasonal transport of organic and inorganic materials on the benthos will not be discussed in detail. The effects of these materials upon the benthos range from being positive by providing food and shelter, to destructive, through scouring, smothering and altering chemical conditions to such an extent that existence is stressed.

Much of the research conducted to date has concentrated on the woodland stream system. As we continue to alter our wooded regions, through industrialization, urbanization and agriculture, it becomes obvious that in order to manage streams more effectively, we need to know more about the dynamics of energy flows in "modified" streams which often become progressively more autotrophic. Consequently some recent work on organic matter movement in modified streams will be discussed.

Definition of terms and assumptions in this paper: the majority
of the streams in which movement of coarse fractions of organic
matter has been studied are first to fourth order systems. Further,
these are typically largely unmodified, woodland streams. However,
before proceeding further it is imperative to define the types of
organic matter which will be discussed in the paper. Organic
matter in aquatic systems has been partitioned in the past on the
basis of particle size. As Wetzel and Rich (1973) indicate, a
basic separation can be made between the dissolved/colloidal and
particulate fractions by filtration at the 0.5 µm level. While
there has been a proliferation of terms used to describe categories
of material within the particulate phase, Boling et al. (1975a)
proposed a system consisting of six fractions for particles in the
range 0.45 µm to 16 mm. Boling et al. (1975b) proposed eight size
categories which also included dissolved organic matter (Table 1).
The latter system (Boling et al. 1975b) is useful, as the dominant
constituents of each fractions are suggested by the name. The scale
in terms of particle sizes, however, does not form a continuum, for

Table 1.   Investigators who have commented on stream transport
of organic particles in the size ranges shown.

Particle size range*	Descriptor and Dominant Constituents*	Acronym*	Authors
> 16 mm	Aggregated whole organic material (2 or more leaves, pieces of bark, branches and masses of twigs)	AWOM	1,3,4,5,7,8, 9,12,13,14,15 19,21
> 16 mm	Small whole organic matter (individual leaves)	SWOM	
1 mm – 16 mm	Large particulate organic matter (leaf, twig and bark fragments)	LPOM	1,3,4,5,7,8,9, 12,13,14,15,17,18
250 µm – 1 mm	Medium particulate organic matter (detrital fragments)	MPOM	1,2,3,4,5,7,8, 9,12,13,15
75 µm – 250 µm	Small particulate organic matter (small detrital and fecal fragments)	SPOM	1,2,3,4,6,7,9, 10,11,12,13,15, 16,20
0.45 µm – 75 µm	Fine particulate organic matter (fine detrital and fecal fragments)	FPOM	

Authors

1. Bilby and Likens, 1979	8. Hodkinson, 1975	15. Naiman and Sibert, 1978
2. Boling et al., 1975a	9. Likens et al., 1977	16. Post and De la Cruz, 1977
3. Boling et al., 1975b	10. Lush and Hynes, 1978	17. Rau, 1976
4. Brinson, 1976	11. Manny and Wetzel, 1973	18. Sedell et al., 1974
5. Dance et al., 1979	12. McIntire et al., 1975	19. Swanson et al., 1976
6. De la Cruz and Post, 1977	13. McIntire and Colby, 1978	20. Wetzel and Manny, 1977
7. Fisher and Likens, 1973	14. Naiman, 1976	21. Young et al., 1978

* After Boling et al. (1975a)

example, the leaf fragment organic matter category contains 4-8 mm
diameter particles and the next smallest fraction is large
particulate organic matter 1-2 mm in diameter. Further, eight
separate terms may be too much subdivision for convenient use. In
Table 1, I have provided a list of workers who have reported
studying organic particles in each of the six size categories
proposed by Boling et al. (1975a). Note that the drift of aquatic
invertebrates which falls into these size categories is excluded
from this tabulation, hence, organic particles are of plant origin.

        The final term to be defined is transport (Figure 1). For the
purposes of this paper, transport is movement either into a
stretch of stream being studied or export from a study reach.
Thus transport includes the subcomponents - import and export.
Input components include leaf fall onto stream surfaces and lateral
movement of organic materials, whether caused by wind, water,
humans or animals.

COMPONENTS OF SEASONAL ORGANIC MATTER TRANSPORT

        Mann (1979) makes the point that, functionally, woodland
streams resemble pipes. Particulate, allochthonous organic energy
falls into the stream and is carried downstream at times of high
flow. Such a description points toward several of the integral
components which come into play in stream transport. First, there
must be a channel through which water and particulate organic
energy flow, there must be a source of particulate organic
energy and a supply of water which pushes it through the vessel and,

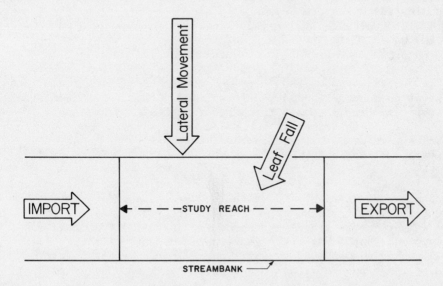

Figure 1. Major components of stream transport as described in this
           paper.

according to Mann's suggestion, most of the transport occurs
during periods when there are large volumes of water flowing down
the channel. Further analysis reveals that the transport of
organic material is seasonal for several reasons: (1) the inputs
of leaf and litter energy are seasonal in many geographic regions,
(2) the conversion of large organic bits to smaller ones that can
be transported under conditions other than high flow is also
seasonal, and (3) the volumes of precipitation, which supply water
to surface drainage systems, vary seasonally in many regions of
the world; that is, there are dry seasons in which stream flows
may be minimal and there are rainy or melt-water seasons when flows
are considerably greater. These factors which contribute to the
total seasonal transport effect will be discussed in detail below.

Seasonality of leaf litter inputs: Papers published up until 1977
indicated that generally across Europe and North America 500 g/m^2
dry weight of leaves and other debris falls to the deciduous forest
floor per year. It therefore seems reasonable to assume that the
same quantity would fall into narrow streams which have treed banks
(Dance, 1977). While in areas of more shrubby vegetation or in
less densely forested regions, the quantities may be less than half
this amount (Kaushik et al., 1975; Dance, 1977; Post and De la Cruz,
1977).

     The types of vegetation which dominate a region are controlled
by geographic location and thus climate. Each species of tree, in
turn, possesses its own life cycle and, thus, sprouts and drops it
leaves, needles and twigs in an annual pattern. Several papers
describe the times and amounts of leaf fall in relation to stream
dynamics and there are undoubtedly many more papers in the forestry
literature which deal with this subject *per se* , but I have not
attempted to review these as the stream-related papers illustrate
the point to be made here sufficiently well.

     On a tropical island, Harrison and Rankin (1975) found that
vegetation composition in the forest and along their study streams
changed with elevation. Since inputs of leaf energy were
controlled by the time of abscission which varied among the plant
species, in effect, there may be a seasonality of input on a
longitudinal basis in the stream that tumbles down a mountain side.

     In the Loblolly pine-oak forest of southern Mississippi the
greatest quantities of litter fall occur in December (De la Cruz
and Post, 1977). These workers found also that wind and
precipitation drastically affected the rate of litter fall even to
the extent that local differences were evident. Although litter
fall peaked in December in this forest, considerable quantities of
energy input to the stream occurred in April and May as well.

Gosz et al. (1972) reported the results of their northern
hardwood forest study in which litter traps were emptied at weekly
intervals. Catches of falling leaf, branch, bark, fruit, bud scale,
flower and insect frass materials were weighed. As would be
expected, the largest quantities of litter fall occurred in the
autumn when leaves and needles drop. During summer storms,
considerable quantities of material fell to the forest floor. This
finding reinforces the later comment of De la Cruz and Post (1977)
that wind and precipitation play important roles in knocking material
from trees.

It is also important to look at the ratio of leaf fall to
total litter input in the northern hardwood forest. Fisher and
Likens (1973) found that autumnal leaf fall accounted for 90% of
the total annual leaf input and annual leaf fall constitutes two-
thirds of the total litter input to Bear Creek. This supports the
idea that the leaf input to the woodland streams of eastern North
America is in a large pulse over a relatively short period of time -
in autumn.

Boling et al. (1975a) in their model of an eastern deciduous
forest stream, recognize the significance of timing of inputs
(Figure 2). Their model incorporates major energy inputs at
appropriate seasons. In these streams daily inputs of particles
> 1 mm are generally greater than 2.5 g/m^2.

In the Pacific Northwest of the U.S. a slightly different
situation exists in that the litter fall to streams studied by

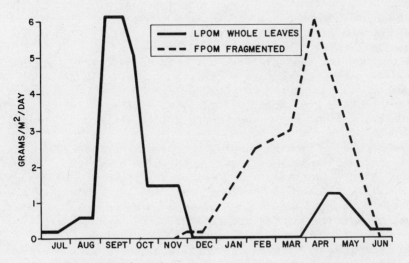

Figure 2.   Input schedule for whole leaves and for fragmented FPOM.
Modified after Boling et al. (1975a).

Sedell et al. (1974) consists primarily of Douglas fir and hemlock
needles (65%). Timing of organic debris input into these streams
varies depending upon the species source. That is, deciduous trees
provide inputs primarily from mid-October through November. The
conifers, however, drop needles and other debris throughout the
year but again most of the needle fall is between September and
December. Generally, the inputs are about 2.5 $g/m^2/day$ in these
streams.

McIntire and Colby (1978) have incorporated data from field
measurements on organic matter input to streams into their Oregon
stream model. Inputs range from a minimum in April (0.70 $g/m^2/day$)
to a maximum in November of 3.42 $g/m^2/day$.

In summary, the timing of organic matter inputs from forests
to streams varies on both a seasonal and geographic basis.
Anything that affects vegetation species composition, e.g.
altitude, latitude, climate, in turn affects the timing of leaf
and needle fall, while individual precipitation events and wind
storms may produce litter fall at times other than during the peak
leaf fall season. In Canada and the U.S., deciduous trees primarily
tend to lose their leaves between September and December, while
coniferous trees lose their needles throughout the year but to the
greatest degree in the autumn.

Changes in particle size which affect transport: once in the
stream certain of the large particles of organic matter undergo
various changes which affect the probablity of being transported,
primarily their conversion to small particles. I do not intend to
discuss the mechanisms and broader implications of this process.
However is it, necessary to discuss the conversion of LPOM to MPOM
or SPOM in general terms since smaller leaf particles are usually
more easily transported by running water systems than are large
pieces of material and during much of the annual cycle, organic
matter is present in relatively small particles. In terms of
seasonal transport, then, one would expect LPOM to be a major form
of organic matter available for transport in the autumn with MPOM
and SPOM being of greater importance in late winter and spring.

The idea of litter processing in streams has been described
by Boling et al. (1975a) as a system in which there is a sequential
series of bins (Figure 3). Large-sized particles are processed
until they are reduced to a certain size (medium) then they are
sieved out into the medium-sized bin where the process is repeated.
The rate of turnover of resistant particulate organic matter (> 16
mm in size) is in the order of several years (Boling et al., 1975b).
That is, limbs, whole trees and roots, etc. decay slowly. If
such a large unit of organic matter fell into a small stream, the
majority of it would not really be available for transport, since

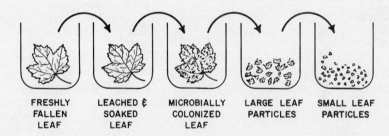

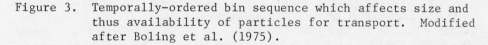

Figure 3.   Temporally-ordered bin sequence which affects size and
             thus availability of particles for transport. Modified
             after Boling et al. (1975).

the flows in that stream could not move it downstream. Swanson et
al. (1976) report debris dams consisting of logs and trees which
have remained in one location for more than twenty years.

    A number of workers have published results which indicate that
organic inputs in autumn consisting of whole leaves (SWOM) are
reduced to FPOM during the period of one year (Fisher and Likens,
1973; Sedell et al., 1974; Naiman and Sibert, 1978; Bilby and Likens,
1979). Petersen and Cummins (1974) found that leaves of certain
species are reduced to smaller particles more rapidly than others
because of innate characteristics of the leaves. Reice (1974)
found that the rate of leaf breakdown was mediated by water
temperature, and composition of the leaf pack, in terms of the
species which contributed to the accumulation, also affected the
rate at which smaller sized particles were formed. Coniferous
needles are generally slow to decompose (Sedell et al., 1974).

    Recently, Gelroth and Marzolf (1978) found that stream channel
form and type played an important role in determining whether leaf
packs were formed or whether leaf material was swept into pools and
buried by sediment. In "natural" physically complex stream
channels leaf packs form and processing can occur. However, in
artificially channelized streams, the homogeneous bottom provides
little in the way of obstructions for leaf packs to form on. Thus
in channelized streams, leaf packs are absent and leaves are
exported from these riffles into the pools. Additionally, these
workers found that leaf processing occurred more rapidly in the pools
of natural streams than in those of channelized streams.

    In-stream processing of organic inputs determines to what size
the particles are reduced during each season and particle size clearly
affects the availability for transport. Refractory POM may not move
for twenty or more years, but as soon as a leaf has been processed

into a number of small bits it can be transported by even the slowest
current in areas where there are few obstructions to impede its
movement.

Seasonality of precipitation and stream flows: above we have accounted
for inputs of organic matter to be transported and have shown how,
at different times of the year, OM is present in various sized
packages which require different latent energies to move them.

Now we turn to a discussion of the principal transporting agent,
water. In streams of order four or less, precipitation events and
subsequent run-off volumes considerably affect stream flows. Large
quantities of water are capable of carrying a total OM transport
load that exceeds that at lower flows. At high flows the stream
channel width is increased and OM that was lying adjacent to the
stream is then washed downstream.

Obviously, the season of maximum precipitation and stream flows
varies according to geographic region. In the humid tropics
(Guatemala), Brinson (1976) found that the wet season begins in
May and tapers off to the end of December. The dry season lasts
from December to April or May. In this region maximum flows occur
during the rainy season. British workers (Dawson, 1976; Moore, 1978)
report that discharge follows the rain pattern. Precipitation
quantities increase rapidly in autumn to a maximum in winter and
steadily decrease until late summer, thus maximum river velocity
(and undoubtedly flow as well) usually occurs in the period between
December and March.

Figure 4 shows the seasonality in flow of streams located in
five regions of North America. Anderson et al. (1976) provide a
synopsis of regional precipitation and flow regimes in the United
States and it is instructive to examine their findings as a variety
of factors which determine local peak flows are outlined. The
occurrence of highest flows in early spring is the distinguishing
characteristic of northeastern coniferous and hardwood forests.
Thirteen of the 23 inches of annual stream flow from New England
forests occur during spring months. Occasional autumn hurricane
storms produce the highest local peak flows. In the north central
area of southeastern hardwood forests winter is the principal season
of high flows. Similarly in the southeastern pine forests,
January through March and hurricane season produce highest flows.
In west coast forests highest stream discharges result from heavy
winter rains falling on wet soils. Rain falling on snow and pack
melt in the mountains can produce snowmelt floods as late as June.

Local climate generally controls the timing of maximum flows
which are associated with peak precipitation periods, snowmelt
season or storms such as hurricanes or electrical storms. Peak

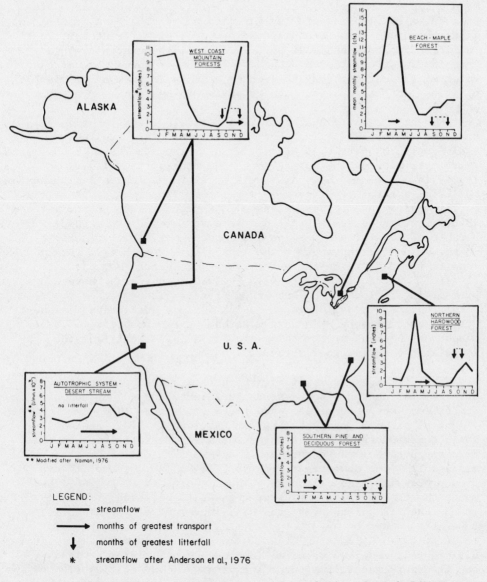

Figure 4.   Timing of stream flow, litter fall and peak POM
            export in streams of certain North American regions.

flows are obviously of importance since theoretically they have
the potential to transport the greatest quantities of available OM.
Furthermore, the duration of substantial flows on a seasonal basis,
also need to be considered.

Mechanisms of transport: Leaves may fall directly onto the water
surface, also wind and gravity may combine to provide a lateral
input which has been recognized as making a significant contribution
to litter input (Kaushik et al., 1975; Phillipson et al., 1975).
Once in the stream the OM packet may be pushed downstream by the
force of the current, may join other packets to form a leaf pack,
or it may settle in a pool and be buried in sediment.

Recent work by a number of investigators (Young et al., 1978
Dance et al., 1979) indicates that the distances travelled by
individual LPOM particles are not great. The studies of Young et al.
raise some interesting points about the transport of individual
leaves in a stream. Mean drift distance of wet leaves was found to
be 192 m, but for dry leaves it was 226 m. The mean distance for
entrainment of all leaves released as a group was 886 m for wet and
1040 m for dry leaves. Drift distance varied with tree species
because of leaf size and shape and depended on stream discharge
on the test day, and whether the leaves were wet or dry. Dry leaves
tend to float along on the stream surface, whereas wet ones are
partially or totally submerged in the water and, thus, have a greater
probability of lodging on submerged stones. Stream order may also
affect the percentage of leaves which becomes entrained in a short
distance after entry. In narrow streams the probability of becoming
caught on the bank or something growing along or out from the bank
is greater than in wider areas.

The distances over which OM packets are transported in a single
move are influenced also by stream gradient, the number and size of
obstacles, bottom substrate and stream velocity. In Woodcock Creek,
(Young et al., 1978), there are fluctuations in flows seasonally
which function to dislodge leaves which become entrained, so that
by early December the number of leaves left in the study reach was
negligible. The downstream transport mechanism does not remove OM
from processing areas altogether, but exports it to the next stream
reach where processing begins or resumes (Sedell and Triska, 1976).
In the Pacific Northwest of the U.S., movement of large debris is
thought to occur by a variety of mechanisms including episodic
flushing during floods and the less obvious daily release of finely
divided and dissolved material (Swanson et al., 1976). Clearly,
the transport mechanisms vary depending upon the initial size of the
OM particle to be moved.

Still other workers have commented on factors which affect
timing of transport and the quantities of OM involved. Lush and
Hynes (1978) found that local physiography, past history of rainfall
and damping effects of lakes and marshes were of importance in
determining quantities of smaller particles of OM transported.
De la Cruz and Post (1977) suggest that since such a large number of
factors control downstream transport of OM, comparisons of export
quantities among different streams should be done cautiously or not

at all. Factors of importance which they mention include: stream size, configuration, current, drainage basin topography, litter decomposition rate and local weather.

There are a number of important points which arise from the above discussion. The first is that wind, water, gravity, animals and man may all play a role in the export of leaf energy into streams. The role of man and animals is probably quite small and incidental. The second important point is that the timing of leaf fall during the annual cycle obviously has a great deal of bearing on the season at which large quantities of fresh leaf material enter streams. It must be stressed though that in many locales precipitation and wind storms are responsible for causing considerable amounts of forest litter fall activity in the so-called "off season".

Once leaf, twig or bark energy enters the stream, it begins a variety of physical and chemical transformations. The critical point here is that the size of the packets begins to reduce. This can result in greater availability of organic matter for transport at a particular time. Large particles may only move under conditions of maximum discharge even though they are "available" to be transported for many months of the year. Conversely, smaller particles may slowly move downstream on an almost daily basis. The other major factor is the agent which provides the energy for OM movement. It is reasonable to assume that the greatest amount of organic matter transport will occur during the seasons of peak stream flow providing that this occurs shortly after or during leaf fall. Also, the timing of leaf fall, and precipitation events varies from region to region in North America and probably throughout the Northern and Southern Hemispheres.

Finally, drift distances for leaves have been found to be a function of leaf size and shape, stream discharge and whether the leaf is water soaked or dry. As well, stream gradient, bottom substrate, water velocity and the number and size of in-stream obstacles affect the distance that leaves are transported. Fluctuations in stream flow, past history of rainfall and effects of lakes and marshes have been found to determine the quantities of small OM particles transported (Lush and Hynes, 1978).

CASE STUDIES ON COARSE FRACTION OM TRANSPORT

Woodland streams: in eastern North America three major groups of workers have studied OM dynamics in woodland streams. These are the Cornell, Michigan State and Waterloo schools.

Cornell has concentrated its work in the Hubbard Brook Experimental Forest of New Hampshire. Likens and his colleagues have published numerous papers on their findings and have summarized

results to the mid-seventies in a recent book (Likens et al., 1977).
The seasonal chemical composition of litter fall components in the
Hubbard Brook hardwood forest was reported in 1972 (Gosz et al.,
1972). It was found that nitrogen and phosphorus concentrations in
leaves and branches falling from trees were greater in October than
in June, August or February. As a result, the original food value
of litter varies seasonally.

Gosz et al. (1976) provided an estimate of losses from the
forest ecosystem via hydrologic export. Only 10 kg/ha/yr were
thought to be lost as particulate organics with another 17 kg/ha/yr
being lost as dissolved organics from a woodland in which the
forest floor biomass was estimated to be 46,800 kg/ha. Likens et al.
(1977) report annual export of materials during the period 1965 to
1973. Of most importance to the present review is the finding that
annual export of larger organic particles as trapped in the
ponding basin ranged from 2.12 kg/ha/yr to greater than 17.88 kg/ha/yr.
The eight-year average was 7.44 kg/ha/yr. The explanation for the
great range in quantities is to be found in a more recent paper
published by these workers (Bilby and Likens, 1979). This paper
outlines the influence of discharge on POM transport and it is
probable that during years when extremes of discharge occur so
also do extremes of OM transport arise. It was previously found
that one large summer or autumn storm could contribute as much or
more to the annual total than the remaining days of the year
(Fisher and Likens, 1973). Bilby and Likens (1979) report that the
small amounts of FPOM being transported at the start of the
snowmelt season result from the OM being blanketed and held in place
by ice and snow and by low temperatures hindering the processing
of CPOM into FPOM. Extremely high concentrations of FPOM result
during the peak of snowmelt from vigorous scouring of available OM
from the streambed and from the washing in of OM originating in
areas adjacent to the stream bank. They also found that FPOM
concentrations during summer storms are dependent on the length of
time since the last discharge event. If the last event occurred
during the previous 48 hours FPOM concentrations would be expected
to be low.

Workers at Waterloo have published information on OM
transport in southern Ontario streams which flow through beech-
maple or cedar woods interspersed among farm lands. Lush and Hynes
(1978) studied what they called particulate OM (> 0.8 μm)
sampled by obtaining aliquots of stream water. Most of the particles
were found to be < 67.4 μm in diameter. Loads of POM were not
reported but concentrations of POM were generally in the range
2-20 mg/l.

Dance et al. (1979) reported dry weights of drifting OM
captured from adjacent intermittent and permanent streams. During
a thirteen month period (June, 1975-June, 1976) in a spring-fed

Table 2. Monthly contribution of each fraction, in kg dry
weight, to the total amount of non-animal material
drifting past station 2 located on a spring-fed
stream.

Month	Particles 1.3 - 3.0 cm				Particles 253 μm - 1.0 cm			
	Green	Conifer	Decid.	Algae	Green	Conifer	Decid.	Detritus + Algae
June 1975	7.66	6.97	9.56	7.98	1.33	3.39	3.39	353.10
July	0.75	0.25	1.01	0.41	0.32	0.87	1.67	72.72
Aug	1.30	1.23	7.40	3.50	0.50	0.89	2.08	83.64
Sept	0.50	1.32	5.53	1.35	0.24	1.55	1.22	57.49
Oct	0.07	5.61	4.58	0.44	0.04	2.65	1.82	15.29
Nov	0.37	8.35	3.26	1.32	0.16	6.27	1.12	30.62
Dec	0.51	6.42	9.75	0.71	0.22	5.56	2.19	78.07
Jan 1976	0.02	0.28	0.24	0	0.01	0.42	0.07	4.31
Feb	5.85	4.44	11.53	0	0.90	8.48	12.49	250.94
March	0.45	1.60	13.99	0.11	10.05	14.52	52.73	2,156.63
April	0.56	4.04	25.52	0	0.08	0.96	1.35	434.52
May	0.54	2.54	8.99	0	0.14	1.10	1.02	153.37
June	2.20	3.38	7.13	0	3.03	2.57	2.96	344.09
Sub-totals	20.78	46.43	108.49	15.82	17.02	49.23	84.11	4,034.79
Total		191.52				4,185.15		

stream, 4185 kg of particles 253 μm–1.0 cm in size passed one
station while the dry weight of particles in the range 1.3–3.0 cm
amounted to 192 kg (Table 2). The largest quantity of the fine
material (detritus and inorganic particles) was captured in March
of 1976 and accounted for just over half of the thirteen month
catch. Other months of considerable fine particle transport
included April and June. In March, 1976 coarse particle
transport values were quite small with June 1975 and April 1976
being the months during which the larger particles found their
way downstream. During each of these "peak" months, only 15% of
the thirteen month's coarse particle total was captured.

    Leaf fall occurred during the period September to November
(see beech-maple forest, Figure 4). This, then, is the time when
fresh leaf material became "available". Stream flows during the
same period were at about average levels when viewed on an
annual basis. During the leaf fall season, transport of
deciduous material as 1.2-3.0 cm and 253 μm-1.0 cm particles was
similar to that of summer months. The conclusion to be drawn is
that although the leaf material was in the stream and was

82   K. W. DANCE

available to be moved there was no agent of transport.

Table 2 indicates that drifting quantities of fine
deciduous material (253 μm–1.0 cm) were higher during February when
a winter thaw occurred and peaked during March at the time of
spring break-up when the most water was available. Figure 4 shows
the seasonal discharge pattern in this stream. Larger particle
deciduous transport peaked in April, the tail end of the melt and
the time of spring rains. These data support the idea that maximum
transport of leaf energy would be expected to occur during the
season when both fallen leaves and high stream flows are present.

Several workers have studied POM transport in streams in
western North America. Hodkinson (1975) quantified POM transport
in energy units. With drift nets he captured particles in the
range of 210 μm–30 cm (large POM) and also measured fine POM
(< 210 μm). Hodkinson found that 2842 Mj entered the pond system
which he was studying via two streams. This transport of POM was
important to the system since the total energy input to the pond
was 4038 Mj.

According to Sedell et al. (1974), one-third of detrital
input to certain Cascade Mountain streams is exported. Figures
6 and 7 in their paper show that closer to 25% of the 590 kg POM
input is exported. These authors also indicate that in the
White Mountains of New Hampshire, nearly 40% of the total POM
input is exported. In another western U.S. study in which small
coniferous forest streams were studied, less than 5% of the POM
(> 1 mm) present was in the form of leaf material, whereas more
than 60% of POM was present as wood debris (Sedell and Triska,
1976).

Naiman and Sibert (1978) in a study where the quantities of
different sized particles being transported on an annual basis
were measured, found that 3.5×10^5 kg of FPOC (0.5 μm–1 mm) and
4.4×10^3 kg of LPOC (>1 mm) were exported annually, providing
0.7 g $C/m^2/yr$ for the estuary into which the river empties. To
provide a sense of perspective, it is important to note that
discharge in this system was usually in the range of 4–50 m^3/sec,
but during months of maximum discharge flow was in the range of
300–400 m^3/sec.

A study in Mississippi found that for POM (> 0.5 μm) daily
amounts transported ranged from 179 – 2,777 kg dry weight. During
the same sample intervals discharge ranged over an order of
magnitude, 37,000–262,000 m^3 sec (De la Cruz and Post, 1977).

These studies indicate that the quanitities of POM exported
from heterotrophic streams vary from system to system. Such
differences are a function of litter fall, input weights, stream

flow regimes and topography in each drainage basin. For these
reasons the regional differences in POM export quantities that are
evident are to be expected. Indeed, Dance et al. (1979) found
considerable differences in total amounts of POM export in adjacent
streams. It is therefore difficult to compare export quantities
from one stream with those from another. The use of different
sampling apparatus to measure export of the larger sized
particles by each group of workers also complicates this issue.
However, it is recognized, that samplers must be designed to withstand the conditions unique to each stream.

Autotrophic and man-modified streams: a number of authors have
studied POM movement in streams that have been autotrophic for
some time or which as a result of human disturbance are shifting
from a heterotrophic to an autotrophic energy base.

In a British stream where three times as much OM is produced
from aquatic plants than from litter fall, 76 t of suspended OM
were exported and 5 t, or less than 4% of the total export weight,
were found to move as large drift (Dawson, 1976). Wong and Clark
(1976) found that during a six hour period 3.5 kg fresh weight of
the green alga *Cladophora* were caught on a 1 m^2 screen placed in
a southern Ontario river. Dance et al. (1979) found in a ditch-
like stream draining agricultural land that *Cladophora* contributed
to OM drift during the growing season (see "Algae" column, 1.3-3.0
cm particles, Table 3). POM drift quantities in a thermal stream

Table 3.   Monthly contribution of each fraction, in kg dry weight,
to the total amount of non-animal material drifting past
station 3 located on the agriculturally impacted stream.

Month	Particles 1.3 - 3.0 cm			Particles 253 µm - 1.0 cm		
	Green	Decid.	Algae	Green	Decid.	Detritus + Algae
June 1975	0.98	6.10	9.73	0.002	0.04	3.92
July	0.02	0.02	1.10	0.0007	0.002	1.86
Aug	0.15	0.37	1.86	0.03	0.48	90.47
Sept	0.006	0.03	0.12	0.02	0.11	103.06
Oct	0.005	0.06	0.18	0	0.0003	3.64
Nov	0.004	0.03	0.02	2.41	7.14	34.14
Dec	0.05	0.07	0.05	0.78	2.65	149.85
Apr 1976	0.51	4.48	0.005	1.67	1.98	82.83
May	0.96	6.98	37.72	0	0	71.28
June	0.02	0.62	1.91	0.18	0.05	20.24
Sub-totals	2.705	18.76	52.70	5.093	12.45	561.29
Total		74.17			578.83	

in a desert area of the U.S. were found to be controlled by solar
radiation. A 2.1 fold increase in solar radiation was found to
result in a twenty times increase in particulate drift. Only 4%
of annual primary production was lost as POM. Total annual
particulate export amounted to 14,400 g AFDW (Naiman, 1976).

Swanson and Bachmann (1976) found that the efficiency of
transport of algal material in rivers was dependent upon the volume
of flow. The explanation for the occurrence of the greatest
absolute amounts of algal material being transported at high
flows is that these flows result in the sloughing off of algal
growths from stream substrates. Moore (1978) indicates that this
scouring and sloughing off process may be of great importance to
the algal community. He found that virtually all periods
of rapid growth during the summer occurred immediately after
flooding.

Brinson (1976) reports results of OM transport studies in
several Guatemalan rivers. Several of these rivers drained
agricultural lands while portions of some, drained forested areas.
The majority of OM transport occurred during the wet season,
which also represented the season of peak river flows. During times
of high flow, very large packets of OM were transported, e.g. logs
and mats of grass. In contrast, there was little, or no, CPOM
transport during the dry season (November–March). Brinson cites
Hobbie and Likens' (1973) paper which indicates that deforestation
has a large effect on particulate matter losses from a basin.

Gelroth and Marzolf (1978) indicate that human alteration of
streams, e.g. channelization disrupts components of the heterotrophic system processes. In a channel bottom that is quite
homogeneous and lacks obstacles, leaf pack formation is less
probable and leaf material is quickly flushed downstream. Stream-
cleanup in western logging areas has been found to influence
channel morphology. If man removes organic debris dams, the
streams become channelized by stream action and, thereafter, the
stream is perpetually clean and unstable both physically and
biologically (Sedell and Triska, 1976).

Studies by Dance et al. (1979) and Dance and Hynes (1979 and
1980) compare organic matter transport and biological communities
of adjacent streams which possess basins having different amounts
of agricultural land and forest. At the upstream station on a
stream which drains a nearly totally cultivated basin, OM of
coniferous and deciduous origin is virtually absent from the
material transported downstream (Table 3). The contribution of
algae (primarily *Cladophora*) is greater in this more autotrophic
system than it was in an adjacent spring-fed stream (Table 2).
The total weights of POM captured in the "agricultural stream"
are considerably less than that for the Table 2 spring-fed stream.

One reason for the difference in total weights is that the critical late winter-early spring season was not successfully sampled in the "agricultural stream." As has been noted above, this is the period during which a large proportion of the annual POM load is transported.

In Figure 5 I have graphed the quantities of 253 μm to 1.0 cm particles exported each month and have shown how this was related to precipitation quantities and stream flow in an intermittent, agricultural stream. In general, there is a clear relationship between the monthly precipitation and discharge curves. This is to be expected for a stream in which there is virtually no contribution to base flow from ground water. Dry weights of exported POM, in turn, generally increase or decrease from month-to-month in response to an increase or decrease in precipitation and flow.

This stream might typify Mann's (1979) description of a stream as a pipe; the stream channel being basically a vessel for transporting water when it is supplied by precipitation. During snowmelt periods or precipitation events, much water is moved and any POM that is adjacent to or in the channel will be moved downstream as long as the stream flow persists. However, during months of little rain, e.g. October, 1975 (Figure 5) there is little flow and very little POM is exported.

PARTICULATE INORGANIC MATTER TRANSPORT

There are two major mechanisms by which inorganic sediment particles are transported once they enter the stream channel and predominant transport method often depends upon the size of the particle to be moved. Bedload transport is the mechanism responsible for tumbling larger particles along the streambed. Only at high stream flow velocities can relatively large particles be moved. The other principal means of transport is as small particles in suspension.

Bedload generally represents only a small proportion of the total inorganic transport load (Dyvik and Gilchrist, 1977), however, Likens et al. (1977) found that bedload accounted for over 50% of the particulate matter removal at Hubbard Brook. Although it has been found that there is a good correlation between bedload sediment catch quantities and stream discharge under a range of flow conditions, bedload transport in natural systems is highly variable especially in water courses with complex cross-sections (Zimmie et al., 1976a).

There has been some work done which allows some general conclusions to be drawn about the seasons during which most

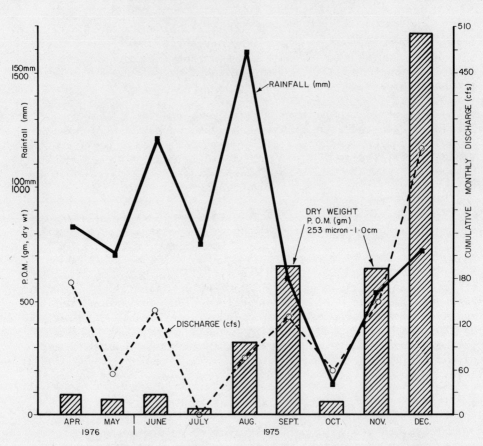

Figure 5.   Relationship between rainfall, discharge and quantities
            of exported POM (253 μm – 1.0 cm). Stream located in
            an intensively farmed basin; data from Dance (1977).

bedload transport occurs. During a 7 month study in New York
state (Dyvik and Gilchrist, 1977) 87% of the total bedload weight
was moved during a five consecutive day period in March. This
indicates that bedload transport in some systems is highly storm-
event related. In this instance, the peak load occurred during
early spring and was probably coincident with snow melt. In
another study Zimmie et al. (1976b) found that late winter/early
spring bedload transport was great but precipitation events in
August and October which produced increased discharge also
resulted in considerable bedload transport. There appears then
to be a similarity between CPOM and bedload transport in that
snow melt related and intense precipitation events can result in

considerable amounts of material being pushed downstream.

The availability of particles which are potentially part of the suspended sediment load in agricultural basins may be affected by management practices. Across certain agricultural areas of Canada and the U.S., only a third to one-half of the year represents the growing season. Under certain cropping practices, ground is left bare for the remaining half to two-thirds of the year; these soils then are exposed to the erosive effects of rain, wind and runoff during a large portion of the year (EPA, 1978). Such soils represent a source of potentially "available" suspended sediments.

Several workers have collected data which indicates that transport of suspended inorganic matter is a seasonal phenomenon, i.e. the greatest quantities of these materials occur during the times of year when flows are greatest.

Decamps and Casanova-Batut (1978) found that in the River Lot, mineral transport was greatest during the winter (November to April), a time coincident with high flows. Work published in 1975 indicated that for several southern Ontario streams, half of the annual suspended sediment load moved downstream during the months of March and April. Ice break-up and snowmelt occur during this period in the study area. Subsequent work in southern Ontario has indicated that, in some streams, more than 50% of the annual suspended sediment transport occurs during the spring months. Van Vliet et al. (1978) found that in certain agricultural basins 70-90% of annual suspended sediment loads were contributed during a single month (based on calculations using their Appendix 4 data). In all but three of the fifteen basins examined, maximum loads occurred during the February-April period. In the three exceptions, maximum values were found in August, a month during which high discharges may occur as a result of cloudbursts associated with thunder storms. Van Vliet et al. (1978) explain that there are seasonal differences in the delivery of sediment from fields to stream systems. During the spring period (January-April) there is a high delivery of sediments from the fields into the stream system due to frozen soil or high antecedent soil moisture conditions and due to a lack of effective ground cover for trapping soil during this transport phase. During the summer (May-August) little or no sediment enters the water course because sediments are trapped by vegetation during transport. The foregoing point is of importance, just as it was in the case of OM transport. If the season of inorganic material availability precedes or occurs at the same time as maximum discharges, then much transport would be expected to occur.

As would be expected from the comments on the important role played by vegetation in reducing erosion and sediment delivery to

streams, land use has been found to have considerable effect on
annual sediment loads (Likens and Bormann, 1974; Walling, 1974;
Hartman et al., 1976), and the way in which the seasonal
distribution of transport events are altered by changes in land use
is important.

Rich and Gottfried (1976) studied changes in stream flow and
sedimentation resulting from a variety of treatments which
manipulated the density of forest vegetation cover. Treatments
included such measures as selection harvest, improvement cuts,
fire, conversion of forest to grass and clearing and planting with
another species. The treatments applied resulted in increased
flows during the historical season of peak flow (winter) and the
month of peak flows remained the same (March). Peak flow volumes
did increase, but there was little apparent seasonal shift in
flows overall. Sedimentation results were reported only as
annual quantities per unit area and thus any seasonal changes are
obscured.

Hornbeck has reported interesting findings on stream and storm
flow from forests in New Hampshire (1973, 1975). In the 1973 paper,
Hornbeck compares storm flow regimes in a drainage basin that was
cleared and maintained essentially unvegetated by application of
herbicides. Since transpiration and interception losses are
reduced after clearing, soils were wetter and had less capacity for
storing water, thus summer quick flow volumes and instantaneous
peak flows were nearly always increased. The absence of the canopy
also affected the timing and rate of snowmelt and changed spring
storm flow events involving snow water. In the later of the two
reports on this project, Hornbeck (1975) explains the seasonality
of flow changes in more detail. When there was no vegetation,
sizeable stream flow increases occurred during June through
September. These increases declined in October and November.
Winter stream flows were small and forest clearing produced an
advance in the initiation of snowmelt. Stream flow increased
during the first month of the major snowmelt period but the effect
of snowmelt on flows was truncated compared with undisturbed
basins. The total volume of snowmelt-produced runoff was unchanged,
it just left the basins in larger volumes over a shorter period.
The application of forest clearing techniques to increase summer
flows should be approached with caution since, under certain
combinations of soil moisture and precipitation, complete forest
clearing can increase floods during summer. Hornbeck (1973)
states that forest clearing can significantly affect flood flows
in basins where a large proportion of the area is cleared, but
that in major forested basins small amounts of timber harvesting
or land development will affect downstream flood potential very
little.

What does this mean for inorganic transport? If we assume that conversion of forested lands to agricultural, industrial or urban uses will result in changes in peak flow volumes and durations and that the majority of sediment transport will occur during the season of peak flows, then the season of peak inorganic transport will remain essentially the same, but it may start earlier and end more quickly than previously in areas where snowmelt is involved. Event-related transport may also become more important since the quantities of storm flows may increase and, thus, have more potential for transporting sediment Total annual inorganic transport would be expected to increase for several reasons if vegetative cover is reduced: (1) more erosion will occur and thus more inorganic material would be available for transport, (2) there would be less vegetation to trap eroded soils, and (3) larger peak flow volume and increased total flow would result in greater potential for carrying loads of inorganic or organic materials.

Finally, it is worthwhile to compare the relative amounts of organic and inorganic materials being transported in streams. Unfortunately, there are few systems for which both types of data are available. Likens et al. (1977) report average particulate matter (> 0.45 μm) outputs (kg/ha oven dry weight) for Watershed 6, and the eight-year average was 33.28 kg/ha of which approximately one-third (11.13 kg/ha) was organic, and two-thirds was inorganic material (22.15 kg/ha).

Conclusions and Directions for Future Research and Management:

1. In the case of organic and inorganic matter transport, as in other processes characteristic of streams, the notion that conditions within the valley play a mediating role in the process (Hynes, 1975) holds true.

2. Vegetation species composition and density, in the drainage basin, affects the amount of organic debris and soil which is available for transport, as well as the runoff regime. Land use, then, plays a role in determining the quantities of organic matter and sediment transported annually.

3. The following factors have been found to affect the concentrations of fine and coarse POM being transported in a stream at any one time: stream order, gradient, flow, presence of lakes or marshes upstream, past history of precipitation and season of the year.

4. It is very difficult to obtain meaningful comparisons of POM loads in different streams because the controlling conditions listed in 3 above may differ tremendously between the systems and at the sampling dates to be compared.

5. The season of maximum in-stream transport occurs at the time
    when much material is available to be transported and when
    a relatively large volume of water (the transporting agent)
    is also present.

6. In many regions there are periods of high stream discharge
    which are responsible for 50-90% of total annual OM and
    sediment transport. These periods are usually associated
    with the rainy season, snowmelt or intense storms.

7. There is presently a considerable knowledge of the dynamics
    of OM and sediment transport and its effects on stream
    energetics and biota. A concerted effort should be made
    to apply the existing knowledge in management schemes
    designed to improve or maintain stream water quality and
    fisheries.

8. Use of a standard terminology for POM fractions would be
    beneficial to all involved in studies of these materials.

9. Standard POM sampling apparatus for each size of material
    would help to reduce the numberof problems associated with
    comparisons of results obtained by different workers.
    Unfortunately, the conditions of current and bottom
    substrate, etc. vary from stream to stream. A "standard"
    apparatus for sampling LPOM may work well in southern
    Ontario, but may not function efficiently in the Cascade
    Mountains of Oregon. Nonetheless, any standardization
    that could be done would be of value.

10. Research directed toward more detailed comparisons of OM
    transport in "altered", autotrophic systems with adjacent
    woodland reaches or streams would be valuable at this
    time. Specifically, the impacts of differences in total
    OM export, differences in composition of size fractions
    exported and shifts in seasonality of export should be
    elucidated. This is necessary because the foregoing
    factors ultimately affect the taxonomic composition and
    density of benthos and fish populations. If land use
    practices and management of stream banks have as great
    an effect on POM and sediment transport as existing
    evidence suggests, the findings of such a study could be
    utilized to develop planning and management guidelines
    directed toward minimizing the impact of land use
    alteration.

11. Efforts directed toward modeling detritus dynamics in
    streams should be developed to the extent that changes in
    land use adjacent to streams could be simulated. The

effects of changes in POM inputs and alterations in
stream discharge regimes which result from land use
impacts could be tested and verified. Again, these
findings could be used to develop effective land use
planning and stream management procedures.

Acknowledgements: I acknowledge the support of Ecologistics
Limited in providing me with secretarial and graphical services,
and Dr. D.J. Coleman and D.R. Cressman for useful comments on a
draft of this review.

REFERENCES

Anderson, H.W., Hoover, M.D. and Reinhart, K.G. 1976. Forests
    and Water: Effects of Forest Management on Floods,
    Sedimentation and Water Supply. U.S.D.A., Forest Service.
    Gen. Tech. Rep. PSW – 18/1976.
Bilby, R.E. and Likens, G.E. 1979. Effect of hydrologic
    fluctuations on the transport of fine particulate organic
    carbon in a small stream. *Limnol. Oceanogr.* 24: 69–75.
Boling, R.H. Jr., Goodman, E.D., Van Sickle, J.A., Zimmer, J.O.,
    Cummins, K.W., Petersen, R.C. and Reice, S.R. 1975a. Towards
    a model of detritus processing in a woodland stream.
    Ecology 56: 141–151.
Boling, R.H. Jr., Petersen, R.C. and Cummings, K.W. 1975b.
    Ecosystem modeling for small woodland streams. *In*:
    B.C. Patten (ed.) Systems Analysis and Simulation in
    Ecology, Vol. 111. 183–204. Academic Press, New York.
Brinson, M.M. 1976. Organic matter losses from four watersheds
    in the humid tropics. *Limnol. Oceanogr.* 21: 572–582.
Dance, K.W. 1977. Seasonal drift of solid organic matter and
    macroinvertebrates in two adjacent agricultural streams.
    M.Sc. Thesis, University of Waterloo.
Dance, K.W., Hynes, H.B.N. and Kaushik, N.K. 1979. Seasonal
    drift of solid organic matter in two adjacent streams.
    Archiv. Hydrobiol. 87: 139–151.
Dance, K.W. and Hynes, H.B.N. 1979. A continuous study of the
    drift in adjacent intermittent and permanent streams.
    Archiv. Hydrobiol. 87: 253–261.
Dance, K.W. and Hynes, H.B.N. 1980. Some effects of
    agricultural land use on stream insect communities.
    Environ. Pollut. 22: 19–28.
Dawson, F.H. 1976. Organic contribution of stream edge
    forest litter fall to the chalk stream ecosystem. *Oikos*
    27: 13–18.
Decamps, H. and Casanova-Batut, T.H. 1978. Les matieres en
    suspension et la turbidite de l'eau dans la Riviere Lot.
    Annls. Limnol. 14: 59–84.
De La Cruz, A.A. and Post, H.A. 1977. Production and transport
    of organic matter in a woodland stream. *Arch. Hydrobiol.*
    80: 227–238.

Dickinson, W.T., Scott, A. and Wall, G. 1975. Fluvial
    sedimentation in southern Ontario. *Can. J. Earth Sci.*
    12: 183–189.
Dyvik, R. and Gilchrist, C. 1977. Comparison and evaluation of
    two bedload samplers in the Hudson River. Unpublished M.
    Eng. Thesis. Rensselaer Polytechnic Institute. Troy, New
    York.
E.P.A. 1978. Nonpoint Source Control Guidance, Agricultural
    Activities. EPA 440/3-78-001. Washington, D.C.
Fisher, S.G. and Likens, G.E. 1973. Energy Flow in Bear Brook,
    New Hampshire: An integrative approach to stream ecosystem
    metabolism. *Ecol. Monogr.* 43: 421–439.
Gelroth, J.V. and Marzolf, G.R. 1978. Primary production and
    leaf litter decomposition in natural and channelized
    portions of a Kansas stream. *Am. Midl. Nat.* 99: 238–243.
Gosz, J.R., Likens, G.E. and Bormann, F.H. 1972. Nutrient
    content of litter fall on the Hubbard Brook Experimental
    Forest, New Hampshire. *Ecology* 53: 769–784.
Gosz, J.R., Likens, G.E. and Bormann, F.H. 1976. Organic
    matter and nutrient dynamics of the forest and forest
    floor in the Hubbard Brook Forest. *Oecologia* 22: 305–320.
Harrison, A.D. and Rankin, J.J. 1975. Forest litter and
    stream fauna on a tropical island, St. Vincent, West
    Indies. *Int. Ver. Theor. Angew. Limnol. Verh.* 19: 1736–1745.
Hartman, J.P., Wanielista, M.P. and Baragona, G.T. 1976.
    Prediction of soil loss in nonpoint-source pollution
    studies. *In*: Soil Erosion - Prediction and Control -
    Proceedings of Conference. Pp. 298–302. Special
    Publication No. 21. Soil Conservation Society of
    America, Ankeny, Iowa.
Hodkinson, I.D. 1975. Energy flow and organic matter
    decomposition in an abandoned beaver pond ecosystem.
    Oecologia 21: 131–139.
Hornbeck, J.W. 1973. Storm flow from hardwood-forested and
    cleared watersheds in New Hampshire. *Water Resour. Res.*
    9: 346–354.
Hornbeck, J.W. 1975. Stream flow response to forest cutting
    and revegetation. *Water Resour. Bull.* 11: 1257–1260.
Hynes, H.B.N. 1975. The stream and its valley. *Int. Ver.
    Theor. Angew. Limnol. Verh.* 19: 1–15.
Kaushik, N.K., Robinson, J.B., Sain, P., Whiteley, H.R. and
    Stammers, W. 1975. A quantitative study of nitrogen
    loss from water of a small, springfed stream. Pp. 110–117.
    Proc. 10th Canadian Symp. Water Poll. Research Canada.
Likens, G.E. and Bormann, F.H. 1974. Linkages between
    terrestrial and aquatic ecosystems. *Bio. Science* 24:
    447–456.
Lush, D.L. and Hynes, H.B.N. 1978. Particulate and dissolved
    organic matter in a small partly forested Ontario stream.
    Hydrobiologia 60: 177–185.

Mann, K.H. 1979. Book Review - Rzoska, J. 1978. On the Nature of Rivers, with Case Stories of the Nile, Zaire and Amazon. *Limnol. Oceanogr.* 24: 1177-1178.

McIntire, C.D., Colgy, J.A. and Hall, J.D. 1975. The dynamics of small lotic ecosystems: a modeling approach. *Int. Ver. Theor. Angew. Limnol. Verh.* 19: 1599-1609.

McIntire, C.D. and Colby, J.A. 1978. A hierarchial model of lotic ecosystems. *Ecol. Monogr.* 48: 167-190.

Moore, J.W. 1978. Seasonal succession of algae in rivers III. Examples from the Wylye, a eutrophic farmland river. *Arch. Hydrobiol.* 83: 367-376.

Naiman, R.J. 1976. Primary production, standing stock, and export of organic matter in a Mohave Desert thermal stream. *Limnol. Oceanogr.* 21: 60-73.

Naiman, R.J. and Sibert, J.R. 1978. Transport of nutrients and carbon from the Nanaimo River to its estuary. *Limnol. Oceanogr.* 23: 1183-1193.

Petersen, R.C. and Cummins, K.W. 1974. Leaf processing in a woodland stream. *Freshwat. Biol.* 4: 343-368.

Phillipson, J., Putnam, R.J., Steel, J. and Woodell, S.R.J. 1975. Litter input, litter decomposition and the evolution of carbon dioxide in a beech woodland - Wytham Woods, Oxford. *Oecologia* 20: 203-217.

Post, H.A. and De La Cruz, A.A. 1977. Litterfall, litter decomposition, and flux of particulate organic material in a coastal plain stream. *Hydrobiologia* 55: 201-207.

Rau, G.H. 1976. Dispersal of terrestrial plant litter into a subalpine lake. *Oikos* 27: 153-160.

Reice, S.R. 1974. Environmental patchiness and the breakdown of leaf litter in a woodland stream. *Ecology* 55: 1271-1282.

Rich, L.R. and Gottfried, G.J. 1976. Water yields from treatments on the Workman Creek Experimental Watersheds in Central Arizona. *Water Resour. Res.* 12: 1053-1060.

Sedell, J.R., Triska, F.J., Hall, J.D., Anderson, N.H. and Lyford, J.H. 1974. Sources and fates of organic inputs in coniferous forest streams. *In*: R.H. Waring and R.L. Edmonds (eds) Integrated Research in the Coniferous Forest Biome. Coniferous Biome Bulletin. Pp. 57-69. No. 5, University of Washington, Seattle.

Sedell, J.R. and Triska, F.J. 1976. Biological consequences of large organic debris in Northwest streams. Logging Debris in Streams Workshop. 10 pp.

Swanson, C.D. and Bachmann, R.W. 1976. A model of algal exports in some Iowa streams. *Ecology* 57: 1076-1080.

Swanson, F.J., Lienkaemper, G.W. and Sedell, J.R. 1976. History, Physical Effects and Management Implications of Large Organic Debris in Western Oregon Streams. U.S.D.A. For Serv., Pac. Northwest For. and Range Expt. Sta. Gen. Tech. Rep. PNW-56.

Van Vliet, L.J.P., Wall, G.J. and Dickinson, W.T. 1978. Soil
    Erosion from Agricultural Land in the Canadian Great Lakes
    Basin. IJC report of PLUARG project, Windsor, Ontario.
Walling, D.E. 1974. Suspended Sediment Production and
    Building Activity in a small British Basin. In: Effects
    of Man on the Interface of the Hydrological Cycle with the
    Physical Environment - Symposium Proceedings. Pp. 137-144.
    Publication No. 113. Internat. Assoc. of Hydrological
    Sciences.
Wetzel, R.G. and Rich, P.H. 1973. Carbon in freshwater
    systems. In: G.W. Woodwell and E.V. Pecan (eds) Carbon
    and the Biosphere. Proc. 24th Brookhaven Symposium in
    Biology. U.S. Atomic Energy Commission Symposium Ser.
    CONF - 720510. Pp. 241-263. NTIS, Springfield, Virginia.
Wetzel, R.G. and Manny, B.A. 1977. Seasonal changes in
    particulate and dissolved organic carbon and nitrogen in
    a hardwater stream. Arch. Hydrobiol. 80: 20-39.
Wong, S.L. and Clark, B. 1976. Field determination of the
    critical nutrient conentrations for Cladophora in streams.
    J. Fish. Res. Bd. Can. 33: 85-92.
Young, S.A., Kovalak, W.P. and Del Signore, K.A. 1978.
    Distance travelled by autumn-shed leaves introduced into
    a woodland stream. Am. Midl. Nat. 100: 217-222.
Zimmie, T.F., Paik, Y.S. and Floess, C.H.L. 1976a. Evaluation
    of a bedload sediment sampler for U.S.A. - Canada
    watershed studies. In: Environmental Aspects of Irrigation
    and Drainage. Pp. 40-55. ASCE, New York.
Zimmie, T.F., Paik, Y.S. and Floess, C.H.L. 1976b. Evaluation
    of a Bogardi T-3 bedload sampler. Civil Engineering
    Department, Rensselaer Polytechnic Institute, Troy, New
    York.

ADDENDUM

Since this paper was written several additional papers have been
published. These have not been reviewed but should be consulted
by interested readers since they contain data and ideas pertinent
to the above discussion.

Dawson, F.H. 1980. The origin, composition and downstream
    transport of plant material in a small chalk stream.
    Freshwat. Biol. 10: 419-435.
Erman, D.C. and Chouteau, W.C. 1979. Fine particulate organic
    carbon output from fens and its effects on benthic macro-
    invertebrates. Oikos 32: 409-415.
Gurtz, M.E., Webster, J.R. and Wallace, J.B. 1980. Seston
    dynamics in southern Appalachian streams: effects of clear-
    cutting. Can. J. Fish. Aquat. Sci. 37: 624-631.
Meyer, J.L. and Likens, G.E. 1979. Transport and transformation
    of phosphorus in a forested stream ecosystem. Ecology
    60: 1255-1269.

Naiman, R.J. and Sedell, J.R.   1979.   Benthic organic matter as
     a function of stream order in Oregon.   *Arch. Hydrobiol.*
     87: 404-422.
Sedell, J.R., Naiman, R.J., Cummins, K.W., Minshall, G.W. and
     Vannote, R.L.   1978.   Transport of particulate organic
     material in streams as a function of physical processes.
     Verh. Internat. Verein. Limnol.   20: 1366-1375.
Webster, J.R.   1977.   Large particulate organic matter
     processing in stream ecosystems.   *In*: D.L. Correll (ed.)
     Watershed Research in Eastern North America.   Volume II.
     Pp. 505-529.   Chesapeake Bay Center for Environmental
     Studies.

THE UPTAKE OF DISSOLVED ORGANIC MATTER IN GROUNDWATER BY STREAM SEDIMENTS - A CASE STUDY

P.M. Wallis

Kananaskis Centre for Environmental Research
University of Calgary
Calgary, Alberta, Canada

INTRODUCTION

The investigation of energy flow in streams has caused considerable interest in the decomposition of dissolved allochthonous organic matter. Dissolved organic matter (DOM) has been shown to be the most important source of organic energy in headwater streams (Wetzel and Manny, 1972; Hobbie and Likens, 1973; Fisher and Likens, 1973) and the pool of DOM in small streams is thought to be in rapid, dynamic equilibrium (Manny and Wetzel, 1973). The source of this organic material is generally considered to be of terrestrial origin (Hynes, 1963; Hynes, 1970; Fisher and Likens, 1973; McDowell and Fisher, 1976; Cummins et al., 1972) but, until recently, it has been assumed that the leaching of dead vegetation which fell into the stream was the only important source of dissolved organic matter. It is known, however, that groundwater is the major component of stream flow throughout the year (Viessman et al. 1977; Freeze, 1974) and that it contains DOM (Kasper and Knickerbocker, 1980; Whitelaw and Edwards, 1980; Leenheer et al., 1974; Wallis et al., in press). The possibility that this DOM contributes large amounts of energy to stream production has not been widely recognized. Fisher and Likens (1973) realized that subsurface inputs of DOM to Bear Brook constituted 21% of the organic energy budget but, because they could find no concentration difference between spring water and stream water, they concluded that groundwater DOM neither diluted nor enriched the stream. This belief was supported by the knowledge that DOM in natural ground and surface waters is refractory in nature, consisting mostly of large molecular weight humic and fulvic acids (Larson, 1978; Otsuki and Wetzel, 1973; de Haan, 1972; McDowell and Fisher, 1976). These substances are not easily degradable and were considered to be chemically and biologically conservative. This

97

paper presents evidence to the contrary and seeks to show that DOM
from groundwater sources can contribute significant amounts of
energy to the detrital pathways in streams. It is important to
note that the "dissolved" organic matter referred to throughout this
study actually consists of molecules in a spectrum of sizes and
states of hydration ranging from truly dissolved through to
colloidal organic species (Lock et al., 1977).

The uptake of DOM in freshwaters is believed to be primarily
a bacterial process (Sepers, 1977; Lock and Hynes, 1976; Wetzel and
Manny, 1972) although fungi, plants, and even benthic invertebrates
may also be involved. Bacteria occur as free-floating organisms in
the water column (planktonic) and as sessile communities attached
to rocks and other objects on the stream bed. Geesey et al. (1978)
have shown that sessile bacteria are numerically much more abundant
than the planktonic forms while Ladd et al. (1979) demonstrated that
sessile bacteria are at least eight times more effective at removing
organic substances from solution in the same stream system (Marmot
Basin; near Calgary, Alberta). Wetzel (1967) found that the uptake
of simple organic substrates increased markedly at the sediment/
water interface of lakes. Lock and Hynes (1976) and Lock and
Wallace (1979) also concluded that sessile benthic bacteria were the
cause of most of the degradation of DOM in several other Canadian
streams. It is also now apparent that these sessile bacteria are
found in a carbohydrate-based slime matrix along with large
populations of algae (Geesey et al., 1978; Lock Chapter 1). If
groundwater is being discharged to a stream from below through the
stream bottom (as it usually is, at least in headwater systems)
then the organic compounds dissolved in it will be exposed to this
sessile microbial community. Groundwaters are usually low in oxygen,
cold, and reducing with respect to Eh (Freeze and Cherry, 1979) and
their emergence into a stream means that they will be exposed to
oxygen and warmer temperatures. Additionally, nutrient concentrations
in the stream may also be higher than those in groundwater, so the
transition from ground to stream water represents a movement toward
more favourable conditions for microbial growth and metabolism.

STUDY AREAS

Field experiments were carried out in the Marmot Basin located
in the Kananaskis Valley near Calgary, Alberta, Canada. (Canadian
Topographic Series Map 82J/14; 115° 10' W., 50° 57' N.) The Marmot
stream system drains the eastern slopes of Mount Allen (2743 m),
a predominantly limestone mountain composed of late Jurassic and
early Cretaceous rock. The catchment basin is approximately 8.2 km^2
in area and is drained by three streams (Twin Creek, Middle Creek,
and Cabin Creek) which are confluent. Stream waters are cold
(average temperature 3.0°C), clear, and of medium hardness (about
110 mg/ℓ as $CaCO_3$). The basin is forested to approximately 2300 m
mainly by Lodgepole Pine (*Pinus contorta*), an Engelmann Spruce hybrid

(*Picea engelmannii* x *Picea glauca*), Fir (*Abies lasiocarpa*), and Alpine Larch (*Larix lyalli*). Below the timberline hybrid willows (*Salix glauca* Linn. x *myrtillifolia* Anderss.), bristly black currant (*Ribes lacustre* Poir), and horsetails (*Equisetum arvense* Linn. and *Equisetum scirpordea* Michx.) dominate the vegetation along the streams (Mutch 1977). Experiments were carried out in the summer of 1977 when the average atmospheric temperature was about 10.0°C and vegetation cover was at a maximum. The streams are all well shaded and the bacterial populations in the sessile slime matrix were about 10^7 cells cm^{-2} (Geesey et al., 1978). Sediment samples for laboratory experiments were taken from Laurel Creek near the University of Waterloo in Southwestern Ontario, a small hardwater stream draining agricultural land. A description of Laurel Creek appeared in Lush and Hynes (1978).

MATERIALS AND METHODS

Field Studies - Marmot Basin: Field uptake experiments were carried out in the Middle Creek Tributary, a small first order stream with an average discharge of about 1.2 ℓ s^{-1} (Figure 1). The Middle Creek Tributary is fed by two small springs at the source and other seepage areas downstream. The stream is only 150 m long and was entirely fed by groundwater during the course of these experiments. By measuring the discharge above and below the experimental reach, the amount of groundwater entering the stream may be calculated by difference. The amount of DOM in the ground water was assumed to be the same as the average concentration found in the piezometers installed around the source springs (Figure 1). The concentration of DOM in groundwaters sampled through these piezometers did not vary by more than 0.3 mg ℓ^{-1} and it is assumed that groundwater discharged evenly over the experimental reach. By knowing the concentration of DOM in groundwater, the amount of groundwater discharged, and the area of the bottom of the stream, it was possible to calculate the uptake of DOM per square metre. The concentration of DOM was measured by the method of Maciolek (1962) after filtration through combusted (550°C for 3 hours) Reeve Angel 934 AH glass fiber filters with a measured standard error of 0.1 mg ℓ^{-1} over 10 replications in the concentration range 0–8 mgC^1.

Discharge was measured at sites A, B, and C by dilution gauging using Cl^- (from NaCl) as a tracer. The salt solution was dripped into the stream at a steady known rate using a constant head bottle. The discharge at each site was calculated using the equation:

$$Q = q \ \frac{(C_1 - C_2)}{(C_2 - C_0)}$$

, where Q is the stream discharge, q is the injection rate of the tracer (in ℓ s^{-1}), C_1 is the concentration of tracer in the bottle, C_2 is the concentration of tracer in the stream

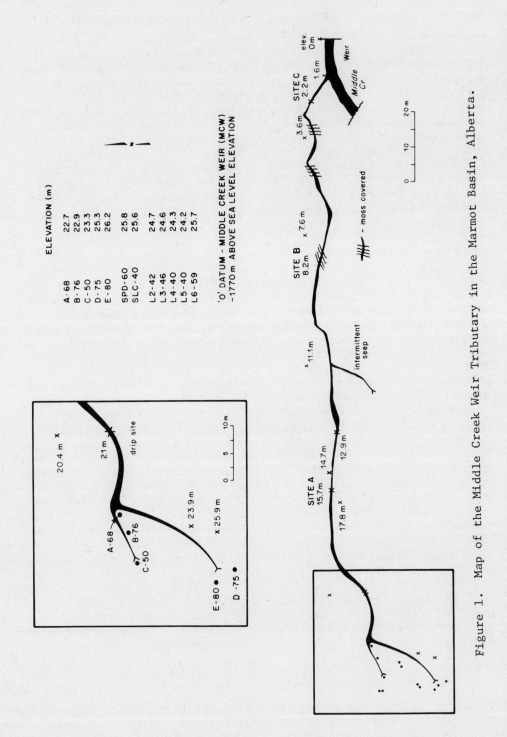

Figure 1. Map of the Middle Creek Weir Tributary in the Marmot Basin, Alberta.

at the measurement site, and C_0 is the background concentration of tracer in the stream.

Measurements of the uptake of groundwater DOM were made on four separate occasions in the summer of 1977. The calculation of the increase in discharge between sites B and C provides a measurement of the amount of groundwater entering the stream. The reach between A and B was not used for experimental measurements because small tributaries enter the stream after passing over marshy ground and an accurate measurement of the total surface area would have been impossible. This groundwater discharge figure may be multiplied by the concentration of groundwater DOM as estimated by averaging the DOM measured in the piezometers on that date. The load of DOM in the stream may be calculated by multiplying the measured discharge by the concentration of DOM at each site. The difference between the amount of DOM which enters the stream in groundwater and that which leaves the experimental reach is considered to represent the net uptake of DOM over the reach from site B to site C. Careful measurements of the area of the stream bed permitted this figure to be converted to uptake per square metre. It should be noted that the amount of DOM measured at each site downstream was constant throughout each experiment.

^{13}C *studies:* ^{12}C is preferentially taken up by living organisms because of the greater amount of energy required to utilize ^{13}C (Deevey and Stuiver, 1964; Degens, 1969). As a result, terrestrial plants are characteristically 'light' with respect to the stable isotope as measured in comparison with the PDB standard (in parts per thousand, $^o/_{oo}$), commonly having a ^{13}C content of $-25\ ^o/_{oo}$ (Deevey and Stuiver, 1964). As vegetation in the basin is fairly uniform on the upper reaches, it is assumed that plant material will have a uniform ^{13}C content. It is further assumed that organic matter which has been reduced to refractory material, such as humic and fulvic acids, should be heavier in ^{13}C than before it was subjected to microbial attack. In other words, the microbial community will selectively take up ^{12}C over ^{13}C leaving the remaining material 'heavier' than when it was first leached from the plant material. By taking samples of throughfall (precipitation which has dripped through the canopy) and stream water and analyzing for ^{13}C it was hoped that some light would be shed on the uptake of dissolved organic matter in the basin.

As the dissolved organic content of Marmot waters is low (average 2.1 mg ℓ^{-1} – Wallis et al., in press) 10 litres of sample must be processed to obtain approximately 20 mg of organic carbon to ensure accurate detection using a mass spectrometer. After filtration (to 1 μm) organic carbon is converted to CO_2 gas by oxidation at low pH and high temperature and is swept out of the sample by a stream of pure O_2. This stream of gas is then bubbled through a solution of $Ba(OH)_2$ which precipitates the CO_2 as $BaCO_3$.

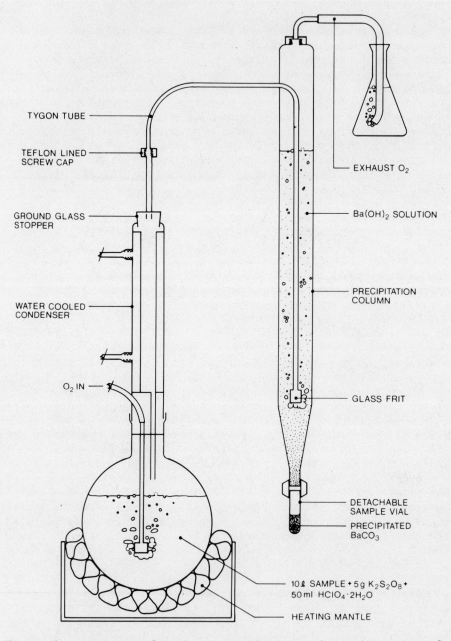

Figure 2.   Apparatus for the wet oxidation and precipitation of
            Dissolved Organic Matter.

The design of the apparatus used to carry this out was adapted from
that of Dr. J. Barker and Dr. P. Fritz of the Department of Earth
Sciences of the University of Waterloo (Figure 2). Wet oxidation
was carried out using 5 g of $K_2S_2O_8$ and 50 ml of $HClO_4 \cdot H_2O$ per 10
litres of water sample, and $Ba(OH)_2$ was prepared by dissolving 100 g
of $BaCl_2$ and 50 g of NaOH in 2.5 litres of water. The wet
oxidation mixture was boiled for one hour while being sparged with
pure O_2. Precipitated $BaCO_3$ was collected in a culture tube with a
teflon lined screw cap containing $Ba(OH)_2$ and stored for analysis.

In order to prepare the sample for the mass spectrometer, it
was decanted and the slurry was poured into a 'quick fit' flask with
a side arm. Dilute HCl was added to convert the $BaCO_3$ to CO_2 which
is then withdrawn by vacuum, water is removed by differential
freezing, and stored in a special tube as CO_2 gas. Sample preparation
and analysis were done using a VG Micromass 602D Twin double collector
mass spectrometer. Samples were taken from the source spring of
Middle Creek in the Marmot Basin, from Middle Creek at a weir about
2 km downstream, from the source of the Middle Creek Tributary used
in the first field experiment, and from throughfall caught by a
plastic sheet. Unfortunately it was not possible to collect enough
groundwater for analysis from the small diameter piezometers available.

Laboratory studies: In order to gain a more direct measurement of
sediment uptake of DOM in groundwater during stream base flow
conditions, a laboratory experiment was undertaken at the University
of Waterloo. A series of glass bowls was modified to contain
sediment on a screen two centimeters from the bottom. A central
core was equipped with an inlet tube which allowed water to seep
under this permeable plate and up through the sediment. The water
was then allowed to overflow into the central core and was carried
to an exhaust container (Figure 3). Fresh sediment was taken from
Laurel Creek about 2 km below Sunfish Lake by chipping a hole
through the ice and scooping it up with a shovel. It was then
returned to the laboratory and equal portions were spread about in
each bowl. The entire apparatus was kept in an environmental chamber
and held at $5^{\circ}C$ in a 12 hour light regime. Groundwater was obtained
on the same day from a nearby seepage well (dug) located on the
North Campus of the University of Waterloo. This well was only about
1 km from Laurel Creek but it was 5 km from the sediment sampling
site. No better groundwater sampling site was available. The
groundwater was allowed to seep through four identical sediment
bowls and was analyzed for DOM before and after passage through
them. Streamwater sampled from Laurel Creek was allowed to percolate
through four identical bowls at the same time for a control. Matched
constant head bottles were employed to facilitate a constant input
rate between the experimental bowls and the control bowls. However,
the resulting rates of seepage between the bowls were not even but,
as this condition also exists in nature, this was tolerated. The

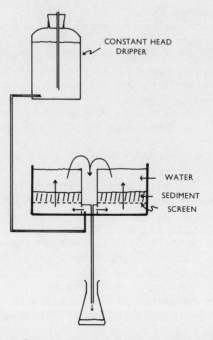

Figure 3. Apparatus for measuring the uptake of Dissolved Organic
           Matter by sediments from Laurel Creek, Ontario.

experiment was repeated twice within 48 hours of sampling so that
eight figures were produced for both the control and the groundwater side. DOM was analyzed using a Beckman 915 Total Organic
Carbon analyzer after filtration through a combusted Reeve Angel
934-AH glass fiber filter. The precision of analysis was
$\pm$ 1.0 mg ℓ^{-1} or 5% in the range of 0-100 mg ℓ^{-1}.

RESULTS

Field studies - Marmot Basin: The results from the calculation of
the uptake of groundwater DOM in the Middle Creek Tributary appear
in Table 1. The average uptake was calculated to be 207 mg m^{-2} h^{-1}
with a standard error of 52.5 mg m^{-2} h^{-1}. This high level of
uptake suggests that stream bacteria are capable of removing
portions of the dissolved organic load which groundwater bacteria
are not despite long periods of exposure to soil micro-organisms
(Wallis et al., 1979). The rocky substrata of the Marmot streams
is covered by a thick layer of microbial slime ($\sim$10^7 cells cm^{-2})
and this community is apparently capable of removing large amounts
of DOM from groundwater percolating upward through the stream
sediments.

Table 1. Uptake of Groundwater DOM in the Middle Creek Tributary

Date 1977	Q-GW[1] (l s^{-1})	DOM in GW (mg l^{-1})	DOM in trib. (mg l^{-1})	Load[2] (mg s^{-1})	Uptake[3] (mg m^{-2} h^{-1})
4/7	1.4	3.4	1.9	2.1	207
10/7	1.7	2.3	1.8	0.9	89
21/7	0.7	5.2	2.5	1.9	187
27/7	1.2	4.8	1.9	3.5	345

Overall average uptake (mg m^{-2} h^{-1}) = 207 ± 52.5 (1 S.E.)

Notes:

[1] Q-GW is calculated as the difference in discharge between sites 'B' and 'C'.

[2] Load is calculated by multiplying the difference between DOM in Groundwater (GW) and DOM in trib. by the discharge (Q-GW).

[3] Uptake is calculated by dividing the Load by the area of the stream between 'B' and 'C' (36.5 m^2) and multiplying by 3600 sec/hr.

^{13}C *studies - uptake:* The results of ^{13}C analysis from organic matter appear in Table 2 and the agreement between replicates is good, indicating that the analytical procedure is not artificially fractionating ^{12}C from ^{13}C. It can be seen that throughfall is significantly lighter in ^{13}C than streamwater and the data are interpreted as follows. Terrestrial plants are light in ^{13}C and it is not surprising that organic matter leached from them contains very little of the isotope ($-31.5^{o}/_{oo}$). As this organic matter enters the soil and is exposed to microbial degradation, ^{12}C is selected from it and ^{13}C tends to accumulate in the remaining organic matter. Groundwater samples taken from the unsaturated zone above the water table contain large quantities of DOM compared with stream water (average 21.2 mg/l in the unsaturated zone compared with an average of only 2.2 mg/l in the stream; Wallis et al., in press) suggesting a great potential for the concentration of ^{13}C by the time groundwater reaches the stream. Because of the difficulty in obtaining enough groundwater for ^{13}C analysis, it is not possible to say how much uptake occurred in the soil and how much took place as the groundwater passed through the stream sediments. It is clear, however, that concentration of ^{13}C is taking place and, by inference, the degradation of dissolved organic matter.

Table 2. ^{13}C in parts per thousand (^{o}/$_{oo}$) in Dissolved Organic
           Matter sampled in the Marmot Basin, Alberta.   Each
           sample was analyzed twice.

Sample	Date 1977	^{13}C ^{o}/$_{oo}$	^{13}C ^{o}/$_{oo}$	Mean ^{o}/$_{oo}$
Throughfall	4/8	-31.5	-31.5	-31.5
Throughfall	6/8	-27.7	-31.5	-29.6
Middle Cr. Source	11/8	-21.2	-17.7	-19.5
Middle Cr. Weir	13/8	-21.6	-21.6	-21.6
Middle Cr. Tributary	7/8	-25.7	-24.4	-25.1

The ^{13}C content of streamwaters sampled at the Middle Creek
Weir and from the mouth of the Middle Creek Tributary are both
lighter in ^{13}C than the source spring for Middle Creek.   This
suggests that additional organic matter which has not been so
efficiently degraded has entered the stream between the two sites.
Presumably this has been leached from terrestrial vegetation which
has fallen into the stream and has not been completely decomposed.
More samples would be required to prove this point but such a
result would not be unexpected.   It would not be surprising to
learn that the complete decomposition of this plant material
might take several weeks.

Laboratory studies - uptake experiments: The results from the
laboratory uptake experiments appear in Table 3.   Although
seepage rates varied from 0.09 to 18.8 ℓ m^{-2} h^{-1}, a linear
regression of seepage rate against uptake yields a correlation
coefficient of only 0.24 so they are not significantly related.
The groundwater half of the experiment showed a definite
decrease in concentration of organic matter while the control
side showed a slight increase in DOM.   The independent t-test
indicated a significant difference between the experiment and the
control at the 99% confidence level (t = 5.0, 15 df).   The average
drop in DOM during Experiment 1 was 3.7 mg ℓ^{-1} (S.D. = 3.1) and
3.4 mg ℓ^{-1} (S.D. = 1.1) during Experiment 2.   This corresponds to a
net uptake of 16.4 ± 3.6 mg m^{-2} h^{-1}, compared with 207 mg m^{-2} h^{-1}
measured in the Marmot Basin.   Although a low level of uptake was
measured by this experiment during winter low flow conditions, a
similar experiment performed in the spring or summer might have
different results.

Table 3. Uptake of groundwater DOM by sediments of Laurel Creek, Ontario.

Bowl No.	Source	DOM Input (mg l^{-1})		Change in DOM (mg l^{-1})		Seepage rate (l m^{-2} h^{-1})		DOM Uptake (mg m^{-2} h^{-1})	
		Expt 1	Expt 2	Expt 1	Expt 2	Expt 1	Expt 2	Expt 1	Expt 2
1	SW[1]	29.3	31.9	+1.3	+4.8	18.8	18.5	–	–
2	SW[1]	29.3	31.9	−1.0	+2.4	10.6	10.9	–	–
3	SW[1]	29.3	31.9	+3.0	+1.8	5.1	3.9	–	–
4	SW[1]	29.3	31.9	+0.5	−0.04	7.9	8.8	–	–
5	GW[2]	40.6	49.6	−8.0	−2.0	5.4	6.0	43.2	12.0
6	GW[2]	40.6	49.6	−0.8	−3.4	4.9	4.6	3.9	15.6
7	GW[2]	40.6	49.6	−2.2	−3.4	0.7	0.09	1.5	0.3
8	GW[2]	40.6	49.6	−3.8	−4.7	6.0	6.7	22.8	31.5

Overall average uptake of DOM (mg m^{-2} h^{-1}) = 16.4 ± 3.6 (1 S.E.)

[1] Streamwater

[2] Groundwater

It is also possible that the bacteria present on the sediments were not pre-adapted to the DOM which passed through the sediment, as the groundwater was taken from a well about 1 km from the stream. Sediment disturbance during transport back to the laboratory may have also influenced the uptake rate but it was not anaerobic and its temperature was not changed by more than 5°C. Finally, it is noteworthy that the DOM levels in the stream were 10 mg l^{-1} lower than the groundwater suggesting that a longer exposure time might have reduced the groundwater DOM to the lower level found in the stream.

SYNTHESIS

The decrease in concentration of DOM in groundwaters as they percolate into the stream through the sediments indicates that a considerable quantity of energy is made available to sessile micro-organisms from this source. This organic matter is mobilized from the humus layer of the soil by infiltrating precipitation and carried through the unsaturated and saturated groundwater zones. Much of this DOM (it is assumed that very little particulate organic matter is transported through soil) is undoubtedly mineralized by soil microbial communities (Wallis et al., in press) but the geochemical conditions in the groundwater zone often do not favour complete decomposition (low oxygen content, temperature, pH, reducing Eh).

Conditions in streams are more suited to microbial growth and it
appears that a further decrease in DOM concentration is possible.
Wallis et al. (1979) found that the concentration of DOM decreased
steadily in groundwater as it moved toward the stream (21.2 mg ℓ^{-1}
in the unsaturated zone, 5.5 mg ℓ^{-1} in saturated zone groundwater,
and 2.2 mg ℓ^{-1} in the stream) in the Marmot Basin. This decrease
in concentration is reflected by the accumulation of ^{13}C in groundwater DOM.

Despite the high uptake rates found during field studies, DOM
in groundwater must still be considered to be refractory in nature
as suggested by McDowell and Fisher (1976). It has been shown,
however, in the literature that even these compounds are at least
partially degradable, usually by a sequence of reaction steps
which involve numerous by-products (Romenenko, 1979; Stabel et al.,
1979; Fleischer and Gahnstroem, 1979; Cappenberg, 1979; Clesceri
et al., 1977). Pitter (1976) has shown that even many aromatic
and toxic organic chemicals can be degraded by sewage bacteria and
it is generally believed that bacteria can break down almost any
organic compound given enough time (Wistreich and Lechtman, 1980).
Reaction rates are slow, however, and a complex assemblage of
microorganisms is required for efficient degradation. If undisturbed
ecosystems are truly more diverse with respect to community structure
then DOM might be expected to accumulate in polluted streams. This
is suggested by the differences in uptake rates found between the
undisturbed Middle Creek Tributary and Laurel Creek which drains
agricultural land. Wallis et al. (1979) also concluded that DOM
disappeared more rapidly in groundwaters under undisturbed forest
soils than from groundwaters beneath an adjacent cultivated field.
This hypothesis remains unproven but further research might lead to
some interesting and useful conclusions.

The mass balance approach used in the Marmot Basin is very
similar to that employed by McDowell and Fisher (1976) in Roaring
Brook. Their work was carried out in the autumn and they have
distinguished between DOM from groundwater sources and from the
leaching of fresh deciduous leaves which had fallen into the stream.
The Marmot experiments were undertaken in the summer in a coniferous
forest and the input of needles (which are more difficult to degrade
in any case) cannot be considered equal to deciduous leaf input in
the fall. As a result, the DOM in the Marmot streams is probably
more refractory than that in Roaring Brook. McDowell and Fisher
(1976) reported that the uptake of DOM over the autumn leaf fall
period (77 days) was 53 g m^{-2} or 29 mg m^{-2} h^{-1}. Although they
considered their results to be high, the uptake rates measured in
the Marmot Basin appear to be almost an order of magnitude higher
(207 mg m^{-2} h^{-1}). The results from Laurel Creek are more comparable
(16.4 mg m^{-2} h^{-1}). Using similar methods, Fisher and Likens (1973)
calculated an uptake rate of 57 mg m^{-2} h^{-1} averaged over a whole
year. The differences between Roaring Brook, Bear Brook, and the

Marmot Basin may have been caused by their differing inorganic water chemistry as the former two are soft water streams compared with the hard water of the Middle Creek Tributary. Lock and Hynes (1975) found that soft waters and their sediments had a lower and more variable potential for taking up DOM leached from leaves than did hard waters and their associated sediments. It is also generally true that benthic communities are less productive in soft water streams than in hard water streams.

Another possible reason for the difference in uptake results between Roaring Brook and the Middle Creek Tributary may lie in the hydrological differences between the two basins. Roaring Brook is described as a "bedrock chute" with numerous exposures of bedrock and thin soils (McDowell and Fisher, 1976) suggesting that there may be no significant inputs of groundwater along the stream channel. This is not the case in the Marmot Basin where groundwater inputs are more continuous. Measurements of the uptake of groundwater DOM by sediments where they did enter Roaring Brook, however, might change the total uptake rate for the stream. In any case, their estimate of ecosystem efficiency (Energy Used/Energy Input = 34%) would not change significantly because an increase would be observed in both the numerator and the denominator.

ACKNOWLEDGEMENTS

This work was carried out while the author was a graduate student of Dr. H.B.N. Hynes, without whose advice and encouragement it would not have been possible. This study was supported financially by a Research Subvention from Environment Canada and an operating grant from the Natural Sciences and Engineering Research Council of Canada.

REFERENCES

Cappenberg, Th. E. 1979. Kinetics of breakdown processes of organic matter in freshwater sediments. *Ergebn. Limnol. H.* 12: 91-4.

Clesceri, L.S., Park, R.A. and Bloomfield, J.A. 1977. General model of microbial growth and decomposition in aquatic ecosystems. *Appl. & Envir. Micro.* 33: 1047-1058.

Cummins, K.W., Klug, J.J., Wetzel, R.G., Petersen, R.C., Suberkropp, K.F., Manny, B.A., Weychuck, J.C. and Howard, F.O. 1972. Organic enrichment with leaf leachate in experimental lotic ecosystems. *Bioscience* 22: 719.

Deevey, E. and Stuiver, M. 1964. Distribution of natural isotopes of carbon in Linsley Pond and other New England lakes. *Limnol. Oceanogr.* 9: 1-11.

Degens, E.T. 1969. Biogeochemistry of stable carbon isotopes *In* Organic Geochemistry. Eglinton, G., and Murphy, M.T.J. (eds.) Springer Verlag, N.Y.

Fisher, S.G. and Likens, G.E. 1973. Energy flow in Bear Brook, New Hampshire: an integrative approach to stream ecosystem metabolism. *Ecol. Monogr.* 43: 421-439.

Fleisher, S. and Gahnstroem, G. 1979. Demonstration of the sequential degradation of organic compounds in aquatic systems. *Ergebn. Limnol. H.* 12: 86-90.

Freeze, R.A. 1974. Streamflow Generation. Review of geophysics and space. *Physics* 12: 627-647.

Freeze, R.A. and Cherry, J.A. 1979. Groundwater. Prentice Hall.

Geesey, G.G., Mutch, R., Costerton, J.W., Green. R.B. 1978. Sessile bacteria: An important component of the microbial population in small mountain streams. *Limnol. Oceanogr.* 23: 1214-1223.

Haan, H. de. 1972. Molecule size distribution of soluble humic compounds from different natural waters. *Freshwat. Biol.* 2: 235-241.

Hobbie, J.C. and Likens, G.C. 1973. Output of phosphorous, DOC and fine particulate carbon from Hubbard Brook watersheds. *Limnol. Oceanogr.* 18: 734-742.

Hynes, H.B.N. 1963. Imported organic matter and secondary productivity in streams. Proc. XVI Int. Congr. Ent. Washington. 4: 324-329.

Hynes, H.B.N. 1970. The Ecology of Running Waters. University of Toronto Press.

Kasper, D.R. and Knickerbocker, K.S. 1980. Organic qualities of groundwaters. Completion Report – Office of Water and Technology A-067-ARIZ ref. 14-34-0001-6003.

Ladd, T.I., Costerton, J.W. and Geesey, G.G. 1979. The determination of the heterotrophic activity of epilithic microbial populations. *In* Native Aquatic Bacteria: Enumeration, Activity, and Ecology, pp. 180-195. ed. Costerton, J.W. and Colwell, R.R. ASTM Press: STP 695.

Larson, R.A. 1978. Dissolved organic matter of a low-coloured stream. *Freshwat. Biol.* 8: 91-104.

Leenheer, J.A., Malcolm, R.L., McKinley, P.W., Eccles, L.A. 1974. Occurrence of dissolved organic carbon in selected ground-water samples in the U.S. *J. Res. U.S.G.S.* 2: 361-369.

Lock, M.A. and Hynes, H.B.N. 1975. The disappearance of four leaf leachates in a hard and soft water stream in S.W. Ontario. *Int. Revue ges. Hydrobiol.* 60: 847-855.

Lock, M.A. and Hynes, H.B.N. 1976. The fate of dissolved organic carbon derived from autumn shed maple leaves (*Acer saccharum*) in a temperate hard water stream. *Limnol. Oceanogr.* 21: 436-443.

Lock, M.A. and Wallace, R.R. 1979. Interim report on ecological studies on the lower trophic levels of muskeg rivers within the Alberta Oil Sands Environmental Research Program study area. Prepared for the Alberta Oil Sands Environmental Research Program by Fisheries and Environment Canada. AOSERP Report 58, Edmonton.

Lush, D.L. and Hynes, H.B.N.   1978.   The uptake of DOM by a small
    spring stream. *Hydrobiologia*   60: 271-275.

Maciolek, J.A.   1962.   Limnological organic analyses by quantitative
    dichromate oxidation.   U.S. Dept. of the Interior. Res. Report
    60.

Manny, B.A. and Wetzel, R.G.   1973.   Diurnal changes in DOC and DIC
    and nitrogen in a hardwater stream. *Freshwat. Biol.*   3: 31-43.

McDowell, W.H. and Fisher, S.G.   1976.   Autumnal processing of
    DOM in a small woodland stream ecosystem. *Ecology*   57: 561-
    569.

Mutch, R.A.   1977.   An ecological study of three sub-alpine streams
    in Alberta.   MSc. thesis, University of Calgary.

Otsuki, A. and Wetzel, R.B.   1973.   Interaction of yellow organic
    acids with $CaCO_3$ in freshwater. *Limnol. Oceanogr.*   18: 490-494.

Lock, M.A., Wallis, P.M. and Hynes, H.B.N.   1977.   Colloidal
    organic carbon in running waters. *Oikos*   29: 1-4.

Pitter, P.   1976.   Determination of biological degradability of
    organic substances. *Wat. Res.*   10: 231-235.

Romanenko, V.I.   1979.   Bacterial growth at natural and low levels
    of organic matter. *Ergebn. Limnol.*   13: 77-84.

Sepers, A.B.J.   1977.   The utilization of dissolved organic
    compounds in aquatic environments. *Hydrobiologia*   52: 39-54.

Stabel, H.H., Moaledj, K. and Overbeck, J.   1979.   On the
    degradation of dissolved organic molecules from Plussee by
    oligocarbophilic`bacteria. *Ergebn. Limnol. H.*   12: 95-104.

Viessman, W., Knapp, J.W., Lewis, G.L. and Harbaugh, T.E.   1977.
    Introduction to Hydrology 2nd edition.   Harper and Row, N.Y.

Wallis, P.M., Hynes, H.B.N. and Fritz, P.   1979.   Sources,
    transportation and utilization of dissolved organic matter
    in groundwater and streams.   Inland Waters Scientific Series
    No. 100.   Ottawa.

Wallis, P.M., Hynes, H.B.N. and Telang, S.A.   (in press)   The
    importance of groundwater in the transport of allochthonous
    dissolved organic matter to the streams draining a small
    mountain basin. *Hydrobiologia*.

Wetzel, R.G.   1967.   Dissolved organic compounds and their
    utilization in two Marl Lakes. *Hidrologiai Kozlony*
    47: 298-303.

Wetzel, R.G. and Manny, B.A.   1972.   Decomposition of DOC and
    nitrogen compounds from leaves in an experimental hardwater
    stream. *Limnol. Oceanogr.*   17(6): 927-931.

Whitelaw, K. and Edwards, R.A.   1980.   Carbohydrates in the
    unsaturated zone of the Chalk, England. *Chem. Geol.*
    29: 281-291.

Wistreich, G.A. and Lechtman, M.D.   1980.   Microbiology.   3rd
    edition.   Glencoe Pub. Co. Inc., U.S.A.

ASPECTS OF NITROGEN TRANSPORT AND TRANSFORMATION IN HEADWATER
STREAMS

N.K. Kaushik*, J.B. Robinson*, W.N. Stammers[†],
and H.R. Whiteley[†]

*Department of Environmental Biology
[†]School of Engineering
University of Guelph
Guelph, Ontario, Canada

INTRODUCTION

Water in a stream and time are alike, both keep moving. It
has been aptly remarked, though in a different context, that today
is yesterday's tomorrow (Vallentyne, 1978). Just as yesterday's
events interact with today's conditions and these together with
tomorrow's, similarly a given stretch of a stream, because of water
flow, is a reflection of its own and upstream conditions and, in
turn, influences what lies downstream from it. Water flow is thus
an important factor that complicates studies on nitrogen transformation in streams. To this, one may add the fact that nitrogen
occurs in various forms ranging from nitrate, the most oxidized
state, to ammonia, the most reduced form. In addition, nitrogen
can enter into streams from point sources as well as from surface
and subsurface diffuse sources. All these factors compound the
situation and make it extremely difficult to carry out nitrogen
budget and transformation studies in streams. It is, therefore,
not surprising that although nitrogen is important to biological
systems, as a constituent of proteins, in the eutrophication
process, in oxygen depletion caused by oxidation of its compounds,
in the development of conditions toxic to fish (caused by ammonia)
and for incidences of methemoglobinemia in humans and animals
(Shuval and Gruener, 1977), the processes that control the
transport and transformation of nitrogen in streams have rarely
been studied in detail.

In order to consolidate available information on nitrogen
transformation in streams, emphasis in the present paper will be
on investigations pertaining to stream environments. A number of

recent and elaborate reviews (Brezonik, 1972; Delwiche and Bryan, 1976; Focht and Verstraete, 1977; Kamp-Nielsen and Andersen, 1977; Keeney, 1972, 1973; Larsen, 1977; Painter, 1977; Wetzel, 1975) deal with nitrogen transformation processes in aquatic environments, mostly lakes. Reference to these and other such papers, will be kept to a minimum in order to emphasize the lack of information on stream environments. Also as this paper deals with headwater streams, reference to a vast body of literature pertaining to streams and rivers receiving sewage effluents and dealing with nitrification and denitrification in such rivers will only be made when essential, and confined to some key publications.

SOURCES OF NITROGEN TO STREAMS

In general, a major portion of the nitrogen reaching lakes is transported through streams and rivers which, in turn, receive nitrogen from a variety of surface and groundwater sources. Brezonik (1972) has listed surface inputs; they include point sources such as sewage treatment plants, large feedlots, industrial waste effluents and non-point or diffuse sources such as runoff from agricultural, forested and undeveloped land. To these must be added the nitrogen contained in natural inputs such as autumn-shed leaves. Land uses such as agriculture can also affect sub-surface inputs to streams. While the magnitude of inputs from point sources can be determined, those from non-point sources are difficult to assess and may be insidious.

Headwater streams seldom receive large-scale sewage and industrial effluents but are frequently affected by agriculture. Animal feedlots and manure storage areas are conspicuous potential sources of nitrogen inputs to rural streams. However, it is difficult to generalize about these sources because site-specific factors such as slope, surface treatment (paved or unpaved), and accumulation of manure all markedly affect the amount and quality of the runoff. In a long-term empirical study of an unpaved feedlot in Ohio, Edwards and McGuinness (1975) showed that multiple regression of runoff volume on rainfall depth and antecedent moisture explained 87% of the variation in runoff volume for one-day periods. Characteristics such as rainfall intensity and feedlot usage did not improve the R^2 values obtained.

A number of studies have been carried out to determine pollutant content, including nitrogen forms, of feedlot runoff. Clark et al. (1975) summarized a number of estimates made in the midwest and western U.S. and gave values of total N ranging from 3000 to 17000 mg l^{-1}. In comparisons of paved and unpaved feedlots in Ontario, Coote and Hore (1977) found that total N loadings to a stream averaged 1.16 and 1.23 mg N per "animal unit" for the paved and unpaved lots respectively. (Animal units are defined as number

of animals which excrete about 150 lbs of N per year, an adult
milk cow for example.) Animal unit densities rarely exceed 1
animal unit per hectare in Ontario and therefore, on a watershed
basis N loads from feedlots (and from manure storage areas) do not
contribute greatly to total stream N loads (Nielsen et al., in
press).

    In recent years a controversy has developed regarding the
agricultural contribution of nitrogen to surface and groundwater.
While some (Commoner, 1968) have called for an outright ban on the
use of chemical fertilizers, others (Viets, 1970, 1971a, 1971b)
have refuted the allegation that agricultural practices cause
significant increases in nitrogen in aquatic environments.
Tomlinson (1970) concluded that an increase in the use of N
fertilizer did not cause a marked increase in NO_3-N in surface
waters. Thomas and Crutchfield (1974) did not find a good
relationship between land use and nitrate-N concentrations in
eight streams in Kentucky draining from almost completely forested
to mostly cultivated lands. Nitrate-N varied from 6 mg l^{-1} to
zero, with the highest value being found in a stream draining a
watershed which was 98% in bluegrass pasture. They also noted that
although the fertilizer use had increased since 1921, there was no
corresponding increase in the nitrate-N in the surface waters of
Kentucky streams. Hill and Wylie (1977) determined that based on
nitrate-N concentrations at 48 stream sites associated with
fertilized fields in southern Ontario, there was little evidence
of any significant contribution from fertilizers to nitrate levels
in adjacent rivers. Similar observations have been made by
Caporali et al. (1981) who compared nitrogen contents of streams
draining an agricultural and a forested watershed in central Italy.
From the agricultural watershed annual NO_3-N losses were nearly
double the amount lost as organic-N, whereas from the forested
watershed losses of these fractions were similar.   During
the winter months concentrations of NO_3-N in the stream draining
the agricultural watershed were nearly twice that in the stream in
the forested watershed. In a year 10.44 kg ha^{-1} were lost from the
former but 27.08 kg ha^{-1} from the forested watershed which also
received more rainfall. Caporali (1981) concluded that N losses
are more dependent upon total flow than upon concentration and
land use.

    In contrast to the above papers, others have found that
fertilizer use in agriculture has the potential to increase nitrate
levels in streams. Hill and Wylie (1977) recorded high nitrate
concentrations (10.0 - 34.0 mg l^{-1}) in groundwater underlying
heavily fertilized fields. Based on analyses of weekly water
samples, Casey and Clarke (1979) concluded that the nitrate
concentration in the River Frome, England has increased by 0.11
mg $NO_3 l^{-1}$ per year during the period 1965-75. Earlier it was

concluded (Casey, 1977) that rainfall and sewage effluents could
not account for most of the increase. The observed increase was
attributed to the increase in fertilizer usage, or growth of more
legume crops or more nitrogen imported as livestock food. Miller
(1979) reported that the average annual concentrations of NO_3-N in
tile drainage water from some mineral soils in Ontario which were
fertilized at or below recommended rates ranged between 4.0 and
8.2 mg l^{-1}, whereas in samples from sites fertilized at rates
greater than recommended, the corresponding values were between
11.3 and 20.4 mg l^{-1}. Similarly, Nicholls and MacCrimmon (1974)
observed that the mean concentration of 0.75 mg l^{-1} of inorganic
N in subsurface water under a cultivated plot was about 10 times
higher than under an uncultivated area during the growing season.
In another study of a small agricultural watershed, Whitely et al.
(1980) recorded that watertable-region samples from wells in
abandoned field areas gave NO_3-N concentrations consistently less
than 2 mg l^{-1}, whereas in samples from beneath cultivated fields the
concentration was never less than 5 mg l^{-1}, and had a mean value of
6.7 Mg l^{-1} of NO_3-N. A more detailed, three-year investigation
dealing with the nitrogen content of the runoff and percolate water
from fields under different cropping and fertilizer application
regimes is reported by Chichester (1977). Nitrogen concentration
in the percolate under meadow did not exceed 10 ppm whereas in
corn the corresponding value was 70 ppm - far in excess of
recommended levels of NO_3-N for potable waters. Similar findings
are reported by other workers (Baker et al., 1975; Burwell et al.,
1976; Calvert, 1975; Jackson et al., 1973; Miller and Nap, 1971;
Pratt et al., 1972).

     To understand the role of agriculture in nitrogen inputs to
streams it is necessary very briefly to discuss some of the
important nitrogen transformation steps in soil (for details see
Kolenbrander, 1977). The bulk of the nitrogen in soil is in the
soil organic matter and is not available to the crop unless it is
mineralized by microorganisms which use organic matter as an energy
source. As a result, organic nitrogen is first converted to
ammonium-nitrogen, a cation that is strongly absorbed to the cation-
exchange complex of the soil and is thus relatively immobile.
This form of nitrogen as well as organic N may be lost to streams
as a result of surface sediment erosion and surface runoff. In
alkaline soils, ammonium nitrogen may become volatile as a result
of ammonia formation. This also occurs when ammonium fertilizer
or farmyard manure is applied without drilling it into the soil
and without plowing it down as soon as possible. Ammonia may be
dissolved in surface waters.

     In the next step of mineralization, ammonium nitrogen is
nitrified by nitrifying bacteria first to nitrite and then from
nitrite to nitrate. As the nitrate formed is anionic, there is
little adsorption to clay and humus particles and thus nitrate is

highly mobile. Under wet conditions, nitrate percolates into groundwater, and may find its way into streams. It is obvious that a
number of factors govern these N transformations and eventual loss
of nitrogen to streams. Such factors have been discussed by Nelson
(1972).

Direct input of N into streams from agricultural land is
possible because of runoff and erosion. Factors affecting this
input include sloping lands receiving high levels of fertilizers
and intensive rainfall; events of heavy rainfall soon after surface
application of fertilizers; inadequate plant cover on the fields;
and soils having low rates of infiltration either because of the
nature of soil, impervious conditions or frozen layers. Percolation
of nitrates to subsurface water depends upon the presence of
excessive nitrates in the soil, rainfall exceeding evapotranspiration,
soils with low water holding capacity and high infiltration rates
(Nelson, 1972). Thus, except under exceptional circumstances
mentioned above and direct runoff from feedlots, streams are not
likely to have surface inputs of large quantities of nitrogen.
However, nitrate in high-production agricultural systems is likely
to lead downward (Robinson, 1976) and therefore possibilities of
nitrates entering streams through subsurface waters always exist.
Nitrogen inputs, recorded by different authors, in surface runoff
(or as noted in streams) from land under agriculture or other uses
are shown in Table 1. Observations made on subsurface waters are
given in Table 2. It can be clearly seen that subsurface inputs
are likely to reach much higher concentrations than values noted
in surface runoff.

It would appear from the above review that under many
situations agriculture does have the potential of adding substantial
quantities of nitrogen into streams through subsurface flow.
However, it does not imply that streams simply act as conduits
and transport this nitrogen to the downstream regions or to lakes.
As will be discussed in the following pages, substantial
quantities of nitrogen in transport may be lost from streams. For
this reason, the accuracy of estimation of nitrogen output from an
agricultural area by analysing stream water samples depends
largely upon the distance of the sampling sites from the source.
In addition, Lee and Hynes (1978) suspected denitrification to
occur as the groundwater with high nitrate + nitrite concentrations
seeped into a stream.

The relative importance of different sources of nitrogen to
streams has been estimated only rarely. In the Great Ouse, a
river that drains a rural area in England, sewage inputs contribute
only about 20% of the total inorganic nitrogen input (Owens et al.,
1972) while in the upper Roanoke River basin 58% of the nitrate-
nitrogen originated in agricultural and rural runoff (Grizzard
and Jennele, 1972). In studies on a number of lakes in Florida

Table 1. Discharge of nitrogen from rural watersheds based on analyses of surface water samples.

Authors	Land use and nature of soil	Discharge of NO_3-N — Total N (kg ha^{-1} yr^{-1})		Remarks
Jackson et al. 1973	Agriculture (corn); loamy sand	(1969) 0.125	–	Surface runoff collected in channels around the perimeter; NO_3-N concentration in most of the runoff in 3 years was less than 1 mg L^{-1}, maximum was 3 mg L^{-1}
		(1970) 0.432	–	
		(1971) 0.340	–	
Schuman et al. 1973	Watershed 1, corn;silt loam	(Average) 1.69	39.64	High fertilizer rate
	Watershed 2, corn;silt loam	(1969-71) 0.97	23.16	Recommended fertilizer rate
	Watershed 3, Bromegrass, silt loam	0.76	1.21	" " "
	Watershed 4, corn, silt loam	0.18	2.62	High fertilizer rate
Thomas and Crutchfield, 1974	Bluegrass pasture, limestone (Cave Creek)	22.2	–	Samples collected from a stream
	94% wooded forest, sandstone and shales (Helton Creek)	2.1	–	" " "
Burwell et al. 1976	Watershed 1, corn; silt loam	1.26	–	Average value for 1969-74
	Watershed 2, corn; silt loam	0.59	–	" " "
	Watershed 3, Bromegrass; silt loam	0.56	–	" " "
	Watershed 4, corn, silt loam	1.26	–	" " "
Johnson et al. 1976	Watershed 1; 34% forest, 40% agriculture (dairying), rest abandoned agriculture	5.5		
Hill 1978	21 watersheds; 49% agriculture, 21% forest and rest abandoned land	range 1.41-7.31		Values include NO_3- and NO_2-N
Caporali et al. 1981	71% agriculture, mostly calcareous clay	10.44	15.65	
	97% forest, sandstone	27.08	58.64	

Table 2. Discharge of nitrogen from rural watersheds based on analyses of subsurface water samples.

Authors	Land use and nature of soil	Discharge of NO$_3$-N kg ha^{-1} yr^{-1}	Remarks
Jackson et al. 1973	Corn, loamy sand	(1969) 34.13 (1970) 46.89 (1971) 22.01	Average monthly concentrations in the three years ranged between 5.80 and 12.57 mg L^{-1}
Baker et al. 1975	Tile-drained cropland, silt loam	30.6	Average for four years. NH$_4$-N included in some samples. Values ranged between 0.0 and 93.0 kg ha^{-1} yr^{-1}
Burwell et al. 1976	Watershed 1, corn, silt loam Watershed 2, corn, silt loam Watershed 3, Bromegrass, silt loam Watershed 4, corn, silt loam	20.72 6.84 11.14 35.12	Average value for 1969-74 " " " " " " " " "
Miller, 1979	Site 1, organic soil, onions Site 2, organic soil, onions Site 3, organic soil, onions/carrots Site 5, clay surface, corn/soybeans/vegetables Site 6, fine clay, soybeans/wheat/barley/corn Site 7, high organic matter, fine sand with clay at 90 to 100 cm, corn/beets/beans Site 8, fine sandy loam, corn Site 9, fine clay, corn Site 10, high organic matter, fine sand, corn	84.35 141.15 135.67 57.00 16.50 15.00 3.50 8.00 55.00	Average of four years " " " Average of three years Average of two years More than recommended fertilizer Average of two years, less than recommended fertilizer Average of two years, less than recommended fertilizer Average of two years, fertilizer as recommended Average of two years, less than recommended fertilizer Average of two years, more than recommended fertilizer
Whiteley et al. 1980	Sandy loam, uncropped areas area with mostly corn, fertilized area with crops, without fertilizer	3.50 80.0-135.0 35.0- 60.0	Values are for total N (not NO$_3$-N) " " " " " "

(Shannon and Brezonik, 1972), agricultural runoff was a major source
of nitrogen in most cases. Similarly, Loehr (1974) observed that
waste waters contributed the bulk of the phosphorus to surface
waters but little nitrogen.

Analyses of samples from 11 small agricultural watersheds in
the Canadian Great Lakes Basin showed that unit area loads of
total N ranged from 3.5 to 29.2 kg N ha^{-1}yr^{-1}. Nitrate plus NO_2
loads, ranging from 2.4 to 25.9 kg N ha^{-1}, averaged 75% of total
N loadings. All the above values were correlated with manure and
fertilizer applications, and the percentage of the watershed area
cultivated in row crops, in corn and tile drained (Neilsen et al.,
in press). In an earlier report Neilsen et al. (1978) estimated
that agriculture was found to account for 98 and 95% of the nitrate
input and 94 and 83% of the total nitrogen loading to the Saugeen
and Grand rivers respectively.

Based on many detailed studies in Sweden, Landner (1977)
estimated that of the total 132,000 tons of nitrogen annually
transported to surface waters in Sweden, 45,000 tons (34%) is from
agricultural drainage, 27,000 tons (21%) from municipal and
industrial wastes, 2,000 tons (1.5%) from forest drainage and
58,000 tons (44%) from background or natural sources.

Headwater streams passing through wooded areas are known to
receive about 400 g m^{-2}yr^{-1} to 600 g m^{-2}yr^{-1} (see Bird and Kaushik,
Chapter 2) of particulate allochthonous organic matter, mostly in
the forms of autumn-shed leaves. As mentioned in the following
pages, this organic matter not only serves as an energy source for
denitrifiers but is also a source of organic nitrogen; on
average, autumn-shed leaves contain about 1% nitrogen (Kaushik and
Hynes, 1971).

NITROGEN TRANSFORMATIONS

The nitrogen transformation processes that can occur in
streams (Fig. 1) are similar to those occurring in other aquatic
systems and in soils and the reader is referred to Wetzel (1975)
who has provided an excellent discussion for lake environments,
to Patrick et al. (1976) and Kamp-Nielson and Andersen (1977) who
have emphasized the processes at the mud-water interface, and to
Painter (1977) for general microbiological aspects. Important forms
of nitrogen in freshwater are: organic nitrogen (organic compounds
such as amino acids, amines, proteins), ammonium (NH_4^+), nitrate
(NO_2^-) and nitrate (NO_3^-).

There is continuous input, conversion and loss of nitrogen in
stream systems. Although most processes are microbial in nature,
stream benthos may not only be important directly (e.g., processing

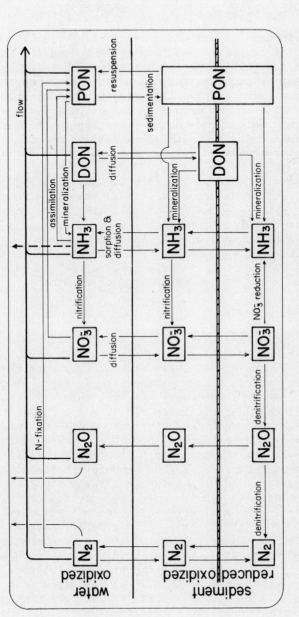

Figure 1. Nitrogen transformation processes (modified from Kamp-Nielsen and Andersen, 1977).

of organic nitrogen) but may also affect microbial activity. In addition, through the uptake by macrophytes, algae and microorganims, nitrate or ammonium can also change to the organic form in the plant tissue - a process referred to as *immobilization*.

Organic nitrogen in sediment, originating from leaves, other allochthonous inputs, stream macrophytes and dead animal tissue is mineralized as a result of the activity of heterotrophic bacteria and the ammonium produced is changed to nitrate through a process called *nitrification*. In the first step of nitrification, ammonium is changed in the aerobic zone near the sediment surface to nitrite by microorganisms mostly belonging to the *Nitrosomonas* group. In the second step, nitrite is changed to nitrate primarily by microorganisms belonging to the genus *Nitrobacter*.

Nitrate resulting from nitrification, or entering directly into a stream from groundwater or surface runoff, diffuses into the sediment. Here, eventually, it will likely encounter an area of active heterotrophic microbial activity at low redox potential and will be converted to N_2O or N_2 both of which are lost to the atmosphere. This process, *denitrification*, is brought about by facultative anaerobic bacteria belonging to the genera *Pseudomonas*, *Achromobacter*, *Bacillus* and *Micrococcus* among others.

Nitrogen Immobilization:

Nutrient uptake and distribution of plants in relation to nutrients has recently been reviewed by Westlake (1975) and Haslam (1978). Water plants obtain their mineral nutrients from the sediment and/or water. Free-floating plants (e.g., *Lemna minor*), which are not common in headwater streams, obtain nutrients from the water, tall emergent plants on banks (e.g., *Phragmites communis*) from the sediment and soil, and others from both the sediment and water. Nutrients can be taken up by roots, stems and leaves. It has been shown that different plants grow best at different concentrations of nitrate-nitrogen, ammonia-nitrogen and nitrite-nitrogen (see Haslam, 1978). Although a few workers (McColl, 1974; Vollenweider, 1968) have pointed out the importance of plants in nitrogen uptake in streams, very little quantitative information is available. Crisp (1970) prepared a nutrient budget of a bed of watercress (*Rorippa nasturtium-aquaticum*) grown commercially. During a period of about 14 months, uptake of nitrogen by watercress accounted for only about 8% of the input to the system. In a spring-fed stream, Kaushik et al. (1975) estimated that nearly 100 kg of N (based on 1974 data) entered the stream from the spring and about 46.7 kg were lost as the water moved about 2 km downstream. The biomass of macrophytes, mostly watercress, in the stream during the summer months ranged from 42.7 to 75.4 kg. The nitrogen content was between 1.4 and 2.6 kg. It was estimated that the macrophytes accounted for slightly more than 5% of the nitrogen lost, or about

3% of the nitrogen input in the stream from the spring water.
Similarly, after detailed considerations, Casey (1977) concluded
that the utilization of the nitrate in solution by macrophytes in
the River Frome and Bere stream is a very low percentage of the
throughput — approximately 2% of the former and even less for the
latter. In contrast to these studies, Stake (1967) recorded that
in a Swedish stream, plants removed 15% of the inorganic nitrogen.
In a more elaborate study, Vincent and Downes (1980) examined the
role of watercress in the removal of nitrate and ammonium from
a low-order stream in New Zealand. They also determined, by
performing *in situ* assays, that the rate of nitrate uptake by
watercress roots was 150 µg N g root tissue^{-1}h^{-1}. Based on the
estimated watercress-root biomass, it was calculated that during
the late summer season watercress growth was responsible for
removing 72% of the nitrogen loss observed between the upstream
and downstream sites. However, these authors did not consider the
possiblity that denitrification may have accounted for some of
this loss. It may be pointed out here that the values for the
percentage loss in this study or in similar studies (e.g. Kaushik
et al., 1975) are based on the apparent loss of nitrogen between
the upstream and downstream stations. The actual loss of nitrogen
from the system because of nitrate produced through nitrification
and then lost possibly through denitrification, is probably greater
than is recorded in all of these studies.

Mineralization:

     Organic nitrogen present in the organic matter produced
within streams or accruing from the adjoining land enters the
nitrogen cycle through degradation and mineralization. Although
bacteria and fungi are the main decomposers, the fact that a
substantial amount of leaf material, which is a major component
of organic matter input, is processed by stream invertebrates (see
Bird and Kaushik, Chapter 2) suggests that mineralization rates
may be affected through the activity of invertebrates. A number
of papers have dealt with the processing of leaf material by
invertebrates, however, work on mineralization has been ignored.
Also, availability of leaf nitrogen in streams will depend upon
leaf retention and the rate of mineralization which, in turn, are
governed by the nature of the stream (water flow, pool and riffle
areas, substrate, etc.) and the quality of input. Quality relates
to decomposibility of the material and its C:N ratio. Since the C:N
ratio of freshly-fallen autumn-shed leaves is about 40 to 50, the
process initially is one of nitrogen immobilization, until the
ratio decreases and nitrogen is in excess. The C:N ratio of
organic matter originating from aquatic plants is generally less
than that for autumn-shed leaves but here also immobilization
preceeds mineralization. Nichols and Keeney (1973) assessed
release of nitrogen from water milfoil decomposed in water. Organic

N appeared in the water in the early stages of the experiment, but was depleted rapidly and inorganic N was apparently immobilized as soon as it was formed.

Nitrification:

In aquatic environments, nitrification is a source of bio-chemical oxygen demand (BOD). The oxygen demand of nitrogenous matter is about 4.57 times, by weight, of ammonia nitrogen and 1.14 times, by weight, of nitrite nitrogen, compared with 2.67 times, by weight, of carbonaceous material (Bansal, 1977). Thus the nitrification process could substantially affect oxygen levels of streams and rivers receiving reduced nitrogen compounds. For this reason this process has been studied extensively by sanitary engineers and biologists dealing with rivers receiving sewage inputs (for references see Bansal, 1977; Gujer, 1976). Available literature on aquatic nitrification, mostly on lakes, has been reviewed by Keeney (1973) and Kamp-Nielsen and Andersen (1977). The latter authors have cited results that show that the rate of nitrification in Lake Mendota sediment ranged from 540 to 900 mg $N\ m^{-2}\ day^{-1}$, while for a sewage-loaded stream the values were between 0 and 150 mg $N\ m^{-2}\ d^{-1}$. Work on nitrification in headwater streams is negligible. In a laboratory investigation (Chatarpaul et al., 1980), sediment from Canagagigue Creek, a headwater stream in southern Ontario, was placed in plexiglass columns and topped with 500 ml of nitrate- N as $^{15}KNO_3$ (90 atom-percent ^{15}N, 9 ppm N). The supernatant was aerated constantly and the columns, incubated for 48 days at $15^{\circ}C$, were sacrificed at different time intervals to determine various N-forms in the supernatant and the sediment. As discussed in the following pages, labelled N was lost from the system indicating occurrence of denitrification. Also it was noticed that ^{15}N abundance of the nitrate-N in the supernatant changed from the initial value of about 90 atom-percent at the end of the experiment suggesting production of nitrate because of nitrification. From this, on the basis of sediment surface area, they (Chatarpaul et al., 1980) estimated that the average rate of nitrate-N production was 29 mg $m^{-2}\ d^{-1}$. In some columns tubificid worms (*Tubifex tubifex* and *Limnodrilus hoffmeisteri*) were introduced and in such columns the rate of nitrate-N production, of 69 mg $N\ m^{-2}\ d^{-1}$ was significantly higher compared with that in columns without worms (Fig. 2).

It was speculated that the increase in nitrification because of worms could be due to increased rate of sediment-water exchange of ammonium-N caused by the burrowing activity of the worms and because of ammonia excretion by the worms and its subsequent oxidation to nitrate. Preliminary investigations in our laboratory indicate that nitrate production in sediment columns is also significantly increased in the presence of the larvae of the crane fly (*Tipula ignobilis*).

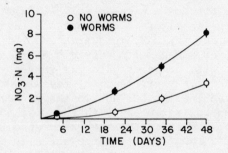

Figure 2. Effect of tubificid worms on the rate of nitrate-N
           production in sediment columns (redrawn from
           Chatarpaul et al., 1980).

For *in situ* measurement of nitrification in streams,
Schwert and White (1974) and White et al. (1977) have used a
method in which a plywood chamber, with holes in the upstream and
downstream ends, is placed in a stream and changes in NO_2 and NO_3
concentrations of water recorded. Since denitrification and
nitrification may occur simultaneously (see above), this
technique could underestimate nitrate production.

Denitrification:

Microorganisms can of course assimilate nitrate as a source
of nitrogen thereby immobilizing the nitrogen as a part of the
microbial cell. Nitrate can also be used by microorganisms as a
terminal electron acceptor instead of oxygen. In this case, it
does not become part of the cell and the process (nitrate
dissimilation or respiration) is called denitrification. In this
process nitrate and nitrite are reduced as a result of bacterial
metabolism to gaseous end-products such as dinitrogen (N_2) and
nitrous oxide (N_2O). These end-products may be lost from the
system and thus the process of denitrification represents a
permanent nitrogen sink in terrestrial and aquatic ecosystems.
Because of the direct implications of this process in the
availability of nitrogen to crops, this process has been studied
quite extensively by agronomists and microbiologists (Painter,
1977; Payne, 1973). Also extensive research on this process has
been carried out in wetlands (Patrick et al., 1976) and in lake
ecosystems (Keeney, 1973; Brezonik, 1977). Fewer papers pertain
to denitrification in running waters and most of them have
appeared during the last decade.

Field observations strongly indicate the occurrence of

denitrification in streams. Owens et al. (1972) estimated nitrogen
loadings to two English rivers, the Great Ouse and the Trent, by
estimating inputs from the catchment area and adding to these the
values determined for sewage effluents discharged in the rivers.
These input values were compared with the values actually observed
in the river discharge. Their results, revealed that in both the
rivers the estimated loading for the months of November through
April, showed a general agreement with the discharge observed.
However, during the summer months the calculated loadings were
considerably greater than the discharge actually observed. In
fact, observed values were even less than the values that were
expected as a result of sewage inputs alone. At a maximum, the
deficit during the summer months was approximately 1200 kg d^{-1} in
the Great Ouse and 25,000 kg d^{-1} in the River Trent or, on a river-
bed area basis, 0.75 g m^{-2} d^{-1} and 1.4 g m^{-2} d^{-1} respectively.
Owens et al. (1972) estimated that the uptake by aquatic plants
accounted for only 200 kg N d^{-1} during the summer and that a similar
amount was immobilized by algal growth during the months of June
and July. Since nitrogen transported during the other months did
not exceed the expected values, it was concluded that the differences
observed during the summer months were actually because of nitrogen
lost from the system and it was suggested that denitrification was
responsible. In addition to their own study, Owens et al. (1972)
discussed other reports indicating losses of inorganic nitrogen
during the summer months from the Sangamon River, Illinois and the
rivers Chelmer, Blackwater and Stour in England. Indirect evidence
for the importance of stream denitrification is contained in the
study by Tomlinson (1970) who tried to correlate nitrate concentrations in some English rivers with the agricultural use of nitrogenous
fertilizers in adjacent counties. Although during the 15-year-
period (1952/53 to 1966/67) the usage of nitrogen fertilizer in
18 counties had increased to approximately two to four times the
initial level, no general relationship existed between nitrate
concentration of river waters and nitrogen fertilizer usage in
adjacent areas, possibly because of denitrification in the streams.
Van Kessel (1977a) reported that 56% of the nitrate from sewage
effluents discharged in canals disappeared over a 800 m long reach.
During the 20-day period of study the average nitrate loss was 913
mg NO_3-N m^{-2} d^{-1}.

Records of nitrogen loss are not confined to running waters
receiving sewage effluents. During summer in a 2 km stretch of a
spring-fed Ontario stream, Swift's Brook, NO_3-N concentration of
about 5 mg l^{-1} at the spring gradually decreased to about 2.0
mg l^{-1} at the most downstream station (Kaushik and Robinson, 1976;
Kaushik et al., 1975). Figure 3 shows that decrease in NO_3-N
concentration were observed (Robinson et al., 1979) even under
extreme low flow conditions, indicating that the decrease did not
result because of dilution with surface runoff. Nitrogen transport
in Swift's Brook was investigated extensively (Robinson et al.,

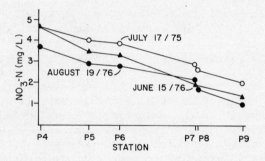

Figure 3.   Downstream decrease in NO_3-N concentration in Swift's
            Brook (redrawn from Robinson et al., 1979).

1979) by continuous gauging at the spring and the most downstream
point and by measuring flow from the intermittent channels. Water
samples were collected as frequently as every 6 hours by an
automatic sampler located at the most downstream location and at
weekly intervals from other locations. This was supplemented with
extensive sampling during thaw and storm events. Samples were
analysed for NH_4-N, Organic-N and NO_3-N. Based on these detailed
investigations it was estimated that during 1975 and 1976 about
250 kg N and 450 kg N, respectively, were lost from the stream
water. On an average this works out to 350 kg N yr^{-1}, or about
480 mg $m^{-2}d^{-1}$ based on streambed area. The bulk of this loss was
considered to be the outcome of denitrification because uptake by
macrophytes and nitrogen immobilization accounted for a very small
fraction of the observed loss (Kaushik et al., 1975).

     Nitrogen export from 21 watersheds and subsequent transport in
Duffin Creek, Ontario was investigated by Hill (1978). During the
summer months he recorded nitrogen loss from various reaches of
the stream. A detailed analysis of data (Hill, 1979) for six
reaches of the stream for 18 days of low flow conditions during the
summer showed a consistent nitrate-N loss. The mean daily loss for
the six reaches ranged from 1.7 kg (40 mg m^{-2} of stream bed) to
20.3 kg (300 mg m^{-2}) of nitrate-N. Based on other measurements he
concluded that this loss was because of denitrification. During
the summer months this loss represented 50% of the average daily
input of nitrogen and 75% of the nitrate-N input.

     Although some of the detailed field studies pertaining to
nitrogen budgets of streams and running waters (e.g., Robinson et
al., 1979; Hill, 1979) strongly suggest that the process of
denitrification is the causal factor for most of the nitrogen losses
observed during the summer months, no definitive field investigations

have been published. In lake ecosystems, *in situ* experiments in
which $^{15}NO_3$-N, added to lake sediments samples in plastic bottles
and returned to the lake bottom, has been monitored to evaluate
N-transformations, including denitrification (Chen et al., 1972).
Because of the flow, such experiments cannot be carried out in
streams and rivers. However, a number of researchers (Hill, 1979;
Robinson et al., 1979; Sain et al., 1977; Toms, 1975; Van Kessel,
1977b) have performed column experiments in which sediments taken
from a stream or a ditch are topped with nitrate solution and, to
determine denitrification, nitrate concentration is routinely
monitored. It is interesting to note that even when the supernatant
nitrate solution is bubbled with oxygen to simulate turbulence in
streams and to maintain an oxygen-saturated supernatant, a decrease
in nitrate concentration is observed. Depending upon temperature
and sediment depth, Sain et al. (1977) recorded losses ranging
from 61.25 to 165.50 mg N m^{-2}d^{-1}. Hill (1979) confirmed these
findings by using sediment from three different sites and recorded
that the average daily losses of N ranged between 100 and 251 mg m^{-2}
of sediment surface for silt-rich deposits and between 20 and 60
mg m^{-2} for sediments with high sand and gravel content. Losses of
nitrate-N from water overlying stream sediment were presumed to be
because of denitrification. However, this needed to be confirmed
by using labelled NO_3-N, showing that nitrogen has been lost from
the system. This has been done by Van Kessel (1977b), Chatarpaul
and Robinson (1979) and Chatarpaul et al. (1980). They
conclusively showed that the bulk of the nitrate was lost from
the system as a result of denitrification.

In streams, denitrification in the reduced sediment and
nitrification in the oxygenated water layer may proceed
simultaneously. Therefore in estimating the actual amount of
nitrate-N lost via denitrification, the nitrate which is added
via nitrification and possibly lost, subsequently must also be
included. Estimation of the rate of nitrification is possible
through the use of ^{15}N (Chatarpaul et al., 1980). In many of the
above cited studies where the extent of nitrification was not
determined, the rate of denitrification was underestimated.

Factors affecting denitrification in sediments from a lake
and a reservoir have been discussed by Terry and Nelson (1975).
Like most biological processes, microbial denitrification is also
affected by temperature and Terry and Nelson (1975) did record
that the rate of denitrification increased with increasing
temperature in the range of 5-23°C. Similar results have been
reported by Sain et al. (1977) and Van Kessel (1977b). Besides
directly affecting activity of the denitrifying bacteria, higher
temperatures accelerates other microbial processes, thus
accentuating anoxic conditions which, in turn, increase sites for
denitrification.

In lake situations, denitrification can occur in the hypolimnion
as water in this zone does become anoxic. In headwater streams,
because of flow, water is mostly saturated with oxygen. Denitrification in such streams, therefore, can occur only in the sediment.
Sain et al. (1977) conducted column experiments in which columns
containing 1.0, 2.5, 5.0 or 10.0 cm deep sediment from a stream
were overlain with NO_3-N solution. Such columns were incubated at
10, 15 or $22^{\circ}C$ and the supernatant monitored for nitrate loss.
The results showed that at 10 and $15^{\circ}C$, there was a significant
difference in loss of nitrate between 1- and 5-cm columns.
However, at $22^{\circ}C$, there was no significant difference with sediment
depth. Such results indicate that at higher temperatures,
because of high denitrification rates, the nitrate diffusing in
the sediment gets denitrified rapidly and is eliminated from the
system even before it diffuses up to about 1 cm in the sediment.
So under such circumstances of high temperature, whether a column
has 2.5, 5.0 or 10.0 cm deep sediment is of no significance.
Under conditions of low temperatures, because of low denitrification
rates, nitrate may diffuse into deeper layers of sediment. Obviously,
if a column has only about 1.0 cm deep sediment, sediment depth
will be a limiting factor for denitrification at usual stream
temperatures (for details see Sain et al., 1977). It should be
obvious that the nature of stream sediment, in terms of its
organic matter and particle size, will also affect the sediment
depth that could be operative in denitrification. Van Kessel
(1977b) experimented with two types of sediment, one with 20.1 mg
and the other with 11.6 mg of organic matter per g of wet sediment.
He determined that at $25^{\circ}C$ with nitrate content of 25.2 mg l^{-1}
the supernatant only about a 7 mm thick layer in the former and a
14 mm thick layer of sediment in the latter case appeared to be
involved with denitrification. Measurement of Eh in the study by
Sain et al. (1977), who used stream sediment containing about 30%
organic carbon, indicated that the oxidizing zone below the
sediment-water interface was less than 1 cm thick despite the
constant aeration of the water. Similar results are reported by
Engler and Patrick (1974) for flooded soils. These findings
indicate that quite shallow (< 1 cm) deposits of organic sediments
in streams can support denitrification.

It should be intuitively obvious that denitrification in
stream sediment also depends upon the availability of a metabolizable
energy source to denitrifiers. It has been repeatedly shown in our
laboratory that columns prepared from Swift's Brook sediment
containing about 60% organic matter could be repeatedly overlain
with about 10 mg l^{-1} NO_3-N solution without much change in the rate
of NO_3-N loss. In contrast, when columns containing sediment,
with only about 10% organic matter, from Canagagigue Creek, are
topped a second or third time with NO_3-N solution, the rate of loss
of NO_3-N declines. In another experiment, Canagagigue Creek
sediment, with organic matter content of about 10% and C:N ratio

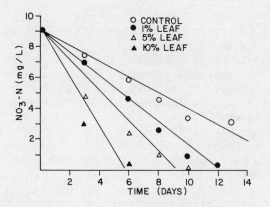

Figure 4.   Nitrate loss from sediment columns with and without
            added organic matter (unpublished Ph.D. thesis, L.
            Chatarpaul, 1978, Univ. of Guelph, Guelph, Ontario).

of 16:1 was ammended with varying amounts of dried ground leaves
(3:1 mixture of maple and water cress).   Figure 4 shows the results of
nitrate loss from columns without added organic matter and from
those with 1, 5 and 10% (w/w, air-dried basis) added organic matter.
Differences in the loss of nitrate due to the added leaf material
were significant.   Engler and Patrick (1974) have shown that soil
with only 0.7% organic C content overlain with floodwater has a
shallow oxidized zone at the surface.   This zone is much thinner
when the soil had 0.5% rice straw added to it, and is almost
eliminated when it contained 2% rice straw.   It is obvious that
allochthonous organic matter in streams is not only important as
an energy source for stream invertebrates (see Bird and Kaushik,
Chapter 2) but also for microbial processes.

    We have already alluded to the fact that the presence of
tubificid worms in the sediment accelerates the rate of nitrification.
Recently, Chatarpaul et al. (1979, 1980) confirmed that these worms
also enhance the rate of denitrification (Fig. 5).   Using ^{15}N as
a tracer, they (Chatarpaul et al., 1980) showed that when worms
were present in the sediment the rate of denitrification was
increased by 80%, from 50 to 90 mg N m^{-2}d^{-1}.   Their experiments
also indicated that besides indirect effects of worms such as
enhancing diffusion of nitrate into the sediment, denitrification
may have occurred in and/or on the worms as the denitrifying
bacteria could be isolated from both worm exteriors and gut contents.
Working with rice-field soil submerged in water, Kikuchi and
Kurihara (1977) observed that in the presence of tubificids the
concentration of nitrite and nitrate in the overlying water
increased rapidly after 45 days of incubation.   The absence of
nitrate in the earlier stages was inferred as inhibition of

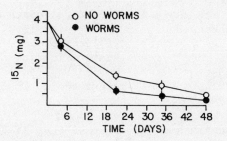

Figure 5. Enhancement of rate of denitrification of sediment by tubificid worms (redrawn from Chatarpaul et al., 1980).

nitrification because of worms. However, Chatarpaul et al. (1980) attributed this to the process of denitrification. In a subsequent work (Fukuhara et al., 1980) it was noticed that the presence of worms decreased the number of heterotrophic aerobes, heterotrophic anaerobes and nitrifiers in the sediment-water system. In an earlier work, Edwards (1958) showed that introduction of larvae of *Chironomus riparius* in settled activated sludge increased the depth of oxidized zone of the sediment and also decreased the concentration of nitrate-N in water overlying sludge. Similar results have been obtained by Andersen (1976) when he added *C. plumosus* larvae to lake sediment. Chatarpaul et al. (1980) believe that diffusion and mechanical transfer of both nitrate and oxygen to the deeper zones of sediment are enhanced by the burrowing activities of these invertebrates. However, this effect is not likely to be uniform and microsites with reducing conditions are likely to occur. Increased availability of nitrate at such microsites can result in increased denitrification. Obviously more detailed investigations are needed to elucidate invertebrate-microbial interactions relating to nitrogen transformations.

Kinetics of the Observed Denitrification Process:

Patrick and Reddy (1976), working with flooded soils, first inferred the kinetics of the denitrification process in water overlying soil by observing the temporal reduction of nitrate concentration of the water. Their conclusion of first-order kinetics was based on the assumption that the flood-water nitrate concentration was identical to the nitrate concentration in the soil water. This approach is equivalent to ignoring the diffusive transport of nitrate from the supernatant solution into and within the soil. At the same time, Toms and coworkers (1975) reported similar conclusions in their studies on nitrate removal

by river sediments. An analysis of Patrick's data by Phillips et
al. (1978), using a continuous model for nitrate diffusion and
reduction, provided no further justifications for either a zero or
first order kinetics model of the denitrification process as the
approach to rate constant determination required an *a priori*
specification of a kinetic model before the optimal selection of
rate constants could be affected.

Eckenfelder and Argaman (1978) state that since both carbonaceous matter and nitrate are involved, the rate of denitrification
is expressed by a Monod-type equation descriptive of saturation
kinetics. They also indicate that since saturation constants for
nitrate are very low, the analytical form of the Monod equation
corresponds to a zero-order reaction with respect to nitrate down
to very low concentrations. The above authors quote denitrification
rates for a number of different carbon sources.

Stammers et al. (in press), employing results from sediment
column experiments reported by Sain et al. (1977) and Robinson
et al. (1977) demonstrated that the denitrification rate constants
could not be obtained using either a lumped or continuous model of
nitrate diffusion and reduction into and in the sediment from the
transient behaviour of the supernatant solution nitrate concentration,
but could be obtained from the steady-state behaviour. The
analysis of these data also lead to the conclusion of zero-order
kinetics.

*Dynamic Models of Stream Water Quality and Denitrification Rate
and Kinetics:*

Two main objectives are generally served by the development
of dynamic models of stream and river water quality. The first is
the management of water quality and, in this context, a dynamic
water quality model may be either a component of on-line quality
regulation or a simulation tool for the generation of scenarios
resulting from different operating rules for water quality
management. The second is a research objective. Here, the dynamic
model is employed to gain insight into interrelationships that
exist among biochemical processes and between biochemical and
hydrologic and/or hydraulic processes and variables.

An excellent introduction to the principles of dynamic
modeling of stream water quality is provided by Thomann (1972)
and a recent, more detailed, account of model structure, state
and parameter estimation is given by Rinaldi et al. (1979). Both
of the above works provide information on the coupling of
biochemical, hydrologic and hydraulic sub-models to generate
transport models for dissolved and suspended components including
nitrogen. The emphasis is on aerobic processes present in the
stream water and their utilization as a component of the waste

treatment process. The objective of this modeling is clearly in
support of the first of the above stated objectives. Review
articles by Hendricks (1979), Krenkel and Navotny (1979), and
Krenkel and Ruane (1979) are also illustrations of this
perspective.

Little published work is evident which related strongly to
the second objective and, in particular, which is concerned with
anaerobic biochemical processes in stream sediments which impact
strongly on nitrogen transport. Najarian and Harleman (1977),
although concerned with a model of an estuarine system involving
nitrogen-limited ecosystems, demonstrated the influence of
hydrodynamic variables on the spatial and temporal distribution
of nitrogen forms. Their work provides insight into sampling-
time-interval specification for estuarine systems. A dynamic
model for nitrate transport involving the anaerobic reduction of
nitrate in stream sediments in given by Stammers et al. (1978).
The model provides a characterization of the nitrate transformation
and transport processes in which denitrification is the dominant
biochemical process in terms of dimensionless groups or combinations
of variables representing stream geometry and hydraulics, sediment
depth and gross denitrification kinetics and the interface
transport of nitrate from stream to sediment. The model provides
a means of quantifying the nitrate-removal-capability of small
streams in terms of the above dimensionless groups.

From the foregoing it is evident that headwater streams draining
agricultural watersheds can have substantial inputs of nitrogen depending upon the proportion of cropland, numbers of livestock, nature
of drainage area and agricultural practices. Quite apart from understanding the theory behind recommended agricultural practices which
may lead to less waste of energy from agricultural operations, through
reduced nitrogen output, it is useful to understand nitrogen transport
and transformation in streams. Such studies can increase the
possibilities of successful management of nitrogen in these environments. However, in comparison with those pertaining to lake and
terrestrial environments, studies on nitrogen transformation processes
in headwater streams are generally lacking. More research is needed
on such processes as uptake of nitrogen by plants, nitrogen release
from decomposing organic matter, nitrification, denitrification, etc.
Benthic invertebrates constitute an important component of the stream
environment and, as a few studies indicate, some of them profoundly
influence the above processes and it will be interesting to fully
elucidate their role in mineralization, nitrification and denitrification. Besides recording direct effects, such studies would entail
investigations of invertebrate-microbial interactions. Also, it
appears that allochthonous organic matter, which in many headwater
streams is an important food source for invertebrates, is an energy
source for microbes involved in nitrogen transformation. As our
guru, Dr. Hynes, has often remarked it is important to study stream
processes. This applies to those relating to nitrogen also.

REFERENCES

Andersen, J.M.   1976.   Importance of the denitrification process
    for the rate of degradation of organic matter in lake
    sediments.   *In:* Interactions between sediments and freshwater.
    Golterman, H.L. (ed.)   Junk, Amsterdam.
Baker, J.L., Campbell, K.L., Johnson, H.P. and Hanway, J.J.
    1975.   Nitrate, phosphorus, and sulfate in subsurface
    drainage water.   *J. Environ. Qual.*   2: 406-412.
Bansal, M.K.   1977.   Nitrification in natural streams.   *J.*
    Wat. Pollut. Control Fed.   48: 2380-2393.
Brezonik, P.L.   1972.   Nitrogen.   Sources and transformations
    in natural waters.   *In:* Nutrients in natural waters.
    Allen, H.E. and Kramer, J.R. (eds.).   John Wiley & Sons.
    New York.
Brezonik, P.L.   1977.   Denitrification in natural waters.
    Prog. Wat. Tech.   8 (4/5): 373-392.
Burwell, R.E., Schuman, G.E., Saxton, K.E. and Heinemann, H.G.
    1976.   Nitrogen in subsurface discharge from agricultural
    watersheds.   *J. Environ. Qual.*   5: 325-329.
Calvert, D.V.   1975.   Nitrate, phosphate and potassium movement
    into drainage lines under three soil management systems.
    J. Environ. Qual.   4: 183-186.
Caporali, F., Nannipieri and Pedrazzini, F.   1981.   Nitrogen
    contents of streams draining an agricultural and a
    forested watershed in Central Italy.   *J. Environ. Qual.*
    10: 72-76.
Casey, H.   1977.   Origin and variation of nitrate-nitrogen in
    the chalk springs, streams and rivers in Dorset, and its
    utilization by higher plants.   *Prog. Wat. Tech.*   8: (4/5):
    225-235.
Casey, H. and Clarke, R.T.   1979.   Statistical analysis of nitrate
    concentrations from the River Frome (Dorset) for the period
    1965-76.   *Freshwat. Biol.*   9: 91-97.
Chatarpaul, L. and Robinson, J.B.   1979.   Nitrogen transformations
    in stream sediment: ^{15}N studies.   *In:* Methodology for biomass
    determinations and microbial activities in sediments.
    Litchfield, C.D. and Seyfried, P.L. (eds.).   ASTM STP 673,
    American Society for Testing and Materials.
Chatarpaul, L., Robinson, J.B. and Kaushik, N.K.   1979.
    Observations on the role of tubificid worms in nitrogen
    transformations in stream sediment.   *J. Fish. Res. Board Can.*
    36: 673-678.
Chatarpaul, L., Robinson, J.B. and Kaushik, N.K.   1980.   Effects
    of tubificid worms on denitrification and nitrification in
    stream sediment.   *Can. J. Fish Aquat. Sci.*   37: 656-663.
Chen, R.L., Keeney, D.R., Graetz. D.A. and Holding. A.J.   1972.
    Denitrification and nitrate reduction in Wisconsin lake
    sediments.   *J. Environ. Qual.*   1: 158-162.

Chichester, F.W. 1977. Effects of increased fertilizer rates
    on nitrogen content of runoff and percolate from monolith
    lysimeters. *J. Environ. Qual.* 6: 211-217.
Clark, R.N., Gilbertson, C.B. and Duke, H.R. 1975. Quantity and
    quality of beef feedyard runoff in the Great Plains.
    *Proceedings 3rd International Symp. on Livestock Wastes --
    Managing Livestock Wastes. Amer. Soc. Agric. Engg.
    St. Joseph, MI.*
Commoner, B. 1968. Threats to the integrity of the nitrogen
    cycle: nitrogen compounds in soil, water, atmosphere and
    precipitation. *In:* Global effects of environmental
    pollution. Singer, F.S. (ed.). Springer-Verlag, New York.
Coote, D.R. and Hore, F.R. 1977. Runoff from feedlots and
    maure storages in Southern Ontario. *Can. Agric. Eng.*
    19: 116-121.
Crisp, D.T. 1970. Input and output of minerals for a small
    watercress bed fed by chalk water. *J. Appl. Ecol.* 7:
    117-140.
Eckenfelder, W.W. Jr. and Argaman, Y. 1979. Kinetics of
    nitrogen removal for municipal and industrial applications.
    In: Advances in Water and Wastewater Treatment, Biological
    Nutrient Removal. Wanielista, M.P. and Eckenfelder, W.W.
    (ed.). Ann Arbor Science Pub. Ann Arbor, MI.
Edwards, R.W. 1958. The effect of larvae of *Chironomus riparium*
    Meigen on the redox potentials of settled activated sludge.
    Ann. Appl. Biol. 46: 457-464.
Edwards, W.M. and McGuinness, J.L. 1975. Estimating quality and
    quality of runoff from eastern beef barn lots. *Proceedings
    3rd International Symp. on Livestock Wastes -- Managing
    Livestock Wastes. Amer. Soc. Agric. Engg. St. Joseph, MI.*
Engler, R.M. and Patrick, W.H. Jr. 1974. Nitrate removal from
    flood water overlying flooded soils and sediments. *J.
    Environ. Qual.* 3: 409-413.
Focht, D.D. and Verstraete, W. 1977. Biochemical ecology of
    nitrification and denitrification, p. 135-199. *In:*
    Advances in microbial ecology. Alexander, M. (ed.).
    Plenum, N.Y.
Fukuhara, H., Kikuchi, E. and Kurihara, Y. 1980. The effect
    of *Branchiura sowerbyi* (Tubificidae) on bacterial populations
    in submerged ricefield soil. *Oikos*   34: 88-93.
Grizzard, T.J. and Jennelle, E.M. 1972. Will wastewater
    treatment stop eutrophication of impoundments? *27th
    Purdue Industrial Waste Conf. Lafayette, Indiana.*
Gujer, Von W. 1976. Nitrifikation in Fliessgewassern --
    Fallstudie Glatt. *Schweiz Z. Hydrol.* 38: 171-189.
Haslam, S.M. 1978. River plants. Cambridge Univ. Press.
    London.
Hendricks, D.W. 1979. Application of water-quality models.
    Ch. 19. *In:* Modeling of Rivers. Shen, Hsieh Wen (ed.).
    John Wiley, N.Y.

Hill, A.R. 1978. Factors affecting the export of nitrate-
    nitrogen from drainage basins in Southern Ontario. *Water
    Res.* 12: 1045-1057.
Hill, A.R. 1979. Denitrification in the nitrogen budget of
    a river ecosystem. *Nature* 281: 291-292.
Hill, A.R. and Wylie, N. 1977. The influence of nitrogen
    fertilizers on stream nitrate concentrations near Alliston,
    Ontario, Canada. *Prog. Wat. Techn.* 8: (4/5): 91-100.
Jackson, W.A., Asmussen, L.E., Hauser, E.W. and White, A.W.
    1973. Nitrate in surface and subsurface flow from a small
    agricultural watershed. *J. Environ. Qual.* 2: 480-482.
Johnson, A.H., Bouldin, D.R., Goyette, E.A. and Hedges, A.M. 1976.
    Nitrate dynamics in Fall Creek, N.Y. *J. Environ. Qual.* 5:
    386-391.
Kamp-Nielsen, L. and Andersen, J.M. 1977. A review of the
    literature on sediment-water exchange of nitrogen compounds.
    Prog. Wat. Tech. 8: (4/5): 393-418.
Kaushik, N.K. and Hynes, H.B.N. 1971. The fate of the dead
    leaves that fall into streams. *Arch. Hydrobiol.* 68: 465-515.
Kaushik, N.K., Robinson, J.B., Sain, P., Whiteley, H.R. and
    Stammers, W.N. 1975. A quantitative study of nitrogen loss
    from water of a small, spring-fed stream. *In:* Water Pollut.
    Res. in Canada. Proc. 10th Canadian Symp. Toronto.
    110-117.
Kaushik, N.K. and Robinson, J.B. 1976. Preliminary observations
    on nitrogen transport during summer in a small spring-fed
    Ontario stream. *Hydrobiologia* 49: 59-63.
Keeney, D.R. 1972. The fate of nitrogen in aquatic ecosystems.
    Literature Review No. 3. Water Resources Centre, Univ. of
    Wisconsin. p. 59.
Keeney, D.R. 1973. The nitrogen cycle in sediment-water systems.
    J. Environ. Qual. 2: 15-28.
Kikuchi, E. and Kurihara, Y. 1977. *In vitro* studies on the
    effect of tubificids on the biological, chemical and physical
    characteristics of submerged ricefield soil and overlying
    water. *Oikos* 29: 348-356.
Kolenbrander, G.J. 1977. Nitrogen in organic matter and
    fertilizer as a source of pollution. *Prog. Wat. Tech.*
    8: (4/5): 67-84.
Krenkel, P.A. and Navotny, V. 1979. River water quality model
    construction. Ch. 17. *In:* Modelling of rivers. Shen, Hsieh
    Wen (ed.). John Wiley, New York.
Krenkel, P.A. and Ruane, R.J. 1979. Basic approach to water
    quality modeling. Ch. 18. *In:* Modeling of rivers. Shen,
    Hsieh Wen (ed.). John Wiley, New York.
Landner, L. 1977. Sources of nitrogen as a water pollutant:
    inudstrial waste water. *Prog. Wat. Tech.* 8: (4/5): 55-65.
Larsen, V. 1977. Nitrogen transformation in lakes. *Prog.
    Wat. Tech.* 8: (4/5): 419-431.
Lee, D.R. and Hynes, H.B.N. 1978. Identification of groundwater
    discharge zones in a reach of Hillman Creek in Southern
    Ontario. *Water Pollut. Res. Canada* 13: 121-133.

Loehr, R.C. 1974. Characteristics and comparative magnitude of non-point sources. *J. Water Poll. Control. Fed.* 46: 1849-1872.

McColl, R.H.S. 1974. Self-purification of small freshwater streams: phosphate, nitrate and ammonia removal. *N.Z.J. Mar. Freshwater Res.* 8: 375-388.

Miller, M.H. 1979. Contribution of nitrogen and phosphorus to subsurface drainage water from intensively cropped mineral and organic soils in Ontario. *J. Environ. Qual.* 8: 42-48.

Miller, M.H. and Nap, W. 1971. Fertilizer use and environmental quality. Report prepared for The Advisory Fertilizer Board of Ontario. p. 40.

Najarian, T.O. and Harleman, D.R.F. 1977. A real time model of nitrogen-cycle dynamics in an estuarine system. *Prog. Wat. Tech.* 8: (4/5): 323-345.

Neilsen, G.H., Culley, J.L. and Cameron, D.R. 1978. Nitrogen loadings from agricultural activities in the Great Lakes Basin. Final Report Task Group C, PLUARG, International Joint Commission. p. 103.

Neilsen, G.H., Cameron, D.R. and Culley, J.L. (In press). Agriculture and water quality in the Canadian Great Lakes Basin, IV: Nitrogen. *J. Environ. Qual.*

Nelson, L.B. 1972. Agricultural chemicals in relation to environmental quality: chemical fertilizers, present and future. *J. Environ. Qual.* 1: 2-6.

Nicholls, K.H. and MacCrimmon, H.R. 1974. Nutrients in subsurface and runoff waters of the Holland Marsh, Ontario. *J. Environ. Qual.* 3: 31-35.

Nichols, D.S. and Keeney, D.R. 1973. Nitrogen and phosphorus release from decaying water milfoil. *Hydrobiologia* 42: 509-525.

Owens, M., Garland, J.H.N., Hart, I.C. and Wood, G. 1972. Nutrient budgets in rivers. *Symp. Zool. Soc. Lond.* 29: 21-40.

Painter, H.A. 1977. Microbial transformations of inorganic nitrogen. *Prog. Wat. Tech.* 8: (4/5):3-29.

Patrick, W.H. Jr. and Reddy, K.R. 1976. Nitrification-denitrification reaction in flooded soils and water bottoms: dependence on oxygen supply and ammonium diffusion. *J. Environ. Qual.* 5: 469-472.

Patrick, W.H. Jr., Delaune, R.D., Engler, R.M. and Gotoh, S. 1976. Nitrate removal from water at the water-mud interface in wetlands. *Technical Report No. EPA-600/3-76-042.*

Payne, W.J. 1973. Reduction of nitrogenous oxides by micro-organisms. *Bacteriol. Rev.* 37: 409-452.

Phillips, R.E., Reddy, K.R. and Patrick, W.H. Jr. 1978. The role of nitrate diffusion in determining the order and rate of denitrification in flooded soil: II. Theoretical analysis and interpretation. *J. Soil Sci. Soc. Am.* 42: 272-278.

Pratt, A.F., Jones, W.W. and Hunsaker, V.E. 1972. Nitrate in
    deep soil profiles in relation to fertilizer rates and
    leaching volume. *J. Environ. Qual.* 1: 97-102.
Rinaldi, S., Concini-sessa, R., Stehfest, H. and Tamura, H.
    1979. Modeling and control of river quality. McGraw-Hill
    International, N.Y.
Robinson, J.B. 1976. Integrating food production into nature's
    biogeochemical cycles. *J. Milk Food Technol.* 39: 297-300.
Robinson, J.B. et al. 1977. Modelling nitrogen transport
    and cycling in natural flow system. Final Report,
    Inland Water Directorate, Environment Canada, Water
    Resources Res. Program. p. 324.
Robinson, J.B., Whitely, H.R., Stammers, W.N., Kaushik, N.K.
    and Sain, P. 1979. The fate of nitrate in small streams
    and its management implications. *In:* Best management
    practices for agriculture and silviculture: Proceedings of
    the 10th Annual Cornell Agricultural Waste Management
    Conference. R.C. Loehr, D.A. Haith, M.F., Walter and C.S. Martin
    (eds.) 1978. Ann Arbor Sci. Publishers Inc., Ann Arbor, MI.
Sain, P., Robinson, J.B., Stammers, W.N., Kaushik, N.K. and
    Whiteley, H.R. 1977. A laboratory study on the role of
    stream sediment in nitrogen loss from water. *J. Environ.
    Qual.* 6: 274-278.
Schuman, G.E., Burwell, R.E., Piest, R.F. and Spomer, R.G.
    1973. Nitrogen losses in surface runoff from
    agricultural watersheds in Missouri Valley, Loess.
    J. Environ. Qual. 2: 299-302.
Schwert, D.P. and White, J.P. 1974. Method for *in situ*
    measurement of nitrification in a stream. *Appl. Microbiol.*
    28: 1082-1083.
Shannon, E.E. and Brezonik, P.L. 1972. Relationship between
    lake trophic state and nitrogen and phosphorus loading
    rates. *Environ. Sci. and Technol.* 6: 719-725.
Shuval, H.I. and Gruener, N. 1977. Infant methemoglobinemia and
    other health effects of nitrates in drinking water.
    Prog. Wat. Tech. 8: (4/5): 183-193.
Stake, E. 1967. Higher vegetation and nitrogen in a rivulet
    in central Sweden. *Schweiz. Z. Hydrol.* 29: 107-124.
Stammers, W.N., Robinson, J.B., Whitely, H.R., Kaushik, N.K.
    and Sain, P. 1978. Modeling nitrate transport in a
    small upland stream. *Water Poll. Res. Canada* 13: 161-173.
Stammers, W.N., Robinson, J.B., Whiteley, H.R. and Kaushik, N.K.
    (In press). Characterization of the kinetics of
    'denitrification' in stream sediments. Proc. Symp.
    Dynamics of Lotic Ecosystems., Augusta, Georgia. 1980.
Thomann, R.V. 1972. Systems Analysis and Water Quality
    Management. Environmental Res. and Applications, Inc.,
    New York.

Thomas, G.W. and Crutchfield, J.D. 1974. Nitrate-nitrogen and phosphorus contents of streams draining small agricultural watersheds in Kentucky. *J. Environ. Qual.* 3: 46-49.

Tomlinson, T.E. 1970. Trends in nitrate concentrations in English rivers in relation to fertilizer use. *Wat. Treat. Exam.* 19: 277-295.

Toms, I.P., Mindenhall, M.J., and Harman, M.M.I. 1975. Factors affecting the removal of nitrate by sediments from rivers, lagoons and lakes. Technical Report TR 14. Wat. Res. Centre, Stevenage, Herts, England.

Terry, R.E., and Nelson, D.W. 1975. Factors influencing nitrate transformations in sediments. *J. Environ. Qual.* 4: 549-554.

Vallentyne, J.R. 1978. Today is yesterday's tomorrow. *Verh. Internat. Verein. Limnol.* 20: 1-12.

Van Kessel, J.F. 1977a. Removal of nitrate from effluent following discharge on surface water. *Water Res.* 11: 533-537.

Van Kessel, J.F. 1977b. Factors affecting the denitrification rate in two water-sediment systems. *Water. Res.* 11: 259-267.

Viets, F.G. 1970. Soil use and water quality -- a look into the future. *J. Agr. Food Chem.* 18: 789-792.

Viets, F.G. 1971a. Water quality in relation to farm use of fertilizer. BioScience. 21: 460-467.

Viets, F.G. 1971b. Fertilizers. *J. Soil and Water Conserv.* 26: 51-53.

Vincent, W.F. and Downes, M.T. 1980. Variation in nutrient removal from a stream by watercress (*Nasturtium officinale* R. Br.). *Aquat. Bot.* 9: 221-235.

Vollenweider, R.A. 1968. Scientific fundamentals of the eutrophication of lakes and flowing waters with particular reference to nitrogen and phosphorus as factors in eutrophication. Organization of Economic Co-operation and Development, Paris. DAS/CSI/ 68-27. p. 192.

Westlake, D.F. 1975. Macrophytes. *In:* River Ecology. Whitton, B.A. (ed.). Blackwell Scientific Publications, London.

Wetzel, R.G. 1975. Limnology, W.B. Saunders Co., Toronto.

White, J.P., Schwert, D.P., Ondrake, J. and Morgan, L.L. 1977. Factors affecting nitrification *in situ* in a heated stream. *Appl. Environ. Microbiol.* 33: 918-925.

Whiteley, H.R., Robinson, J.B., Stiebel, W.H., Kaushik, N.K. and Stammers, W.N. (In press). Factors affecting nitrogen fluxes in a small Ontario agricultural watershed. *In:* Proc. Nutrient Cycling in Agricultural Systems. An International Symposium Univ. of Georgia, Athens, Georgia, 1980. Ann. Arbor Sci. Publishers Inc., Ann Arbor, MI.

ORGANIC CARBON IN AQUATIC ECOSYSTEMS: BEYOND ENERGY – CONTROL

Donald L. Lush

Beak Consultants Limited
Mississauga, Ontario, Canada

For several years now the limnological literature has carried
a wide variety of articles dealing with descriptions of the
aquatic organic carbon pool. Organic carbon in aquatic
ecosystems has been broken down into a series of components
ranging from DOM (dissolved organic matter) through to POM
(particulate organic matter) with a wide variety of sub-
descriptors such as FPOM (fine particulate organic matter) and
CPOM (coarse particulate organic matter), principally for
descriptive convenience. It should be recognized at the outset
that in all natural freshwater systems, dissolved organic matter
exists as a continuum of sizes from very small molecular weight
amino acids and simple sugars to very large molecular weight
humic materials. In a similar context, although this material is
called 'organic', in many instances its inorganic component of
sorbed or complexed metal silicates or hydroxides may be considerable. The fact that particulate organic matter also spans a
continuum from colloidal substances and bacteria through to fish
is perhaps more obvious.

Within the aquatic sciences the study of organic carbon is
changing rapidly from a descriptive science to a mechanistic one,
where the descriptive questions are being replaced by "what is it?",
"why is it present?" and "what role does it play in the ecosystem?".
In this brief review of selected literature, spanning a variety of
fields ranging from geochemistry to soil science through aquatic
ecology and cellular physiology, certain trends may be seen as
developing. Organic matter in aquatic systems is dynamic. It is
continually changing, being modified by both geochemical and
biological processes. Although biologically produced, it plays an
important role in geochemical surficial weathering processes, at

the same time controlling the supply of essential trace elements
as organic matter complexes for the community responsible for its
production. Certain organic fractions are soluble which, within
a given geochemical environment and time frame, are mobile, while
others are insoluble and thus restrict the rate of flow of organic
energy and any complexed essential elements. Beyond the soluble
vs. particulate control systems, control of 'soluble' elements is
mediated through restrictions on bioavailability (the state of a
substance relating to its suitability for biological uptake) which,
again, are demonstrated to be dynamic processes shifting
temporally and spatially with shifting chemical environments
related to parameters such as pH and eh. As many of the more
complex and higher molecular weight organic materials pass through
aquatic systems, their energy contributions are relatively
insignificant but their effect on energy flow through their ability
to transport and control the bioavailability of a wide range of
trace nutrients may be very great. At present, we have but a few
glimpses into what promises to be a very important field of aquatic
science.

NATURE OF FRESHWATER ORGANIC MATTER

    This organic component of freshwater ecosystems is believed to
exert both direct and indirect controls over biological systems.
There seems to be little doubt that the simpler, small molecular
weight organic molecules can be used directly as soluble energy
sources and that both fine and coarse particulate organic matter
is ingested and digested to varying degrees by filter feeders
found in both stream and lake systems. In this role it serves as
a transporter of energy between various spatial and trophic
components of the aquatic ecosystem.

    In addition to the organic component which appears to be
utilized as a nutrient and energy source, some other organic
components, principally the humic and fulvic acid solubles and
aggregated particulates, appear to be very resistant to biological
degradation and are known to have carbon content half lives on the
order of 600 years (Schnitzer, 1971). The role of this organic
component is more subtle and appears to be more concerned with the
control of aquatic biota through the regulation of energy flow and
trace element distribution.

    Within most natural fresh waters the concentration of organic
carbon commonly falls in the range of from 0.1 to 10 mg ℓ^{-1} (Stumm
and Morgan, 1970). It is derived principally from two major
sources: 1) allochthonous sources through the direct input of
organic materials such as autumn-shed leaves (Kaushik and Hynes,
1971), and indirect inputs of soil organic material from overland
flow and shallow groundwater inputs; and 2) from autochthonous
sources such as the cellular debris and extracellular excretion

products of aquatic organisms (Fogg, 1966; Leppard et al., 1977).
The chemistry of this organic material is varied and complex. The
nature of simple soluble exudates such as glycolic compounds from
algae has been described by Fogg (1966), while others such as
Leppard et al. (1977) have described the morphology and, to a
limited degree, the chemistry of algal and bacterially-derived
fine particulate organic matter.

The chemistry of direct particulate inputs of organic matter
from terrestrial sources to freshwater systems has been described
from the standpoint of animal nutrition with relation principally
to components such as carbohydrates and protein (Kaushik and
Hynes, 1971). Within the soil systems that supply organic
leachates to lake and stream systems, humic substances constitute
the bulk of the organic matter (Bordovskiy, 1965). These
materials have been described (Schnitzer, 1971) as acidic, dark
coloured, partially aromatic, and chemically complex substances
with molecular weights ranging from a few hundred to several
thousand. They lack specific chemical and physical characteristics
(e.g., melting point, refractive index, exact elementary
composition, etc.) usually associated with well-defined organic
compounds. Based upon their solubility, Schnitzer (1971) has
suggested humic materials may be broken down into: 1) fulvic acids,
which are soluble in both alkali and acid and are considered to
have the lowest molecular weights; 2) humic acid, which are
soluble in alkali but insoluble in acid and are of intermediate
molecular weight; and 3) humins, which are insoluble in both
alkali and acid and have the highest molecular weight.

Despite the general categories of these compounds mentioned
above, it is worth noting that there are no sharp chemical
divisions between the three broad classes of fulvic acids, humic
acids, and humins. They are all part of an extremely heterogeneous polymer system and the differences noted by Schnitzer
(1971) and others (Gjessing, 1976) are due to some degree of
variation in elemental composition, acidity, degree of polymerization and molecular weight among other factors. It should
also be noted that the composition of these materials, despite
their apparent ^{14}C age, is unlikely to be static either in
terrestrial soils or aquatic sediment. Evidence collected by
investigators such as Kuznetosov (1975) has shown that microorganisms in sediment are responsible for the decomposition of some
humic materials and possibly, either directly or indirectly, for
the polymerization or synthesis of others. It is also very likely
that, within the water column, microbial populations are also
responsible for this continuous process of degradation and
resynthesis. Thus, within any component of the aquatic ecosystem
there is likely to be a continuum of humic materals, although the
percentage composition will likely vary between aquatic compartments
such as stream waters and lake sediments.

At any point in time a freshwater aquatic system is likely to
be receiving a wide variety of organic inputs. Many of the
smaller molecular weight materials such as soluble algal exudates
or material released from autumn-shed leaves are likely to feed
directly into secondary production within the aquatic ecosystem.
Additionally, the release and uptake of small molecular weight
organic materials, while in itself a complex and difficult topic
in the area of secondary production, is further complicated by the
role of larger molecular weight humic compounds in controlling
many aspects of aquatic secondary production. This control may
range from a regulation of weathering rates of trace elements to
the restriction of bioavailability of many essential nutrients.
The larger molecular weight soluble fulvic and humic acids,
particulate humins, or microfibrillar materials (Leppard et al.,
1977) are likely to fulfill a different role, related more to the
long-term stabilization of the aquatic system, if not the entire
watershed.

ROLE OF ORGANIC MATERIALS IN REGULATION OF AQUATIC SYSTEMS

It is axiomatic that nothing in nature is wasted and that
those things soon disappear which no longer serve a role in the
efficient functioning of the ecosystem. Thus, it is easy to ponder
the role of a ubiquitous substance like aquatic organic matter
but it is another matter to generate appropriate paradigms which
are capable of being tested. To truly understand the nature of
this phenomenon we must understand where it fits into the patterns
of life responsible for its origin.

During the past several decades, soil scientists have been
studying the nature of organic materials in the soil which are
responsible, to a large degree, for that characteristic of soil
known as fertility. This research has sprung in large part from
man's need to manage his farms and forests for the betterment of
society. By comparison, municipal engineers have historically
regarded the corresponding organic compounds occurring in aquatic
systems simply as a nuisance. They have tried to characterize the
organic materials solely with the purpose of ridding the water of
them, through coagulation, bleaching, filtration, etc., thereby
making it more potable (Black and Willems, 1961; Gjessing, 1976;
Klyachko, 1964). More recently the geochemical disciplines have
been studying the nature and behaviour of water soluble organic
materials (Jackson et al., 1978). In this discipline it has been
recognized that these organic substances, because of their ability
to trap and hold metals, provide valuable clues both to the mode
of creation of certain types of ore bodies and to the location of
potentially valuable commercial deposits (Schenek, 1968; Breger,
1963; Manskaya and Drozdova, 1968). Within the area of aquatic
biology, tantalizing bits of information have also been appearing
relating to the possible role many naturally-occurring organic

substances play in controlling trace metal availability both from
a restrictive standpoint (Sakamoto, 1971) and from a bioconcentra-
tion point of view (Leppard et al., 1977). Other mechanisms of
control exerted by humic materials have also been discussed by
Jackson et al. (1980) and Jackson and Hecky (1980). Through the
interrelationships that are developing between the sciences of
geochemistry, soil science and aquatic ecology, it is becoming
clear that soluble and particulate organic materials act as a
transporting agent in watersheds. They control to a significant
degree the rate of erosion of many elements as they move in
solution or particulate form through streams, rivers and lakes.
And they control the short-term geologic sink for many elements
found in the deep sedimentary basins of many lakes. The cycling
of organic materials in oligotrophic and dystrophic lakes has been
investigated recently by Seki et al. (1980).

Dissolved organic materials first begin to influence the
biotic nature of the watershed in the headwater areas where rains
and melting snow provide the water necessary to transport
terrestrial organic matter from the soils to streams. In the
soils and headwater wetlands of the watershed, the vegetation
releases soluble and insoluble, relatively simple organic
compounds, many of which are used as an energy source by a wide
variety of soil and sediment micro-organisms. This microbial
utilization of organic materials in terrestrial (Mathur and Paul,
1967a) and aquatic environments (Nissenbaum and Kaplan, 1972)
results in the synthesis of more chemically complex and chemically
diverse fulvic and humic materials (Martin and Haider, 1971;
Sundman, 1965). It appears that the character of terrestrial and
aquatic humic material is dependent upon a very complex series
of interactions involving not only vegetation and microbial
populations but also the parent soil minerals (Flaig et al., 1971).
Once these more complex humic materials are synthesized, their rate
of decomposition appears to be slowed but they are still sources
of energy and are continually metabolized by microbial
populations (Mathur and Paul, 1967b; Neunylov and Khaukina, 1968).
As these soluble organic materials move out of the shallow soil
horizons and organic rich aquatic sediments, and as they are
carried through shallow groundwater and surface stream systems,
they come in contact with the minerals of the parent – sedimentary
or bedrock deposits.

As early as 1929 Harrar recognized the ability of organic
acids to leach, reduce and dissolve minerals. Since that time, the
role of naturally-occurring organic acids in the dispersion of
many elements has been widely recognized (Jackson et al., 1978).
Some marked effects of organic acids can be shown in their
influence on the rates of dissolution of some metals. When fulvic
acid was present in water, lead sulphides were found to dissolve

10 to 60 times faster than in control waters at the same pH (Bondarenko, 1968). Even such relatively stable elements as gold (Fisher et al., 1974), and highly refractory silicate minerals (Singer and Naurot, 1976) are found to react with fulvic acid, forming very stable complexes. Whether or not these trace metal scavenging reactions are direct reactions between fulvic acids and inorganic minerals in nature is unknown. It may be that in some cases the reactions are mediated by bacterial exoenzymes which serve to "pluck" the trace element from the host mineral, after which fulvic acid complexes the trace element. Regardless of the mechanism, the net result is the same - a biologically accelerated weathering of trace elements and their transport to the aquatic system. Ramamoorthy and Kushner (1975 a and b) and others (Pauli, 1975) have determined the stability constants associated with a variety of weathering trace elements, but how these relate to the evolution of trace element biogeochemistry in freshwater systems is at present unknown.

The resulting stable soluble and colloidal organic complexes are transported throughout the watershed carrying with them a wide variety of trace elements and nutrients. The nature of the bioavailability of many of the elements, complexed or otherwise bound to the soluble and colloidal organic materials in water, is not well known but may be expected to fluctuate seasonally. In the case of iron, some evidence is available to suggest that within oxidized systems, such as occur in the majority of lakes and streams, the precipitation of iron as inorganic hydroxide is strongly inhibited by humic compounds (Rashid and Leonard, 1973). Other evidence (Sakamoto, 1971; Poldoski, 1979) suggests that the stabilization of iron and other elements by some organic acids may restrict its bioavailability to phytoplankton. The degree of bioavailability of trace elements and micro and macronutrients (Lean, 1973) undoubtedly relates to the form in which it is bound (Jackson and Schindler, 1975) and may not relate to whether it is in true solution or not. Iron has been found to readily cause the co-precipitation of the complex microbial degradation products of algal cells, whereas the simpler pigment and lipid materials appear to stabilize iron in solution (Akiyama, 1973). Thus, the more chemically complex organic acids under certain circumstances may be responsible for the co-precipitation and removal of many elements to stream and lake sediments, whereas the simpler organic compounds may stabilize many elements in solution or colloidal suspension. Within natural waters containing significant amounts of natural organic materials, several studies have shown that a very high percentage of elements were in some manner associated with the organic matter (Beck et al., 1974; Ramamoorthy and Kushner, 1975 a and b).

The mechanisms by which the algae and microflora of streams and lakes obtain trace metals and nutrients from soluble and

particulate organic materials is not well understood. One
possible mechanism may involve contact cation exchange between
the soluble or colloidal organic material and microfibrillar
material produced by algal and microbial cells (Leppard et al.,
1977). These carboxylated polysaccharide polymers with micro-
fibrilar characteristics are known, in soil systems, to bind cations
and hold them in exchangeable form (Ramamoorthy and Leppard, 1977)
and thus make the cations available for direct cellular uptake.
Rorem, in 1955, showed that bacterial cells increase their
production of extracellular carboxylated polysaccharides in
response to a reduction in mineral nutrients in the culture medium.
Other investigators (Haug et al., 1969) have suggested that in
blue-green algal cells the synthesis of these extracellular
polysaccharides is under strict genetic control and that cells can
secrete any one of a variety of different acid polysaccharides
depending upon the nature of their external chemical environment.
Based upon the abundance of these carboxylated polysaccharide
materials in natural lake waters, whether they are attached to
algal cells, suspended in the water column, or on the surface of
sediments, it has been suggested (Leppard et al., 1977) that
these materials act as a carrier system having a major role in
the binding and redistribution of biologically available cations.
It is thus possible to speculate that the humic and fulvic acid
materials may act as a more long-term transport system, tying up
many trace elements and micronutrients in a refractory pool that
is only accessible through the mediation of extracellular cation
exchange materials (microbially derived carboxylated polysaccharides)
produced in direct response to cellular requirements.

     Leppard et al. (1977) also noted that the intestinal
epithelium of vertebrates contains microfibrillar material
similar in structure to that produced by algal cells and that this
material is presumed to play a role in ion binding. In freshwater systems much of this organic material having strong cation
exchange properties is known to aggregate and settle to the
sediments, where it may facilitate nutrient uptake by detritivores.
Within the water column itself the secretion of selective cation
exchange material by planktonic cells may also be active in
cellular protection when toxic materials are found at high levels.
In this instance, genetically-controlled production of extracellular material with selective cation exchange properties may
facilitate the adsorption of required elements, while the toxic
elements would be left in the more refractory organic acid pool.
The alternative hypothesis has been suggested by Hardstedt-Romeo
and Gnassia-Barelli (1980) in which the extracellular products of
phytoplankton are responsible for decreasing the rate of uptake
of some metals, while other elements are in turn continuously
removed from this pool through coagulation of the organic acid
fraction and its subsequent precipitation to lake and stream
sediments.

It is interesting to note that humic acids and other organic
detrital materials have the capacity to adsorb significant
quantities of hydrophobic organic compounds, including many of the
herbicides and pesticides applied to watersheds (Khan, 1973; Odum
and Drifmoyer, 1978). One of the major consequences of this is
the relatively rapid shifting of many pesticides and herbicides
from the aqueous phase to the particulate phase in natural freshwater systems with their consequent potential for uptake by
filter feeders or loss to the sediments. Deposition in the sediments
of many potentially hazardous organic materials adsorbed to humic
materials effectively removes much of this material from
circulation within food chains to which man is directly exposed
(Hakanson, 1980). The fact that precipitated humic materials
have a low density ensures their relatively rapid transport to
deeper lake basins (Emerson and Hesslein, 1973; Hesslein et al.,
1980) where they are not subject to resuspension by currents.
Within these deeper sediments, which contain relatively refractory
organic materials, it is suspected that microbial competition for
available energy is intense and biological breakdown of
decomposable materials relatively rapid.

The diverse nature of the microflora found in most lake and
stream sediments, and their ability to metabolize a wide variety of
natural and synthetic organic materials can also have unpleasant
consequences for man. Many metals which are toxic in elevated
concentrations are suspected of being stripped from natural organic
materials and removed from the microfloral environment by being
solubilized, principally as methylated metal complexes (Ferguson
and Gavis, 1972; Wood, 1974). These methylated materials such as
methyl mercury enter the water column and depending on environmental
conditions may become available for uptake by higher trophic levels
(Jernelov and Lann, 1973). In the case of trace nutrients this
microbial and possibly higher plant mobilization may serve as a
very valuable recycling system (Lovelock, 1979). However, under
polluted conditions, such a mobilization could be harmful.

Under normal conditions this recycling of trace elements from
sediments is a diffusion controlled process aided by bioperturbation
and it is possible that a unique biofeedback control mechanism
exists here. As trace metal or toxic organic materials rise
through sediments and become bioavailable through microbial
mobilization, toxic levels of these substances may be exceeded
for the macroinvertebrates responsible for bioperturbation. The
macroinvertebrate populations would then decline, and thus
bioperturbation processes would be reduced with a simultaneous
reduction in the flox of toxic substances to the overlying water.

Coupled with this bioperturbation-related feedback for some
metals is a complex chemical system involving sulphides.

Polysulphide ions (Bouleque, 1974) are strong complexing ions and act as good acidity and redox buffers in reduced organic sediments. Polysulphides are also thought to be responsible for the stabilization of humic materials through the incorporation of polysulphide chains into the structure of the humic material. Other studies (Timperley and Allan, 1974) suggest that, in some cases, the metal sulphide may be directly responsible for trace element immobilization (e.g., copper sulphide), whereas in other cases adsorption to the sulphide mineral may be the controlling factor (mercury on various sulphide minerals). Under changed environmental conditions these controlling mechanisms can change rapidly. In one instance, a beaver dam washed-out, exposing reduced sediments to the atmospheric oxygen (Jonasson and Timperley, 1973). Within a few weeks of being exposed to atmospheric oxygen the sediments had lost more than half their original content of copper, nickel, zinc, lead and arsenic, and more than 80% of their sulphur and mercury, although the carbon content remained constant. It is mobilization processes similar to these that are of concern when reduced muds are dredged for either onshore disposal or dumping into oxygenated deeper waters.

In freshwater systems, one can easily visualize that when deep organic sediments are exposed to oxygen with resultant redox potential changes, either on a seasonal basis during lake turnover or on a longer-term cycle, that trace nutrients may be released. Longer-term cycles may relate to decreases in sediment oxygen demand which in turn may be related to lower water column productivity. Accordingly, this controlling mechanism may be expected to release essential trace nutrients, thus contributing to higher water column productivity and a consequent slowdown in nutrient release causing the cycle to repeat.

It is thus possible to visualize sediments rich in humic materials acting as a large trace element reservoir. The chemical properties of these allochthonous and autochthonous organic materials are responsible both for the removal of excess levels of metals from the water column (Jackson, 1980) under conditions of high organic input and the release of these metals back to the water column under conditions of low organic input.

The adsorption of trace elements to organic materials is also a function of pH (Schindler et al., 1980) and, despite the pH and redox buffering capacity of the humic and polysulphide materials, some evidence is emerging of trace element release from sediments related to changes in pH (Bourg, 1979). It is suspected that much more research will be done in this area in the future as concern about acid rain affecting aquatic systems mounts.

CONCLUSIONS

Evidence has been emerging over the last several decades from
a variety of disciplines, principally those of soil science,
geology and aquatic ecology, that dissolved and particulate
organic matter plays a vital role in the functioning of aquatic
ecosystems. At present, we are just beginning to understand the
physical, chemical and biological nature of many of these materials.
It appears that, through the wide variety of properties they
exhibit, they are involved in processes ranging from a microscopic
scale in regulating micronutrient supply to phytoplankton
(Leppard et al., 1977) to a global scale in regulating the level
of oxygen abundance in our atmosphere (Lovelock, 1979). There are
a variety of new research thrusts focussed toward the understanding
of the nature and function of organic materials in natural waters.
With the advent of rapid, automated and inexpensive analytical
techniques, sedimentary organic materials are being examined in
great detail by the geochemical prospecting community and a large
and useful database is being developed. At the other end of the
spectrum, many resource development companies and government
agencies are becoming more aware of the need to understand the
mechanisms associated with "pollution" assimilation in freshwater
ecosystems. In both of these areas organic material dominated by
the humic materials plays a key role. As more and more data
become available from a wide range of disciplines, new horizons
are being opened up for ecologists and an opportunity is emerging
to better understand those factors which control our freshwater
ecosystems.

REFERENCES

Akiyama, T. 1973. Interactions of ferric, and ferrous irons and
    organic matter in water environment. *Geochem. J.* 7: 167-177.
Beck, K.C., Reuter, J.H. and Perdue, E.M. 1974. Organic and
    inorganic geochemistry of some coastal plain rivers of the
    southeastern United States. *Geochim. cosmochime acta.*
    38: 341-364.
Black, A.P. and Willems, G.D. 1961. Electrophoretic studies of
    coagulation for removal of organic colour. *J. Amer. Water
    Works Assoc.* 53: 589-604.
Bondarenko, G.P. 1968. An experimental study on the solubility
    of galena in the presence of fulvic acids. (Trans. from)
    Geokhimiya. 5: 631-636 (from Jackson et al., 1978).
Bordovskiy, O.K. 1965. Accumulation and transformation of organic
    substances in marine sediments. *Marine Geol.* 3: 1-14.
Boulegue, J. and Michard, G. 1974. Interactions between sulphur,
    polysulphide system and organic material in reducing media.
    C.R. Acad. Sci., Ser. D, 279 (1): 13-15.

Bourg, A.C.M. 1979. Effect of ligands at the solid-solution
    interface upon the speciation of heavy metals in aquatic
    systems. *In*: Proc. Int. Conf. on Man and Concentration of
    Heavy Metals in the Environment. Long, Sept., 1979.
    pp. 446-449.
Berger, I.A. (ed.). 1963. Organic Geochemistry. Pergamon Press,
    New York, N.Y.
Emerson, S. and Hesslein, R. 1973. Distribution and uptake of
    Radium-226 in a small lake. *J. Fish. Res. Bd. Canada*
    30: 1485-1490.
Ferguson, J.F. and Gavis, J. 1972. A review of the arsenic
    cycle in natural waters. *Water Research* 6: 1259-1274.
Fisher, E.I., Fisher, V.L. and Miller, A.D. 1974. Nature of
    interaction of natural organic acids with gold. *Soc. Geol.*
    7: 142-146.
Flaig, W., Kuster, E., Haider, K., Bautelspacher, G., Filip, Z.
    and Martin, J.P. 1971. Influence of clay minerals on the
    formation of humic substances by some soil fungi. *Soviet Soil
    Sci.* 4: 322-330.
Fogg, G.E. 1966. Algal Cultures and Phytoplankton Ecology.
    Univ. of Wisconsin Press, Madison, Wisconsin.
Gjessing, E.T. 1976. Physical and Chemical Characteristics
    of Aquatic Humus. Ann Arbor Sci. Publ. Inc., Ann Arbor,
    Michigan.
Hakanson, L. 1980. An ecologist risk index for aquatic
    pollution control: a sedimentological approach.
    Water Res. 14: 975-1001.
Hardstedt-Romeo, M. and Gnassia-Barelli, M. 1980. Effect of
    complexation by natural phytoplankton exudates on the
    accumulation of cadmium and copper by the Haptophyceae
    Cricosphaera elongota. Marine Biology 59: 79-84.
Harrar, N.J. 1929. Solvent effects of certain organic acids
    upon oxides of iron. *Econ. Geol.* 24: 50-61.
Haug, A., Larsen, B. and Baardseth, E. 1969. Comparison of the
    constitution of alginates from different sources. *In*:
    Margslef, R. (ed.), Proceedings of the Sixth International
    Seaweed Symposium, Madrid, Subsecretaria de la Marina
    Mercante.
Hesslein, R.H., Broecker, W.S. and Schindler, D.W. 1980. Fates
    of metal radiotracers added to a whole lake: sediment water
    interactions. *Can. J. Fish. Aquat. Sci.* 37: 378-386.
Jackson, K.S., Jonasson, I.R. and Skippen, G.B. 1978. The
    nature of metals - sediment - water interactions in freshwater
    bodies, with emphasis on the role of organic matter.
    Earth-Science Reviews 14: 97-146.
Jackson, T.A. and Hecky, R. 1980. Depression of primary
    productivity by humic matter in lake and reservoir waters of
    the boreal forest zone. *Can. J. Fish. Aquat. Sci.* 37:
    2300-2317.

Jackson, T.A., Kipphut, G., Hesslein, R. and Schindler, D.W. 1980.
Experimental study of trace metal chemistry in soft water
lakes at different pH levels. *Can. J. Fish. Aquat. Sci.*
37: 387-402.

Jackson, T.A. and Schindler, D.W. 1975. The biogeochemisty of
phosphorous in an experimental lake environment: evidence for
the formation of humic-metal-phosphate complexes. *Proc. Int.
Assoc. Theor. Appl. Limnol.* 19: 211-221.

Janecek, J. and Chalupa, J. 1969. Biological effects of peat
water humic acids on warm-blooded organisms. *Arch.
Hydrobiol.* 65: 515-522.

Jernelov, A. and Lann, H. 1973. Studies in Sweden on feasibility
of some methods for restoration of mercury-contaminated
bodies on water. *Environ. Sci. Technol.* 7: 712-718.

Jonasson, I.R. and Timperley, M.H. 1973. Field observations on
the transport of heavy metals in sediments" (A.J. de Groot
and E. Allersman). Discussion: *In*: P.A. Krenkel (ed.)
Proceeding Symposium Heavy Metals in Aquatic Enviconment
(Suppl. Progr. Water Technol., Publ. 1975). Pergamon,
Oxford. pp. 97-101.

Kaushik, N.K. and Hynes, H.B.N. 1971. The fate of dead leaves
that fall into streams. *Arch. Hydrobiol.* 68: 465-515.

Khan, S.U. 1973. Equilibrium and kinetic studies on the
adsorption of 2,4-D and picloram on humic acid. *Can. J. Soil
Sci.* 53: 429-434.

Klyachko, V.A. 1964. Oxidation method for the removal of colour
and iron from water. Scientific papers from the Institute of
Chemical Technology, Prague. *Technology of Water* 8: 195-201.

Kuznetzov, S.I. 1975. Role of microorganisms in the formation of
lake bottom deposits and their diogenesis. *Soil Sci.* 119:
81-88.

Lean, D.R.S. 1973. Movements of phosphorous between its biologically
important forms in lake water. *J. Fish. Res. Board Can.*
30: 1525-1536.

Leppard, G.G., Massalski, A. and Lean, D.R.S. 1977. Electron
opaque microscopic fibrils in lakes: their demonstration,
their biological derivation and their potential significance
in the redistribution of cations . *Protoplasma* 92: 289-309.

Lovelock, J.E. 1979. Gaia - A New Look at Life on Earth.
Oxford Univ. Press, Toronto.

Manskaya, S.M. and Drozdova, T.V. 1968. Geochemistry of Organic
Substances. Pergamon Press, Oxford.

Martin, J.P. and Haider, K. 1971. Microbial activity in relation
to soil humus formation. *Soil Sci.* 111: 54-63.

Mathur, S.P. and Paul, E.A. 1967a. Microbial Utilization of
Soil Humic Acids. *Can. J. Microbiol.* 13: 581-586.

Mathur, S.P. and Paul, E.A. 1967b. Partial Characterization of
Soil Humic Acids Through Biodegradation. *Can. J. Microbiol.*
13: 573-580.

Neunylov, B.A. and Khaukina, N.V. 1968. Study of the rate of decomposition and conversion processes of organic matter tagged with ^{14}C in the soil. *Soviet Soil Sci.* 2: 234-239.

Nissenbaum, A. and Kaplan, I.R. 1972. Chemical and Isotopic Evidence for the in situ origin of Marine Humic Substances. *Limnol. Oceanogr.* 17: 570-582.

Odum, W.E. and Drifmoyer, J.E. 1978. Sorption of pollutants by plant detritus: Review. *Environmental Health Perspectives* 27: 133-137.

Pauli, F.W. 1975. Heavy metal humates and their behaviour against hydrogen sulphide. *Soil Sci.* 119: 98-105.

Poldoski, J.E. 1979. Cadmium bioaccumulation assay: their relationship to various ionic equilibria in Lake Superior water. *Environ. Sci. Technol.* 13: 701-706.

Ramamoorthy, S. and Kushner, D.J. 1975a. Heavy metal binding components of river water. *J. Fish. Res. Bd. Canada* 32: 1755-1766.

Ramamoorthy, S. and Kusher, D.J. 1975b. Heavy-metal binding sites in river water. *Nature (London)* 256: 399-401.

Ramamoorthy, S. and Leppard, G.G. 1977. Fibrillar pectin and contact cation exchange at the root surface. *J. Theor. Biol.* 66: 527-540.

Rashid, M.A. and Leonard, J.D. 1973. Modifications of the solubility and precipitation behaviour of various metals as a result of their interactions with sedimentary humic acid. *Chem. Geol.* 11: 89-97.

Rorem, E.S. 1955. Uptake of rubidium and phosphate ions by polysaccharide-producing bacteria. *J. Bacteriol.* 70: 691-701.

Sakamoto, M. 1971. Chemical factors involved in the control of phytoplankton production in the experimental lakes area, northwestern Ontario. *J. Fish. Res. Bd. Canada* 28: 203-213.

Schenek, P.A. and Havenaar, I. (eds.). 1968. Advances in Organic Geochemistry. Pergamon, Oxford.

Schindler, D.W., Hesslein, R.W., Wagemann, R. and Broeckner, W.S. 1980. Effects of acidification on mobilization of heavy metals and radionuclides from the sediments of a freshwater lake. *Can. J. Fish. Aquat. Sci.* 37: 373-377.

Schnitzer, M. 1971. Metal-Organic Matter Interactions in Soils and Waters. *In* Organic Compounds in Aquatic Environments. S.D. Faust and J.V. Hunter, (eds.) Marcel Dekker, Inc., New York.

Seki, H., Shortreed, K.S. and Steckner, J.G. 1980. Turnover rate of dissolved organic materials in glacially-oligotrophic and dystrophic lakes in British Columbia, Canada. *Arch. Hydrobiol.* 90: 210-216.

Singer, A. and Navrot, J. 1976. Extraction of metals from basalt by humic acids. *Nature (Lond.)* 262: 479-481.

Stumm, W. and Morgan, J.J. 1970. Aquatic Chemistry. Wiley-
     Interscience, New York, N.Y.
Sundman, V. 1965. Transformation of lignin-related compounds
     into humic acids. Acta Polytech. Scand., Helsinki. Chapter 40.
Timperley, M.H. and Allan, R.J. 1974. The formation and
     detection of metal dispersion haloes in organic lake
     sediments. *J. Geochem. Explor.* 3: 167-190.
Wood, J.M. 1974. Biological cycles for toxic elements in the
     environment. *Science*   183: 1049-1952.

SECTION II

BENTHOS IN TIME AND SPACE

MIGRATIONS AND DISTRIBUTIONS OF STREAM BENTHOS

D. Dudley Williams

Division of Life Sciences
Scarborough College
University of Toronto
West Hill, Ontario, Canada

INTRODUCTION

Streams are dynamic systems and the organisms that live in
them reflect this in that they move around a great deal. It is
probable that these migrations arose purely as the result of
physical displacement by the current. However, over time,
migrations resulting from behavioural responses have evolved.
Consequently, the types and degrees of migrations shown by
running water inhabitants, today, are extremely varied in their
duration and complexity; the distribution of animals on streambeds is the end product of these movements.

This chapter will review the present knowledge of migrations
and distributions of stream benthos from both descriptive and
experimental studies. It will conclude with a consideration of
the application potential of such knowledge.

We should begin, perhaps, by defining a few terms. The
movement of individual animals into or out of a population or
population area is called dispersal. It may take three forms:
emigration - one-way outward movement; immigration - one-way
inward movement; and migration - periodic departure and return
(Odum, 1971). All forms of dispersal occur at sometime in the
stream habitat, with emigration and immigration being responsible
for spreading species between streams, while most of the dispersal
within a stream is cyclical (Müller, 1954) and therefore
represents true migration. Dispersal between streams is affected
by environmental barriers and the capacity of the organism to
move, that is its vagility. Dispersal within streams is also
affected by these two parameters, but usually on a smaller scale.

155

The stimulus to disperse is probably a combination of population
pressure and environmental change (Dansereau, 1957), and one
particular stage in the life cycle of a species is usually
specifically adapted to disperse, with most insects dispersing as
adults (Lewis and Taylor, 1967). The latter may not be as true for
stream insects, however.

Migrations within streams were earlier studied as individual
events (e.g. Dendy, 1944) although it is now realized that a
single species may migrate in different manners and in different
directions throughout its life (e.g. Dennert et al., 1969). It has
been suggested that under-water migrations of benthic species take
place in the following directions: downstream, upstream, laterally
and within the substrate (Williams and Hynes, 1976a).

DOWNSTREAM MIGRATIONS

These have been studied the most and several reviews on the
topic have appeared (e.g. Waters, 1972; Müller, 1974). Although
animals move downstream along the surface of the substrate (Kureck,
1967) most are carried in the water column, as part of the drift.
Drift is a very important aspect of life in running water and has
been studied from many aspects including its importance as food
for fishes (e.g. McCormack, 1962; Elliott, 1973) and as an escape
mechanism for the benthos (e.g. Hastings et al., 1961; Corkum and
Pointing, 1979). Various authors have underlined the adaptive
significance of drifting in terms of redistribution in response to
competition for food and space (e.g. Townsend and Hildrew, 1976;
Williams, 1977), and it may even play a role in mate-searching in
certain species (e.g. *Gammarus*, Lehmann, 1967). It has been shown,
also, to be the major contributing source of animals to denuded
areas of stream substrate (e.g. 82%, Townsend and Hildrew, 1976;
42%, Williams and Hynes, 1976a). Recolonization studies have shown
that the normal carrying capacity of an area of substrate can be
replenished by drift alone in as little as 10 to 14 days, in some
instances (Waters, 1964), although four weeks seems to be the
recovery time noted by the majority of workers (Mason et al.,
1967; Coleman and Hynes, 1970; Williams and Hynes, 1976).
Undoubtedly, the time required depends on the extent and proximity
of undisturbed areas of substrate that serve as the source of
drifting animals. Even though normal densities may be restored
after this time, the area might not itself produce a normal level
of drifting animals for quite some time (Dimond, 1967).

Certain age or size classes of benthic species drift out of
proportion to their density in the benthos (Ulfstrand et al., 1974).
Their presence high up in the water column, rather than just above
the substrate, suggests an active determination to move (Larimore,
1972), and this may vary seasonally (e.g. Elliott, 1967). Taxa
that exhibit strong diel drift patterns, for example *Gammarus*

and *Baetis*, are readily able to return to the substrate (Ciborowski
and Corkum, 1980). These and many other studies point to a
behavioural component to the drift which may differ in its intensity
between species. This component may be supplemented or, on
occasion, over-shadowed by animals drifting as the result of
abiotic factors, such as water flow, light and temperature; they
constitute the "constant" and "catastrophic" components of Waters
(1965, 1972).

UPSTREAM MIGRATIONS

With the steady displacement of animals downstream in the
drift, it is logical to speculate on the existence of compensatory
upstream migrations; if stream headwaters are not to become totally
devoid of fauna and the seas inundated with freshwater debris.
Theoretically, the numbers of animals moving upstream should
approximate those drifting downstream (this would include the
hatchlings from eggs carried into headwater regions by gravid
females). Few people, however, have been able to come anywhere
near to proving this. Bishop and Hynes (1969) found that upstream
migration in the Speed River, Ontario accounted for only 6.5% by
numbers and 4% by weight of the drift. Elliott (1971) found a
7-10% upstream compensation, by numbers, in spring and summer when
drift was high, but this increased to 30% in the winter when
drift was minimal. The latter study, done on an English Lake
District stream, showed that most of the benthic invertebrates
moved upstream near the banks, where the substrate consisted of
small stones and gravel. Large immature stages of Plecoptera,
Ephemeroptera and Trichoptera moved upstream at various points
across the stream. More invertebrates moved upstream at night and
a distinct nocturnal peak in diel periodicity was seen in
Baetis rhodani and *Gammarus pulex*. In addition to these authors,
many others have pointed to the failure of upstream migrations
in offsetting the loss through drift (e.g. Travers, 1925; Ball
et al., 1963; Schwarz, 1970; Hayden and Clifford, 1974). To
account for this phenomenon, Müller (1954a,b) put forward the
concept of the "colonization cycle" in which he suggested that,
for insects, the number of nymphs moving downstream in the drift
is compensated for by a net upstream flight of the adults prior to
oviposition. These adult migrations have been shown to occur in
the Trichoptera, Plecoptera, Ephemeroptera and Simuliidae (Roos,
1957; Schwarz, 1970; Bengtsson et al., 1972; Madsen et al., 1973;
Svensson, 1974), but in some cases the difference between the
numbers of females moving upstream and downstream were quite
marginal. However, as a single female may lay hundred of eggs,
even a small number ovipositing upstream may suffice. Interestingly
enough, wingless females of the winter stoneflies *Capnia atra* and
Allocapnia pygmaea have been observed walking upstream along the
stream bank for considerable distances before ovipositing (Thomas,

1966; Hynes, 1970).

The colonization cycle might explain the compensating
mechanism in insects, which have an extra-fluvial stage, but what
of those members of the benthos that are confined to the water?
These, too, have been noted to move upstream, but again mostly in
numbers insufficient to counteract the drift, for example *Physa
gyrina* (Noel, 1954), various species of *Gammarus* (Hynes, 1955;
Minckley, 1964; Dennert et al., 1969; Hultin, 1971; Meijering,
1972), and for the crayfishes *Orconectes nais* (Momot, 1966) and
Pacifasticus klamathensis (Black, 1963). Moreover, most of these
migrations were seasonal, probably associated with food,
temperature or reproductive stimuli. However, in a field study
of direction of movements of animals colonizing artificial
substrates, Williams and Hynes (1976a) found that in a permanent
Ontario stream, in July, cladocerans and ostracods predominantly
moved upstream, while nematodes and oligochaetes colonized by
moving upstream and downstream in approximately equal proportions.
At the start of the new lotic phase (November) in a nearby
temporary stream, a greater diversity of taxa moved upstream;
including nematodes, oligochaetes, ostracods, cyclopoids, amphipods
and chironomid larvae (Williams, 1977). It was concluded that
upstream migration might be an important process, along with drift,
in dispersing hatchlings from egg masses and cocoons in these types
of habitats.

Some taxa, notably the Ephemeroptera, exhibit annual upstream
migrations, usually in response to food requirements. For example,
Neave (1930) observed a mass movement of nymphs of *Leptophlebia
cupida* into upstream temporary creeks in the Winnipeg region in
spring. Some of the nymphs moved as much as 1-6 km. Similarly,
Olsson and Söderström (1978) recorded the upstream migration of
hundreds of thousands of *Parameletus chelifer* nymphs into a
temporary tributary of the River Vindelälven, in Sweden, during
nine days in spring. These nymphs grew faster than those in the
river and had emerged before the tributary dried up.

As Hultin et al. (1969) have pointed out, the ecological
significance of upstream and downstream movements probably differs
for different stream communities. Although positive rheotactic
behaviour is widespread in stream animals, it need not automatically
result in extensive upstream migrations. However, the greater a
species' tendency to drift, for whatever reason, the more important
is the development of a compensatory upstream migration.

LATERAL MIGRATIONS

These are known to occur in streams but have been little
studied. Most records involve movements towards the banks prior
to emergence. Verrier (1956), for example, showed that several

mayfly genera gradually move towards the bank, and slower water, as they grow; the preference for either bank may be unequal (Williams, 1981). Müller (1974) recorded that newly emerged adults of *Capnia atra*, in a woodland stream in northern Sweden, crawl away from the banks and into the forest. It appeared that the spatial orientation of the life cycle of this particular population was perpendicular to the stream.

MIGRATIONS WITHIN THE SUBSTRATE

Migrations of benthos within the substrate of a streambed, where interstices occur, are a relatively recent discovery. Kuhtreiber (1934) and Aubert (1959) suggested that stream invertebrates could live elsewhere than on the streambed, as various immature stages of some species (e.g. those of the Leuctridae) disappear from the benthos at certain times of the year. Berthèlemy (1966) showed that these larvae could be found in specialized areas deep within the substrate. Many species are now know to migrate vertically into the substrate to depths of at least up to 100 cm in streams with deep gravel beds (Williams and Hynes, 1974). The existence of these sub-benthic or hyporheic populations has now been substantiated in many countries, for example U.S.A. (Stanford and Gaufin, 1974; Poole and Stewart, 1976), Malaysia (Bishop, 1973), England (Ford, 1962; Gledhill, 1971), Wales (Hynes et al., 1976), Austria (Danielopol, 1976), Germany (Husmann, 1971), Yugoslavia (Mestrov and Tavcar, 1972; Mestrov et al., 1976) and Canada (Coleman and Hynes, 1970; Hynes, 1974).

Williams and Hynes (1974) suggested that this interstitial fauna, termed hyporheos, could be subdivided into an occasional component and a permanent component. The former consists of larvae of most of the surface benthos that live interstitially during part of their lives, while the latter consists of many specialized forms of copepods, mites, ostracods, tardigrades and syncarids that complete their entire life cycle in this zone.

An example of the annual vertical migration of a member of the occasional hyporheos is given in Figure 1. It represents a population of the chironomid *Cladotanytarsus* in the Speed River, Ontario. The species is univoltine with adults appearing from April to early August. Eggs of a new generation begin to hatch in mid-August and early instar larvae are found interstitially at this time. Either this is the result of a downward migration of small animals, coinciding with a drop in the surface water temperature, or the settling of eggs into the interstices shortly after they have been laid; gentle agitation by the current may cause them to passively sink in the same way that the smallest particles in a mixture will be sorted towards the bottom. The

larvae overwinter at various depths in the substrate before an upward
migration in March when the surface water begins to warm up. It
seems that the majority of larval growth takes place in the substrate.
Food, in the form of detritus, is plentiful interstitially, but in
this particular river, detectable dissolved oxygen occurs in only
the top 30 cm of substrate.

There would seem to be many advantages in living interstitially,
particularly for small larvae: the small interstices would prevent
the entry of larger predatory species; the buffering effect of the
substrate would provide a more uniform temperature; and depth would
lessen the severity of spates and droughts. Nymphs of the stonefly
Allocapnia vivipara, for example, have been found diapausing at
depths of from 5 to 25 cm in the substrate during the dry phase of
a southern Ontario temporary stream (Williams and Hynes, 1976b).
This is an adaptation to surviving high summer temperatures
(Harper and Hynes, 1970). A series of core samples taken from this
creek indicated a definite vertical migration taking place. In May,
large numbers of animals were found in the top 20 cm of substrate
in both riffles and pools. This pattern was repeated in June but
there were more animals closer to the surface (0-5 cm). In July,

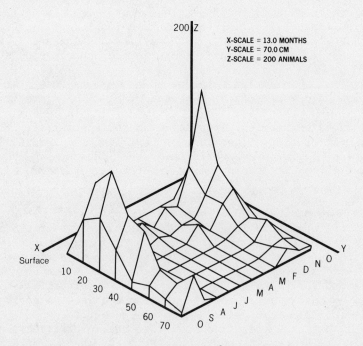

Figure 1.   Annual distribution of *Cladotanytarsus* spp. within the
            substrate of the Speed River, Ontario. (Plot of time
            vs. depth vs. number of animals.)

this upward movement became more evident and the numbers at 10 and
20 cm had declined considerably. This represented mature larvae
moving towards the substrate surface in preparation for pupation
and emergence. By mid-August, when the riffles were dry and the
pools shrinking, comparatively few animals were left in the substrate. There were, however, quite a large number of different
sorts of eggs and cocoons present. As water flow resumed in
October, the numbers of animals in the substrate started to
increase once more as a result of hatching. By early November,
the hyporheos was once again distributed vertically throughout the
substrate in a pattern typical of the lotic phase of the stream.

Clifford (1966) showed that the surface fauna of a stream in
Indiana moved deeper into the substrate during spates - presumably
as a protective mechanism against the increased current. A
similar observation was made by Williams and Hynes (1974). The
fast repopulation of the substrate surface, noted by many authors,
after application of insecticides or chemical spills may well be
from interstitial sources.

Within-substrate migrations do not appear to be restricted
to the vertical plane, as a few recent studies indicate a horizontal
component. Peckarsky (1979) buried metal cages containing known
densities of benthos in the substrate of a Wisconsin stream. Each
cage was open only on the downstream side so that only active
colonization would be recorded. Overall, the animals moved into
and out of the cages in a density-dependent fashion showing them
capable of migrating on an upstream-downstream axis within the
gravel, and to be able to detect one another's presence among the
interstices.

Animals also migrate laterally within the substrate. Figure 2
shows the distribution of hyporheos at a depth of 10 cm on a
transect from midstream to 2 m into the bank of the Speed River.
The data were collected using a combination of core samples and
small colonization pots during November. It shows that some of
what might be termed typical stream taxa - Ephemeroptera, Trichoptera,
Plecoptera, Tricladida and Mollusca (Sphaeriidae) - were entirely
restricted to the stream interstitial environment (although
Schwoerbel, 1967, has recorded movements of nymphs of the mayfly
Habroleptoides modesta into the bank of a stream). On the other
hand, some of the Chironomidae and Elmidae were taken right up to
the stream margin, and some of the Orthocladiinae and Tanypodinae
were found up to 2 m into the interstitial water under the bank.
The Nematoda, Ostracoda, Oligochaeta (Naididae), Cyclopoida,
Harpacticoida and Acarina all showed fairly continuous distributions
from midstream to at least the 2 m bank mark. This supports
Schwoerbel's (1961) idea that the hyporheic zone extends out to a
distance several metres beyond the margin of the stream. Here,

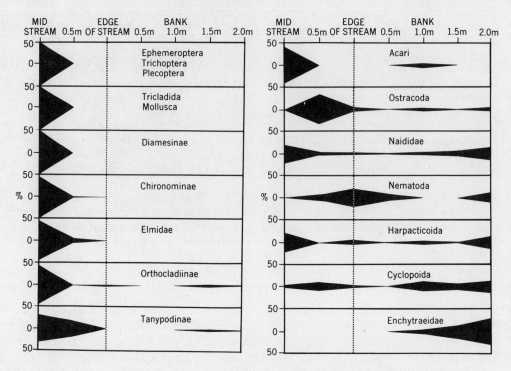

Figure 2.   Distribution of the hyporheos at a depth of 10 cm
into the substrate on a transect from midstream to
2 m into the bank of the Speed River (November).

it presumably merges with the true groundwater and its associated
fauna.   At this point, we begin to detect different species; in
the case of the Speed River, the enchytraeid oligochaetes.

Clearly then, invertebrates do migrate within the hyporheic
zone which, incidentially, can be defined as a middle zone of
interstitial water in streams with deep gravel beds (Orghidan,
1953).   It is bordered by the epigeic, or surface water, of the
stream above, and by the phreatic, or groundwater, below, although
its exact limits may be sometimes difficult to delineate due to
the relative instability of its upper and lower boundaries.   For
example, if after a nearby heavy rainstorm there is a surge of
groundwater, its lower boundary may be pushed nearer the substrate
surface.   Conversely, after spates or during droughts, the vertical
extent of the zone, along with its associated fauna, might
increase as the groundwater table retreats.   In many ways the zone
is analogous to the marine intertidal zone.

BETWEEN-STREAM MIGRATIONS

Movements of animals between streams may be either true
migrations or one-way migrations or emigrations. They are
important in expanding species' distributions and in populating
newly created habitats. For the most part they are by aerial means,
either actively or passively, but terrestrial migrations between
waterbodies are not unknown; Cummins (1921), for example, recorded
inter-pond migrations in the burrowing crayfish *Cambarus argillicola*.

Vagility plays a major role in the successful spreading of a
species. Some stream insects are good fliers, for example many of
the Odonata (Corbet, 1963), some Coleoptera (Balfour-Browne, 1958)
and Hemiptera (Macan, 1939), some Trichoptera (Bouvet, 1977),
Simuliidae (Davies, 1961) and Culicidae (Provost, 1952). Many of
their species are therefore widely distributed in habitats
capable of supporting them. Others, for example most Ephemeroptera
(Edmunds et al., 1976), Trichoptera (Ross, 1956) and Chironomidae
(Davies, 1976), many Plecoptera (Hynes, 1976), and some Coleoptera
(Jackson, 1956) are weak fliers, or even apterous (e.g. some
species of Hemiptera, Plecoptera and Trichoptera) and consequently
may have species that have narrow distributions, with many being
endemic to certain watersheds. Hynes (1970b) is of the opinion
that the ability to fly is of less advantage to stream-dwelling
insects than to lentic species. He states that "a tendency to
disperse widely may be a positive disadvantage because streams in
different valleys are often separated by considerable ecological
barriers, and the chances of a species reaching another suitable
stream by air would appear often to be less than that of reaching
it by water by way of confluences." In many instances this may be
true. Lehmkuhl (1972), for example, has shown that the presence of
Baetisca (Ephemeroptera) in the Saskatchewan River may be the
result of connections between the Saskatchewan and Missouri-
Mississippi River systems in late glacial times. At various times
since the last glaciation, certain species of stoneflies may have
emigrated northwards from the Missouri-Mississippi system or westwards from the St. Lawrence system through drainage connections
(Flannagan, 1978). Such connections may have also led to the
dispersal of many Trichoptera in North America (Ross, 1956).
Nevertheless, many insect species do regularly colonize other
streams through flight. Harrison (1966), for example, observed
that the Munwahuku stream near Salisbury, Zimbabwe, dried up for
several months annually. Recolonization of the benthic insects,
each year, was from eggs laid by flying adults migrating from
nearby permanent rivers. Similar migrations are seen in species of
Coleoptera that colonize temporary water bodies. *Helophorus
orientalis*, for example, typically overwinters in permanent waters
but migrates to temporary ponds or streams in the spring where it
breeds in a more favourable environment. The old and the new

generations of adults then fly back to the permanent waters before
winter (Fernando and Galbraith, 1973).

This brings us to consider how those members of the benthos
that do not possess the power of flight, or cannot walk overland
reach new streams. There have been numerous records of animals
attached to flying waterfowl. Proctor (1964), for example,
recovered viable eggs of Anostraca, Notostraca, Conchostraca,
Cladocera and Ostracoda from the lower digestive tracts of wild
ducks, and other Crustacea and algae from captive killdeer
(Proctor et al., 1967). Similarly, Rosine (1956) found several
dozen live *Hyallela azteca* (Amphipoda) clinging to the feathers
of two dead mallards that had been out of water for several hours;
apparently they had been carried some distance by the ducks (see
also Daborn, 1976). Even fish eggs (pike and perch) have been
found clinging to aquatic birds (Thienemann, 1950).

Smaller hosts may also contribute to the passive dispersal
of certain groups. Stewart et al. (1970) found that algae,
protozoans and fungi could be transported by aquatic Hemiptera,
Trichoptera and other aquatic insects such as tipulids and
chironomids (Revill et al., 1967), aquatic beetles (Milliger and
Schlichting, 1968) and dragonflies (Maguire, 1963). Fryer (1974)
found small clams attached to the setae of corixids, and Lansbury
(1955) saw ostracods similarly attached to notonectids. Fernando
and Galbraith (1970) collected specimens of *Gerris comutus* and
G. buenoi that were heavily infested with the water mite
Limnochares, and Harris and Harrison (1974) found two species of
Hydrachna parasites on corixids. I have seen specimens of the
crayfish *Orconectes p. propinquus*, with sphaeriid clams clamped
tightly on the tips of their pereiopods, walk several metres before
shaking them off; Rees (1952) has shown that amphibians can carry
clams in this manner. Another mollusc, the freshwater limpet,
Ancylus, has been repeatedly found attached to the elytra of water
beetles such as *Colymbetes* (Thienemann, 1950). The major
disadvantage of this type of migration strategy is that the
dispersing organism has little choice in where it is taken by its
host. Presumably, though, specificity to a host species that
selects a suitable habitat might soon evolve.

Some stream organisms may be light enough to be carried by air
currents. Maguire (1963) found this route to be very effective in
the transport of disseminules of some small species of Cladocera,
and some ephippia were shown to be easily picked up by the wind
(Fryer, 1972). Gislen (1948) showed that an "aerial plankton"
exists, and Maguire (1963) studied its colonization of small
water-filled jars. The latter author concluded that the numbers
of different organisms per jar decreased with increased height
above the ground and that they similarly decreased with increasing

distance from the water source. Disseminules were frequently washed into the bottles from nearby vegetation and soil surfaces by rain.

Not only are the smaller aquatic animals passively carried by winds, but larger insects are often carried this way as well. Freeman (1945) studied the insects carried in the air from ground level to 92 m in north Lincolnshire, England. Most of the insects he found were terrestrial, but a few aquatic groups occurred including Ephemeroptera, Coleoptera, Tipulidae, Ephydridae, Chironomidae and Ceratopogonidae. The latter two families of Diptera were most commonly caught in nets close to the ground and it was concluded that they were only likely to be dispersed short distances.

The role of aerial migrants, whether by active or passive means, can be very important in the colonization of new streams, and this importance may vary with time. Williams and Hynes (1977) found that drift was initially the most important source of colonizing animals in a newly created southern Ontario stream. However, its importance gradually declined as more species began to arrive from downstream and via the air. At the end of one year, almost half the species present in the stream were those that must have arrived aerially.

These then are the different means of migration undergone by members of the stream benthos (see summary in Fig. 3), although, of course, no one species may necessarily display all types. For the main part, they are responsible for the benthic distributions that we see today, although the latter may have been influenced by geographical and physiological barriers and habit suitability relative to the species' ability to adapt to the environment in the extended geographical range. Potentially, there are several different ways of looking at stream animal distributions, but for the purpose of this review we shall consider them on a size scale ranging from large scale (world wide distributions) through medium scale (local, within stream distributions) to microdistribution (within a small section of stream), and also in terms of time.

WORLD-WIDE DISTRIBUTIONS

Despite the overall world-wide similarity of stony stream communities (Hynes, 1970a), there are differences in stream and river faunas as one moves from the tropics to the poles. This is best shown in terms of the relative proportions of the two zones of a river known as rithron and potamon (Illies, 1961; Illies and Botosaneanu, 1963). These faunal zones appeared to be distinct ecological entities in the North German streams studied, but the division seems to have world-wide application.

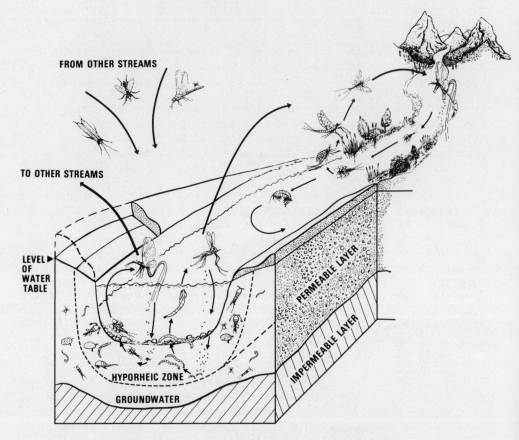

Figure 3. Summary of the different means of migration undergone
by members of the stream benthic community.

The fauna of the rithron is largely cold-stenothermic and
characteristic of flowing water and thus benthic rather than
planktonic. Amongst the insects, it includes Plecoptera (most
families); Ephemeroptera (Baetidae and Heptageniidae); Trichoptera
(most families, but particularly the Rhyacophilidae and the net-
spinning families); Coleoptera (Psephenidae and Elmidae); and
Diptera (Simuliidae, Deuterophlebiidae, Blepharoceridae and
Chironomidae, especially the Diamesinae, Orthocladiinae and
Tanytarsini). The non-insects include Crustacea (particularly
the Pericarida), Gastropoda (Ancylidae), Oligochaeta (Lumbriculidae)
and Tricladida (Planariidae) (Claassen, 1931; Needham et al. 1935;
Ross, 1956; Hynes, 1970a; Macan, 1974; Wiggins, 1977).

The fauna of the potamon is eurythermic or warm-stenothermic
and contains forms that reach their maximum development in lentic
habitats. Planktonic forms may be present. Typical insect
representatives are: Ephemeroptera (Ephemeridae and Caenidae);
Odonata (many families); Hemiptera (Corixidae and Gerridae);
Trichoptera (many case-building families); Coleoptera (Gyrinidae,
Haliplidae, Dytiscidae, Hydrophilidae); and Diptera (Tipulidae,
Dixidae, Culicidae, Stratiomyidae and Chironomidae, especially the
Chironomini). The non-insects include Crustacea (particularly the
Eucarida), Gastropoda (Lymnaeidae and Planorbidae) and Oligochaeta
(Tubificidae) (Shelford, 1913; Needham et al., 1935; Robertson and
Blakeslee, 1948; Pennak, 1953; Hynes, 1960, 1970a; Wiggins, 1977.

At high latitudes, there is a tendency for no true components
of the potamon to show up. Many of the rivers are short and are
typically rithronic throughout their entire length. Streams in
both the Arctic and Antarctic are like this. Conversely, in lower
latitudes, for example the Amazon basin, the rithron is very short
compared with the potamon, and rithronic species may be driven into
high altitude regions (see Hynes and Williams, 1962). In temperate
latitudes, there appears to be more of a balance between the two
zones, but although potamonic animals can survive low temperatures,
they usually do not thrive, whereas rithronic animals cannot
survive at high temperatures.

The migrational abilities of species at various latitudes may
profoundly affect their distribution in sections of rivers where
they can survive. In the tropics, for example, rithronic species
may only occur at high altitude on isolated mountains. Migration
between such habitats may be extremely difficult, particularly for
poor fliers and the resultant isolation may lead to local
speciation. The progressive downstream drift of members of a
rithronic community into the potamon would seem a pointless waste,
although there exists little quantitative data on the extent of
such happenings (see Elliott, 1972 and Williams, 1980a).

So far as the world-wide distribution of species of stream
benthos are concerned, in the northern hemisphere there are great
similarities between Europe and North America because of the various
land connections that stretched into modern (Quaternary) times,
for example the Bering Sea connection (Beringia). However, the
Pleistocene glaciation had an unequal affect on the two areas. In
North America, the ice sheet drove many species south in front of
it, and many returned as it retreated. In northwestern Europe,
however, species were hemmed in between the encroaching northern
ice cap and the one forming on the Alps to the south. This
resulted in the extermination of certain elements of the fauna
(Hynes, 1970b). The Holarctic fauna, in general, is mostly
temperate with minimal tropical forms. Many species are widespread,
for example several species of the caddisfly genera *Apatania* and

Asynarchus (Wiggins, 1977), the many species pairs of chironomids (Saether, 1969), and the oligochaetes *Nais communis, Limnodrilus hoffmeisteri* and *Tubifex tubifex* (Brinkhurst and Jamieson, 1971).

Continental drift may have played an important role in distributing some lotic groups. It is commonly evoked as the distributing mechanism of groups with low vagilities, for example the Turbellaria, Ostracoda, Plecoptera and Chironomidae (see Ball and Fernando, 1969).

The relationships amongst southern hemisphere forms are not well defined, but many species appear to show distant relationships between the land masses. This probably reflects the fact that the southern continent breakup took place earlier, in the Tertiary. Many old insect families characterize these areas, for example certain of the Plecoptera in South Africa and Australia (Hynes, 1964; Harrison, 1965).

From a comparative study of the insects in small headwater streams in temperate and tropical areas, Stout and Vandermeer (1975) showed that species richness was significantly higher in the latter, even though biomass was usually lower. Earlier studies showing no difference between the two area (e.g. Patrick, 1964) were thought to have taken an inadequate number of samples. Stout and Vandermeer theorized that the temperate communities were "regionally" controlled by highly mobile (drifting), opportunistic species that profoundly influenced community structure. On the other hand, the tropical stream communities were thought to be controlled, largely, by local interactions among comparatively sedentary species, resulting in equilibrium communities. This hypothesis was criticized by Fox (1977) who pointed out that the suggestion that temperate species are maintained in a community through the regional factors of migration and local extinction, rather than as the result of local factors such as competition and predation, was based on the premise of higher immigration rates and estimates of drift from temperate streams, although comparative studies of drift in the tropics are very rare (Hynes, 1975). Fox argued that although temperate communities may show more regional effects, this does not necessarily mean that local (biological) interactions are not important also. Reduced benthic biomass in tropical streams may be the result of low energy input and high predator pressures, leading to reduced mobility. In temperate streams, on the other hand, plentiful plant detritus may reduce predation by facultative predators, intensifying competition for space and resulting in higher drift rates.

The composition of a stream community may also be the result of its degree of zoogeographical isolation, although this may not necessarily affect its richness. For example, many common northern hemisphere insect families are absent from New Zealand and it lacks

many plecopteran and chironomid elements. The country does, however,
have a very diverse ephemeropteran and trichopteran fauna, so much
so in fact that in certain streams the species richness approximates
values for comparable habitats in the northern hemisphere (Towns,
1979). Similar situations exist in Australia and South America
where the Leptophlebiidae occupy some niches which in the north are
filled by other families of Ephemeroptera (e.g. Ecdyonuridae)
(Hynes, 1970b).

     Considerable zoogeographical isolation exists between the
stream faunas of the east and west Nearctic. Pennak (1958)
suggested that eastern species may be prevented from naturally
spreading westwards by the prevailing westerly winds, the topography
and headwater drainages of the Continental Divide area, steep
stream gradients, intermittent streams, and extensive desert and
semi-arid regions. Because of these same barriers, he argued, the
West should support a unique population of aquatic invertebrates due
to isolation and speciation, often in habitats of unusual character
(e.g. springs of unusual chemical and thermal regime). Lack of
research in this area results in this prediction remaining
largely unsubstantiated, although examples of western endemism are
to be seen in the lotic gastropods (Pleuroceridae), shrimps
(Atyidae) and caddisflies (Limnephilidae) (Anderson, 1976).

LOCAL DISTRIBUTIONS

     By their very nature, individual streams and rivers change
along their entire lengths. They show one-directional movement
downhill; variable levels of discharge, depending on the nature of
the valley, season and weather conditions; variable levels of
associated parameters such as current velocity, depth, width,
light, turbidity, detritus and chemical content; continual
turbulence and mixing of water layers, except at low altitudes;
and relative instability of the substratum (Coker, 1968; Hynes,
1975). These factors cause longitudinal biotic zonation from the
source area to the point at which the river reaches a lake or the
sea.

     The fact that regions of a single stream appeared different
led to attempts to classify the phenomenon. Various parameters
have been tried including local geology, source of water, size,
current speed, gradient, discharge, substrate composition,
temperature, dissolved oxygen and carbon dioxide, fauna (both
vertebrate and invertebrate) and flora, and productivity – either
singly or in combination (e.g. Carpenter, 1928; Butcher, 1933;
Ricker, 1934; Huet, 1954; Macan, 1961. See also the extensive
review of this topic by Hawkes, 1975). However, many of these
systems apply only locally and run into problems when applied to
a continental or global scale (Hynes, 1960). Perhaps the most

workable system on a global scale is that of Illies and Botosaneanu (1963) with which we have already dealt.

Despite their reluctance to be classified according to their longitudinal characteristics, some streams show very clear examples of the longitudinal zonation of their communities or species within a given taxon. Cummins (1977) produced a general summary of the longitudinal changes seen in the benthic community of a water course (Fig. 4). The headwater region is typically

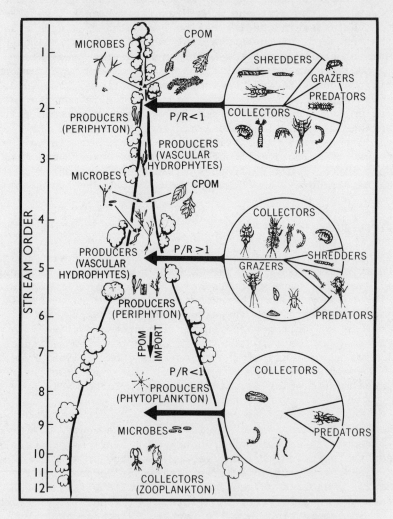

Figure 4.   General summary of the longitudinal changes seen in the benthic community of a watercourse (P/R = production to respiration ratio of benthic community) (redrawn after Cummins, 1977).

dimly-lit (because of dense riparian vegetation and steep valley sides), of high gradient and of reasonably constant water temperature. It is often a region of low aquatic primary production (but see Minshall, 1978), where secondary producers feed on coarse particulate organic matter (CPOM) and its derivatives that mostly come from the breakdown of autumn-shed, riparian leaves (Kaushik and Hynes, 1971). The CPOM is fed on by invertebrate shredders that reduce the material to faecal pellets less than 1 mm in diameter which contribute to the fine particulate organic matter (FPOM) content of the water and sediments. Current theory suggests that the shredder species derive most of their nutrients from microorganisms, such as hyphomycete fungi, growing on and in the leaf tissue (Barlocher and Kendrick, 1973). Alongside the shredder species in the headwater community are the collectors. These either trap the FPOM as it passes in the water column, or collect it directly from the substrate. Again, many of the nutrients appear to come from microorganisms that colonize the particles. The collectors produce faeces that are of much the same size as the particles that they ingest, but minus the microorganisms, and these are returned to the FPOM pool. Because of the oftentimes low primary productivity in headwater regions, the grazer or scraper members of the benthic community may be poorly represented. Predators are present also, particularly in the form of large stoneflies and cool-water species of fishes such as salmonids and sculpins. In general, the headwater community is heterotrophic.

In the midreaches of the watercourse (generally at the level of third to fourth order streams - see Horton, 1945), there is a shift from heterotrophy to autotrophy (Cummins, 1977). This is primarily the result of widening of the river and its valley, reduced shading by riparian vegetation, warmer water temperatures, and high nutrient levels derived from upstream. Here, the ratio of CPOM to FPOM is reduced as the latter increases due to immigration from upstream, from the increased production of periphyton and the rapid breakdown of dead aquatic macrophyte tissues. Predictably, the importance of shredders in the community is reduced while the collector and grazer species dominate. Invertebrate and vertebrate predators are present.

In the lowermost sections of the watercourse, the collector species dominate as a result of the large amounts of FPOM, produced upstream, and locally produced phyto- and zooplankton. The community is heterotrophic as a result of reduced primary productivity through increased turbidity and unstable substrates. Predators are typically odonatan nymphs, aquatic beetles and hemipterans, and bottom-feeding fishes.

The longitudinal zonation of type of food particle, just noted, may result in a linear succession of species groups within a taxon.

Wiggins and Mackay (1978) showed this for the Nearctic subfamilies
of the net-spinning Hydropsychidae (Fig. 5). Three of the four
subfamilies have somewhat restricted distributions in the stream
continuum. The Macronematinae are able to filter very small
particles of FPOM from the water column by means of fine-meshed
nets. They are therefore well suited to large rivers. The
Arctopsychinae, on the other hand, spin nets with coarse meshes
that are more suited to capturing CPOM and small animals in rapid
mountain streams. The Diplectroninae are restricted to cool headwater regions but their feeding biology is not well documented.
Finally, the Hydropsychinae spin nets that have meshes of intermediate size and are more generalized feeders. Consequently, they
can be found throughout the stream continuum.

     Temperature is undoubtedly important in determining the
longitudinal distribution of species (Ide, 1935). It was thought
to play a major role in determining the distribution of
Ephemeroptera species in St. Vrain Creek, Colorado (Ward and Berner,
1980). This stream originates in alpine tundra (3,414 m above sea
level) and drops to plains at 1,870 m over a distance of 54 km. A
few species, for example *Baetis bicaudatus*, were found over almost
the entire range (Fig. 6), while several were found only at one
location (e.g. *B. parvus*). No species were entirely restricted
to the headwaters, but several species found in the headwaters
occurred at significantly higher densities downstream. Species of
the same genus generally had different ranges. The overall

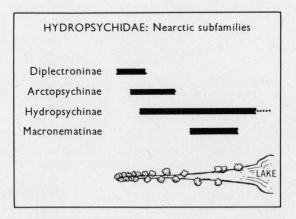

Figure 5. Longitudinal zonation of the Nearctic subfamilies
           of the Hydropsychidae (redrawn after Wiggins and
           Mackay, 1978).

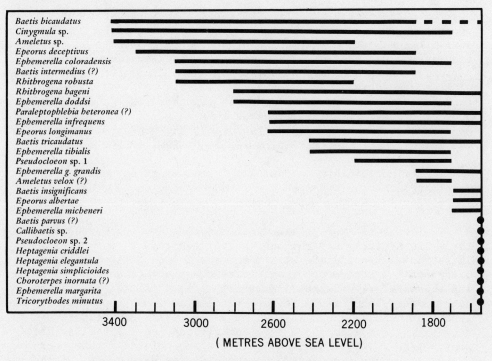

Figure 6. Altitudinal distribution of Ephemeroptera nymphs in St. Vrain Creek, Colorado (solid circle indicates species was found only at that site) (redrawn after Ward and Berner, 1980).

distribution pattern appeared to show selective elimination of species towards higher elevations.

Temperature interacts with dissolved gases, particularly oxygen, and dissolved nutrients and so may act directly or indirectly to control longitudinal zonation. However, a few studies have shown zonation under relatively uniform water temperatures. Hynes (1971), for example, showed a very clear altitudinal zonation of species of the hemipteran *Rhagovelia* in a Trinidadian stream where water temperature varied only 2.2^{o}C. The author concluded that slope and the nature of the streambed were the controlling factors. Thorup (1974) showed a similar succession of three species of Simuliidae downstream from a spring outlet of the stenothermal Rold Kilde in Denmark, but causal factors were not obvious.

    Competition may be a significant factor in determining
zonation. This has best been shown in the Tricladida. *Crenobia*
alpina ranges from the spring source down to a point where water
temperature reaches 14-15°C, when occurring alone in mountain
streams of Europe. *Polycelis felina*, under similar circumstances,
occurs down to the 16-17°C mark, and *Dugesia gonocephala* down
to 23°C. If all three species occur together, they succeed one
another downstream in the order *C. alpina, P. felina* and *D.*
gonocephala. Where *P. felina* is absent, the ranges of the other
two species meet (Thienemann, 1912, in Hawkes, 1975).

    Distribution as the result of factors such as current,
substrate, oxygen, water chemistry and predation will be dealt with
in the section on microdistribution.

    Quite apart from longitudinal zonation of stream faunas,
there are differences from one side of a stream to the other.
These probably correlate with the differences in flow, dissolved
oxygen, accumulation of organic debris, etc., between the middle
and the edge of the water. Most of the insects that live in the
central fast-water area are members of primitive groups (Plecoptera,
Ephemeroptera, Trichoptera). Comparatively few of the more highly
evolved insects occur here (e.g. Simuliidae, some Tabanidae and
Rhagionidae). Insects living near the edge of the stream or in
backwaters include dytiscid and gyrinid beetles and many of the
higher dipterans (e.g. most Tabanidae, Empididae and Dolichopodidae)
(Hynes, 1970b). Many other Diptera are semi-aquatic and live at
the stream margins (e.g. Tipulidae, Ceratopogonidae, Syrphidae,
Tetanoceridae and Ephydridae) (Pennak, 1953). Non-insect groups
may be distributed similarly. Shelford (1913), for example, found
the sequence of molluscs *Sphaerium stamineum* (bivalve),
Campeloma integrum (gastropod) and *Pleurocera elevatum* (gastropod)
from the inside curve (fine sediments and shallow water) to the
outside edge (large stones and deeper water) of a meander of the
North Branch of the Chicago River.

MICRODISTRIBUTION

    This is the sort of distribution of benthic invertebrates seen
within a small section of a stream, often over just a few metres
of the streambed, or even around a single rock. It is the result
of many factors, for example substrate type, food quality and
abundance, light, temperature, water movements and chemistry,
oviposition habits and predation, acting perhaps singly or, more
likely, in combination (Hynes, 1970a).

Substrate

    Many of the above factors are correlated with the nature of
the substrate and this is, therefore, perhaps the primary factor

creating microhabitats, although it may be of secondary
importance to some species (Cummins and Lauff, 1969; Williams,
1980b).

A study done by Linduska (1942) is an example of the
distribution of mayflies controlled by substrate features and
locally induced microcurrents. In a Montana stream, he found only
Ironopsis sp. and *Baetis* spp. on top of a typical 50 cm diameter
boulder in a current of 240 cm/s. Halfway down the sides of the
rock were *Iron longimanus* and *Ephemerella doddsi*. Another species
of *Baetis* occurred anywhere on the rock where the passing water
was slowed to less than 120 cm/s, while *Rithrogena doddsi*
occurred anywhere the current was less than 60 cm/s. Under the
rock, where the current was reduced and especially where debris
had accumulated, *Cinygmula* sp., *Rithrogena virilis*, *Ephemerella
inermis* and *E. yosomite* were found. Many species of *Ephemerella*,
although fairly similar in appearance, have well-defined
substrate preferences (Leonard and Leonard, 1962). Similarly,
Flannagan (1977) showed that emergence densities of Trichoptera
from the Roseau River, Manitoba were directly related to substrate
size, with semi-submerged boulders being the site of emergence of
almost half the caddisflies. Higler (1975), too, found caddisflies
to show distinct preferences for various substrate types.
Limnephilus rhombicus and *Potamophylax rotundipennis* preferred
pebbles over sand, and coarse pebbles over crushed brick.
Sericostoma pedemontanum burrowed under pebbles or stones. Further,
Mackay (1977) showed *Pycnopsyche gentilis* to change its substrate
preference during its life history; avoiding mineral substrates
except during a short period in the final instar when leaf-disc
cases were changed to sand grain cases. *P. scabripennis* similarly
sought out inorganic particles (4-16 mm in diameter), to burrow in,
prior to a three month aestivation period. McLachlan (1969) showed
that substrate selection in the chironomid *Nilodorum brevibucca*
reflected its suitability as tube-building material. Substrate
also affected the rate of migration and settlement of the larvae.
Substrate type was found to determine the distribution of the
burrowing mayfly *Hexagenia limbata* (Eriksen, 1968) as the animal
only burrows successfully in undisturbed fine sediments. McLachlan
and Cantrell (1976) observed that both larval abundance and tube
shape of *Chironomus plumosus* were strongly influenced by sediment
depth.

Sand is often thought to be a rather unsuitable substrate for
macroinvertebrates due to its instability (Hynes, 1970a). However,
Ali and Mulla (1978) found that sandy and muddy substrates in a
Southern California river supported 8-10 times more chironomid
larvae than stony areas. Some species actually appear to prefer
bare sand, for example *Polypedilum breviantennatum* and
Stictochironomus sp. (Chironomini), although the fauna becomes much

richer when mixed with coarse or fine detritus (Tolkamp and Both, 1978). Hildrew and Townsend (1977) and Hildrew et al. (1980) found that susceptibility of stoneflies to predation by the caddisfly *Plectrocnemia conspersa* was highest on sand, persumably as this substrate offered few refuges. Along the same lines, Miller and Bukema (1977) suggested that the availability of adequate-sized interstitial openings in streambed substrate may be a factor in the availability of size ranges of different populations of *Gammarus minus* found by some workers (e.g. Culver, 1971). The above may account for the findings of Williams and Moore (1981) that, under plentiful food supply, a large-sized gravel substrate (31.5 mm diameter) produced significantly lower night-time drift of *Gammarus pseudolimnaeus* than either a medium-sized (11.0 mm diameter) or a small-sized gravel (3.4 mm diameter). Presumably, the larger spaces between the 31.5 mm diameter gravel particles provided maximum protection from predation, while the spaces in the smaller sizes were not adequate and the animals moved around on them more in search of suitable refuges.

Several recent studies have attempted to show which sizes and configurations of substrates support maximal benthic densities; earlier studies had associated the most productive areas of riffles with rubble substrates (e.g. Cummins, 1964; Egglishaw, 1969). William and Mundie (1978), for example, found, by using a variety of uniform substrate sizes (small 11.5 mm diameter, medium 24.2 mm, large 40.8 mm), that the greatest diversity was obtained on the large gravel even though maximum numbers and biomass occurred on the medium gravel. The findings of Wise and Molles (1979) tend to support this in that substrates of between 10 and 25 mm diameter implanted in a New Mexico stream supported more individuals and greater diversity than larger stones (> 75 mm diameter). Complexity (degree of physical distortion) of the surface of the particles has been shown to affect species richness, due to the greater diversity of resources available on such habitats (Hart, 1978). Colonization rates on complex particles may also be greater (Blackburn et al., 1977). Selection of appropriately-sized substrate particles by aquatic insects may involve sensory setae on the tarsi (Keetch and Moran, 1966).

As the result of an experiment designed to test the responses of members of an Ontario stream community to substrates of different heterogeneity, Williams (1980b) suggested that the degree of association with a particular substrate mixture varies from species to species. It may be a primary association in some species (e.g. substrate may provide the necessary physical environment for the species) but secondary (e.g. substrate may provide the necessary physical environment for a predator's prey species), or even tertiary (e.g. substrate may provide the necessary physical environment for the prey of a secondary predator) in others (see also Cummins and Lauff, 1969; Shelly, 1979).

A few studies have shown preferential selection of organic substrates by benthic invertebrates, although in some cases the choice may be directly the result of the organic material's suitability as a food source rather than purely as an attachment site. Egglishaw (1969), for example, found a general increase in the diversity of benthos with an increase in plant detritus in a Scottish river. Mackay and Kalff (1973) showed that two closely related species of caddisfly occurred largely on separate organic substrates in West Creek, Quebec. 90% of the larvae of *Pycnopsyche gentilis* lived in submerged, dead riparian leaves which are used as food and for case-building materials, while the population of *P. luculenta* occurred largely in detritus or in a mixture of leaves and detritus. The distribution of either species from fall to early spring depended on the amount of preferred habitat space available. Barber and Kevern (1973) found larger numbers and biomass of mayflies, caddisflies, chironomids and simuliids associated with macrophyte beds in the Pine River, Michigan. The association was seasonal and it was suggested that as the macrophytes were colonized by those taxa known to be common in the drift, they may act as unoccupied habitats (refugia) thus preventing population losses normally attributed to drift. It is unlikely that many submerged aquatic macrophytes are used directly as food (Westlake, 1975). A similar example of an organic substrate being used purely as an attachment site is seen in the sessile rotifer *Ptygura beauchampi* (Wallace, 1978). Larvae begin substrate selection activities when they chemotactically sense a stimulus associated with the glandular trichomes that cover the trapdoor region of the bladders of *Utricularia vulgaris*. The rotifers do not feed on the plant or its prey.

Current

Many lotic species have an inherent need for current, whether it is because of respiratory demands or because they extract their food from the passing water (Hynes, 1970a). Several studies have shown benthic densities to be related to current speed. Chutter and Noble (1966) showed this for the Simuliidae, while Rabeni and Minshall (1977) showed that the reduction of current velocity alone accounted for reductions in the densities of four or five species in Mink Creek, Idaho. However, the latter authors concluded that, overall, the substrate – detritus interaction was the overriding influence on insect microdistribtion, and current velocity played only a secondary role. Obviously though, at certain times current plays a primary role in animal distribution. In times of flooding, for example, large numbers of animals may be displaced downstream. Anderson and Lehmkuhl (1968) studies the catastrophic drift in an Oregon woodland stream during a fall freshet resulting from less than 2.5 cm of rain. There was a fourfold increase in numbers and a fivefold to eightfold increase in the biomass drifting, with mayflies, stoneflies, dipterans and terrestrial insects constituting the bulk of the increase. Even minor spates might significantly affect the drift patterns of Ephemeroptera (Ciborowski et al., 1977). The ability to hold station on the streambed is influenced by the animal's rheotactic behaviour. Allee (1914), for example, showed that *Asellus* (Isopoda) could withstand approximately twice the current when oriented into the current than when oriented downstream.

Reduced current velocity also may have an affect on micro-distribution. Cessation of flow in creeks has been shown to elicit migration responses in both macropterous and apterous gerrids (Riley, 1920), while regions of slower current are favoured for compensatory upstream migrations in amphipods (e.g. *Gammarus pulex*, Hughes, 1970). Nymphs of the stonefly *Brachyptera risi* normally press their bodies close to the substrate when the water is flowing but swim or hang passively from the surface of the water when flow ceases (Madsen, 1969).

Distribution with respect to currents have perhaps best been studied amongst the net-spinning caddisflies which rely on water movement to bring them food. Edington (1968) found that *Hydropsyche instabilis* preferred water velocities in the range of 15-100 cm/s, while *Plectrocnemia conspersa* was characteristically found in velocities of 0-20 cm/s; the physical nature of the respective capture nets appeared to be the chief factor limiting each species' current range. Philipson (1954) had shown earlier that in the laboratory, *H. instabilis* kept at 30 cm/s built normal feeding nets but in still water, although spinning some kind of structure, failed to make functional nets. In a study of two species of *Hydropsyche* in the North Tyne river system, England, Boon (1978) found *H. siltalai* to prefer fast water, while *H. pellucidula* appeared to tolerate a wide range of current regimes. He suggested that these preferences assist coexistence and avoidance of competition in these two commonly occurring species.

Size and shape may be correlated with current velocity. *Helicopsyche* (Trichoptera) individuals are reportedly larger in faster water than in slow water (Allen, 1951), whereas the shell of the freshwater limpet *Ancylus fluviatilis* is taller in fast water (Hynes, 1970a).

Currents are notoriously difficult to quantify accurately, especially in the field. Alongside difficulties in designing suitable instruments are those of incredibly complex microcurrents which frequently result from the flow of water around just a single uneven rock on the streambed. Complex flow patterns have been shown around even simple shapes placed in a current and the existence of a relatively stagnant boundary layer close to the surface of stones is known (Jaag and Ambuhl, 1964; Trivellato and Decamps, 1968). Many animals apparently living in fast water may, in fact,

be in relatively slow water, either in the boundary layer or in
eddy currents around rocks (Hynes, 1970a). The varying microcurrents
flowing around a rock may favour the growth of different species of
periphyton on different areas (Blum, 1960). This, in turn, may
control the microdistribution of monophagous herbivores such as
mayflies.

Depth

There are relatively few studies linking depth of stream to
microdistribution of the fauna. Harker (1953) found that, in
general, Ephemeroptera nymphs in Broadhead Brook, Lancashire did
not show a distribution correlated with absolute depth, but were
associated with the relative degrees of depth at certain times of
the year. *Ecdyonurus torrentis* and *Heptagenia lateralis*, for example
were present in shallow regions during their emergence periods, but
Rithrogena semicolorata was in deeper water at this time. *Baetis
rhodani*, which emerges from the water column, was randomly distributed at the time of its maximum emergence.

Temperature

Mention has been made of the effects of temperature on
distribution in the earlier section on longitudinal zonation.
However, temperature may also have an effect on a smaller scale.
On the whole, the turbulent flow in streams prevents the
establishment of thermal stratification such as is common in lakes,
although some very deep rivers may show differences between the
surface and the bottom. Even so, small streams may show small
temperature differences between riffle and pool areas, or even from
one side to the other, particularly if shading by riparian
vegetation is unequal (Williams, 1981). Sprules (1947) showed
that the number and diversity of Plecoptera decreased with
increasing average summer water temperature in different sections
of an Ontario stream. This decrease was coincident with a rise in
numbers and diversity of Ephemeroptera and Trichoptera. Hynes
(1970a) cites the example of the odonatan *Agrion virgo* replacing
A. splendens in areas where summer temperatures reach only 16-18°C.
The latter species can be found in water up to 24°C as it is less
sensitive to low dissolved oxygen levels. Hynes concludes that
such distributions probably reflect the direct or indirect effect
of temperature coupled with interspecific competion.

The occurrence of some chironomid larvae deep in the interstices
of streambeds may be the result of their following an optimum
temperature for development, particularly as the interstitial
water is more buffered against temperature change than the stream
water (Williams and Hynes, 1974). Even on the surface of the
streambed animals may seek out less severe temperatures as winter

approaches. Aiken (1968), for example, observed that a stream
population of the crayfish *Orconectes virilis* moved into deeper
water in winter, but thought that this was more associated with
gonadal maturation than with winter survival. In a lake population
of this species, Fast and Momot (1973) observed adult females
migrating into deeper cold water after releasing their young, while
adult males remained in warmer water close to the shore. These
authors hypothesized that under normal conditions of thermal
stratification, the social aggression of the larger males forced
the females into deeper water and that the aggression was temperature
related. Thorp (1978) similarly linked aggression in *Cambarus
latimanus* to water temperature, although in this particular case
the relationship was an inverse one. Gallepp (1977) noted that the
rate of feeding in *Brachycentrus* spp. (Trichoptera) increased with
temperature between 8 and 20°C but declined either side of this
range. The animals seemed more responsive to temperature and food
availability than to current speed.

    Williams (1980c) found that the annual colonization pattern of
artificial substrates by benthos in an Ontario stream was related
to water temperature. Likewise, Muller (1966) showed that an
increase in temperature always increased the drifting rate of
Gammarus pulex, but Keller (1975) found that an increase in
temperature decreased the drift of the mayfly *Ecdyonurus venosus*.
Again, the settling behaviour of the mayfly *Heptagenia* was thought
to be temperature related (Madsen, 1968). Some of the above
apparent contradictions may be explained in terms of the different
adaptive metabolisms seen in members of the benthic community. Some
species, especially those with multivoltine life cycles, grow faster
and move around more when water temperatures are high in the summer,
while others reach their maximal metabolism in the cooler water
during spring and fall. Still others, for example some stoneflies
and mayflies, grow actively during the winter, frequently under ice-
cover (Hynes, 1970b). In fact, Rauser (1962) has suggested that
the threshold temperature for many stoneflies is close to 0°C. At
the other end of the scale, some benthic species are only found in
thermal streams (see review of Castenholz and Wickstrom, 1975).

Dissolved Oxygen

    Most stream animals, particularly those in the rithron, demand
well-aerated water. Because of the nature of this zone (cool
temperatures, rapid flow and turbulence) the water is usually
saturated with oxygen. Only in backwaters, particularly where they
are loaded with decaying debris (Schneller, 1955), or in the
potamon, is the level of oxygen likely to fall. In the latter case,
as we have seen, many inhabitants are derived from lentic stock
and hence are better physiologically suited to such conditions, some
even possessing haemoglobin in their blood. High inputs of organic

pollution may also lead to local oxygen depletion, frequently
eliminating certain elements from the fauna (Hynes, 1960).

Recent studies, however, indicate possible microdistributions
as the result of variation in oxygen content even within the rithron.
Madsen (1968) found that the stonefly *Brachyptera risi*, in a small
Danish brook, occurred mainly on the upper surfaces of rocks where
oxygen levels were always high. Another species, *Nemoura flexuosa*,
conversely, was typically found on the lower surfaces of the rocks
where the oxygen level was much lower. Laboratory experiments on
the respiration of the two species confirmed that *N. flexuosa* could
survive at low oxygen concentrations while *B. risi* died. Kovalak
(1978) thought that nocturnal movements of benthic insects onto
stones and bricks in a Michigan stream were due to the animals
seeking areas of higher oxygen supply. Differing oxygen requirements
were used by Mann (1961) to explain the sharing of the habitat of
two riverine species of *Erpobdella* (Hirudinea). Of course, air
breathing benthic species, for example many of the Culicidae and
other siphonate dipterans, are independent of the dissolved oxygen
content of the water.

Light

A great many benthic species are photophobic, typically
occurring on the undersides of rocks and debris during the day.
Activities such as drift or feeding are frequently confined to the
hours of darkness when the animals are less susceptible to
predation. Manipulations of diurnal light/dark cycles have shown
that organisms typically found in the drift will follow any newly
imposed light regime (Müller, 1974). Kureck (1969), however, has
shown virtually no diel change in the number of drifting simuliid
larvae in a Lapland stream. Hall and Edwards (1978) interpret
this lack of change as a reflection of the decreased influence of
light on the activity of other benthic species at high latitudes.

Most studies have shown no endogenous rhythm to these
activities, although there appear to be exceptions (e.g. in the
caddisfly *Potamophylax luctuosus*, Lehmann, 1972). Laboratory
experiments on *Ephemerella subvaria* have shown that the number of
nymphs in the drift is a linear function of the duration of the
preceding photoperiod, probably acting through an internal
physiological change (Ciborowski, 1979). The size of the
animals in the drift may also vary with the light/dark cycle. For
example, Steine (1972) found that small mayfly nymphs predominantly
drifted during the day, while larger nymphs did so at night. The
relationship between this phenomenon and predation by fishes has
been discussed by Allan (1978). In a tropical stream, Hall and
Edwards (1978) found more larvae settling out of the drift at
night than during the day, with a significant increase in the
settling rate immediately following dusk.

Experiments on the level of light intensity needed to
initiate or terminate drift show considerable variation. Holt
and Waters (1967) found that for *Baetis vagans* and *Gammarus
pseudolimnaeus*, in Minnesota, it was approximately 1.0 lux.
Bishop (1969), however, working only slightly further north, in
Ontario, with different species of mayflies, caddisflies and
stoneflies determined the value at between 0.1 and 0.01 lux. The
latter author tentatively concluded that variations in wavelength
affected the activity patterns little.

Elliott (1968) showed that light and dark stimuli may
interact with other stimuli in controlling the activity response
of mayfly nymphs. Many of the species were negatively phototactic
in running water but in standing water exhibited alternating
periods of high and low activity throughout the 24 hour period
coupled with a loss of negative phototaxis.

Shading of portions of a streambed has been observed to
create noticeable differences in benthos distribution. Hughes
(1966) found *Baetis harrisoni* and *Tricorythus discolor*
(Ephemeroptera) predominantly in areas exposed to sunlight and
shaded by dense riparian vegetation, respectively. He concluded,
from laboratory experiments, that the selection and maintenance of
the microhabitat of *Tricorythus* was the result of the interaction
of its light response, current requirement and marked thigmotaxis,
while that of *Baetis* was the interaction of light response and an
undefined current factor perhaps associated with the boundary layer.
Thorup (1966) similarly found that in a small Danish stream, many
species, particularly those feeding on periphyton, were more common
in unshaded areas.

Food

Food is available to stream species in many different forms
often characteristic of different regions of a stream. Seston,
for example, may be particularly rich at the outflow of lakes and
species that typically feed on this, such as the net-spinning
caddisflies and filter-feeding blackflies, are abundant here
(Williams and Hynes, 1973; Carlsson et al., 1977; Oswood 1976).
In reasonably well-lighted sections of the stream, periphyton
grows on the substratum with local distribution and production
controlled by many environmental parameters, for example light,
current, substrate type and level of nutrients (Blum, 1960; Lowe,
1974; Whitton, 1975; Siver, 1977). Its occurrence may well control
the microdistribution of stream herbivores.

The role of flagellates in establishing pioneer communities
on new substrates has been examined by Yongue and Cairns (1978).
Flagellates become established very quickly (2-3 days) and reach

an equilibrium considerably earlier than other invertebrate groups
(15-28 days, Williams et al., 1977). It may well be that microorganisms like these and bacteria play a role in making bare
substrate surfaces attractive to higher benthic taxa. Bohle (1978)
has shown that food supply in the form of epilithic diatoms, caused
local concentration and lack of drift of nymphs of *Baetis rhodani*.
Depletion of food resulted in a more uniform distribution followed
by increased drift (see also Hildebrand, 1974). Again, Ward and
Cummins (1979) have shown that substrates with higher microbial
activities and biomass produced greater growth rates in the
chironomid *Paratendipes albimanus*. In the marine situation, young
intertidal *Gammarus* appear to benefit from a diet containing
filamentous algae. The life cycles of several species is such that
young are released in the spring and summer when this food is
available in large amounts, but stop in the fall and winter when
the algae are largely absent (Steele, 1967).

Allochthonous plant detritus on the streambed has been shown
to control the microdistribution of some stoneflies, mayflies and
chironomids, although other species (e.g. *Simulium* spp. and
Hydroptilidae) seem unaffected (Egglishaw, 1964). The size
composition of this detritus, as CPOM and FPOM, may also be
important (Malmqvist, 1978). As we noted earlier, Cummins (1975)
classed species feeding on CPOM as "shredders" in his scheme of
functional trophic groups. The distribution of shredder species
on a streambed can therefore be expected to coincide with the
distribution of this food source   at least during times of feeding,
and in particular with the CPOM that has been in the stream
sufficiently long to have become colonized by all-important
microorganisms and aquatic fungi (Kaushik and Hynes,
1971; Barlocher and Kendrick, 1973; Madsen, 1974). The microdistribution of "collector" species may similarly be expected to
correlate with the occurrence of FPOM. Some collectors have a
specific size range of FPOM on which they feed (Chance, 1970) and
the occurrence of these in the water column or amongst the substrate
particles (possibly controlled by relative densities) may result
in a further microdistribution. Other collectors are less selective
and so might be expected to be more widely distributed.

Areas of streams that receive large inputs of dissolved
organic matter, from materials that fall into them and in solution
from the soil, may support high benthic densities through local
formation of FPOM by precipitation and/or microbial uptake (Lush
and Hynes, 1973; Lock and Hynes, 1976).

The microdistribution of predatory species should reflect the
distribution of prey, but might be influenced by factors such as
the degree of specificity of the prey, the period between meals
and other habitat requirements of the predator.

Other animals

Microdistribution of a species may be affected by the density
of the species itself and by other species. The latter may involve
predation or some less drastic influence.

Several studies are reported in the literature where a
relationship between benthic population density of a species and
the amount of emigration of that species from the area has been
sought (e.g. Dimond, 1967: Peckarsky, 1979). The results are mixed
but the majority indicate that movement via drift is a density-
dependent response. Walton et al. (1977), for example found that
drift of the stonefly *Acroneuria abnormis* increased with benthic
density over a range of substrate particle sizes. Hildebrand
(1974) found a similar response in a variety of benthic species
including mayflies, simuliids and hydropsychid caddisflies. The
aggressive nature, including stridulation, of the latter group is
well known (Johnstone, 1964). Glass and Bovbjerg (1969), working
with *Cheumatopsyche*, found that larvae dispersed uniformly in the
absence of refuge and when densities were increased, aggressive
encounters enforced this spacing. When refuges, in the form of
small angular pebbles, were provided, the larvae aggregated at
these sites at all densities. When the refuges were filled to
capacity, excess larvae were forced to the periphery. Working
with the crayfish *Cambarus alleni* Bovbjerg (1959) observed a direct
relationship between initial densities and rate of movement in
populations of 5, 10 and 20 animals. He concluded that aggressive
behaviour was the causal factor and suggested that such dispersal
traits would not be seen in species lacking aggressive behaviour.
In a study of the drift, of *Baetis* and simuliids in Utah and Colorado,
Pearson and Franklin (1968) found that water-level fluctuations were
related to changes in drift rates indirectly through influence on
population density. Reisen and Prins (1972), on the other hand,
found that with the exception of nymphs of *Ephemerella*, drift seemed
only to be initiated by pupation or emergence activity and did not
respond to benthic population increases in a density-dependent
fashion. Corkum (1978) has suggested that there is a gradation
among behavioural groups within the Ephemeroptera in their tendency
to leave the substrate and drift. Studying *Baetis vagans* a species
with a high tendency to enter the water column, and *Paraleptophlebia
mollis*, a more benthic species, she concluded that the propensity
to leave the substrate was related to the species' behavioural
traits rather than density.

Bovbjerg (1964) concluded that population pressure may be one
cause of dispersal for some species, but not for others. His
experiments showed that aquatic snails and amphipods appear to
move independently of density, as they are not aggressive and may
form dense natural populations. Their tendency to disperse is
innate, based on internal factors. Animals like crayfish, on the

other hand, are more aggressive and respond directly to population pressure.

To my knowledge, there have been discovered very few fresh-water counterparts to the phenomena of mutual attraction, stimulation of metamorphosis or other similar intraspecific features in the life cycle that members of some marine bottom communities have upon each other (see Wilson, 1970; Meadows and Campbell, 1972). Brinkhurst et al. (1972) have shown, however, that tubificid oligochaetes can detect the presence of other species (through thermolabile substances carried in the water) and frequently form multispecific clumps in the substrate. Improved rates of growth and respiration of the species involved often result from such aggregations.

Predator pressure may influence movement of benthic species. Corkum and Clifford (1980) showed that the presence of the predatory stonefly *Isogenoides elongatus* affected the drift of two mayfly species, with more than twice as many *Baetis* nymphs drifting than those of *Leptophlebia cupida*. When kept together, neither mayfly affected the drift rate of the other even though *L. cupida* is an active crawler. The authors suggested that a large, actively-foraging, tactile carnivore, having long antennae and cerci, would encounter mayflies more often than either mayfly species would encounter each other. Such tactile predators hunting on the substrate would presumably be effective during both darkness and light. An interesting finding from this study was that the presence of *L. cupida* nymphs significantly influenced the drift of the predator. Peckarsky (1979), conversely, in a study of colonization of substrates buried in the bed of a Wisconsin stream, found no statistically higher colonization rates for invertebrate predators in cages initially inhabited by prey species than in those without prey. She concluded that, in this particular setting, invertebrate predators searched randomly for prey in a space-dependent manner rather than in a density-dependent one.

Predators that use vision rather than tactile senses may be expected to affect the movement and microdistribution of prey species in a different way. For example, Allan (1978) found that predation of *Baetis bicaudatus* by the brook trout *Salvelinus fontinalis* fell disproportionately on the larger nymphs. As a result, large animals drifted mostly at night, whereas smaller instars were aperiodic or day active.

In a recent study, Peckarsky (1980) investigated the predator-prey interactions between stoneflies and mayflies, in the field. Different species of mayfly apparently use different cues for detecting the presence of stonefly predators. For example, *Ephemerella subvaria* and *Stenonema fuscum* were able to avoid *Acroneuria lycorias* through non-contact chemical stimuli, while

Baetis phoebus, Heptagenia hebe and *S. fuscum* showed locomotory evasion (crawling, drifting, swimming) upon direct contact with *A. lycorias*. In a contact situation, *E. subvaria* frequently adopted a "scorpion-like" posture that Peckarsky interpreted as an attempt, by the prey, to alter its size and shape thus confusing a tactile predator. None of the mayfly species tested responded to the presence of stoneflies purely on visual grounds. As the introduction of stoneflies into test chambers statistically decreased the number of mayflies in the region of highest chemical stimulus, they presumably have significant affects on the micro-distribution of mayflies on a streambed.

Similar results have been obtained for the amphipod *Gammarus pseudolimmaeus*. Williams and Moore (1981) found that the introduction of rainbow trout (*Salmo gairdneri*) to artificial streams, in the day or night, caused almost total cessation of drift and upstream activity within minutes. Trials with fish-water suggested that the amphipods detect some form of labile exudate produced by the fishes. In contrast, the amphipods' response to predatory stoneflies appears to be an increase in drifting, again involving detection of some chemical substance produced by the stonefly nymphs (Williams and Moore, in prep.).

Naturally occurring chemicals

The effect of dissolved oxygen content on benthos distribution has already been covered. As for other chemical substances, their effects are not usually seen on the small scale but differences may be evident between distant reaches of a stream. Minshall and Minshall (1978), for example, found that chemical factors directly restricted the activity of benthos in the upper Duddon, England but not in the lower Duddon. In the specific instance of *Gammarus pulex*, they suggested that potassium was a limiting factor. Although pH differed between the two sections, it was not thought to be the primary factor affecting animal distribution. pH has, nevertheless, been shown to control distribution in other studies (e.g. Beck, 1965).

Sutcliffe and Carrick (1973) showed that whereas most benthic insects were not affected by low ionic concentrations, *Gammarus pulex* was, becoming lethargic almost immediately on being immersed in such water (Minshall and Minshall, 1978). Calcium ions are often considered to be of great importance in controlling the distribution of benthic species (e.g. Dittmar, 1953), but apart from good correlations of benthic densities with calcium levels (e.g. Egglishaw, 1968) the relationship has yet to be conclusively proved, particularly as many other ions (e.g. magnesium, sodium, sulphate and chloride) increase with water hardness alongside calcium (Hynes, 1970a).

Areas where the chemical nature of stream water can be seen
to affect the fauna over a short distance are springs. Groundwater emerging in limestone areas, for example, is high in
calcium bicarbonate. As the carbon dioxide in the water is
removed by both equilibrium with the air and photosynthesis around
the boil, calcium carbonate is deposited on the streambed. These
crystalline deposits (marl) are frequently laid down around algal
cells. This substrate, which often becomes covered with periphyton,
may attract certain benthic species such as larvae of the chironomid
Lithotanytarsus, which makes calcareous tubes, and the psychodid
Pericoma (Geijskes, 1935). Spring water in boggy and marshy areas
(low pH) often emerges rich in iron salts (usually ferrous
bicarbonate). Contact with the air releases dissolved carbon
dioxide causing the pH to rise. The ferrous ion is oxidized and a
flocculent brown film of ferric hydroxide is precipitated on the
substrate (Hynes, 1970a); characteristically, this promotes growth
of the iron bacterium *Leptothrix ochracea*. This iron deposit has
been claimed to eliminate certain taxa (e.g. *Gammarus*, Albrecht,
1953) but may not affect others (e.g. Plecoptera, Hynes, 1953).
Hot springs may similarly precipitate sulphur on the surrounding
substrate and this may promote the growth of sulphide-oxidizing
chemoautotrophs such as *Thiobacillus* (Castenholz and Wickstrom,
1975).

Man-made chemicals

Many man-made chemicals are added to sections of a stream or
river and these affect animal distribution either vividly, where
the benthos is locally exterminated, or subtly where only certain
elements of the fauna are eliminated, frequently resulting in a
local zonation of species radiating out from the source of the
pollutant. Pollutants include inorganic and organic suspensions
and sediments, insecticides, oil, pulpmill effluents, road salt
and a host of other compounds the effects of which have been dealt
with elsewhere (e.g. Hynes, 1960; Hart and Fuller, 1974). Coverage
of these topics is beyond the scope of this review.

Internal factors

We have seen that a few benthic species which move periodically do so in response to some sort of endogenous rhythm
(e.g. Harker, 1953; Elliott, 1968). Such internal stimuli may be
advantageous to species living in constant environments, like
caves. The weak endogenous component influencing activity in the
caddisfly *Potamophylax stellatus*, for example, was thought to
initiate locomotion in parts of a stream where there was little
daylight (Thorne, 1969).

Dickson (1977) found that within its natural habitats in
caves in Virginia, populations of *Crangonyx antennatus*

(Amphipoda) exhibited differences in behaviour between pools and
streams. The subterranean stream-dwelling populations were found
to show more cryptic behaviour than those living in the pools, both
under natural conditions and when disturbed. The selective advantage
of these differences was thought to be that the cryptic behaviour
of the stream dwellers made them less likely to be washed out of
their habitat. Whether the behavioural adaptations of the latter
populations were learned or of an inherited nature was not
determined.

Oviposition, controlled by the animal's physiology,
undoubtedly has an important influence (at least initially) on
the microdistribution of the next generation. Even though the
adult stage is frequently brief, it may result in some anomalous
distributions (Hynes, 1970b). Many female insects show distinct
preferences as to oviposition sites and shaded or lighted portions
of a stream may be favoured, as might fast or slow-water areas,
or smooth or rough water surfaces. Again, eggs may be laid singly
or in clumps, they may be deposited below the water surface, above
the water surface or on the water surface using a "dive-bombing"
technique. Some egg masses cling together until the larvae or
nymphs hatch while others explosively fly apart due to a reaction
between the secretion that holds the eggs together and the water
Hynes, 1941). A female may lay all her eggs at one site or
spread them over several sites. Subsequent dispersal of the
immature stages has been shown to be, in many cases, through drift
(Townsend and Hildrew, 1976; Williams, 1977) but this will vary
with the species and its propensity to leave the substrate.

TEMPORAL DISTRIBUTIONS

Several studies have recorded seasonal differences in benthic
faunal densities (e.g. Nelson and Scott, 1962; Logan, 1963).
DeMarch (1976) found that the fauna on certain substrates in the
Rat River, Manitoba was characterized by high variability due to
temporal succession of morphologically related species. She
suggested that this could be the result of seasonal changes in
substrate characteristics working through two mechanisms. First,
if a species is most commonly associated with a certain movable
particle size, the location of its desirable habitat may change
if the particle is moved through normal stream processes. Second,
seasonal influx of fine particles to a substrate area may clog
interstices and ruin the microhabitat for certain species.

Qualitative and quantitative changes in drift are known to
occur seasonally. Clifford (1972), for example, recorded changes
in a brown-water stream in northern Alberta. Drift densities of
entomostracans (cladocerans, cyclopoids and ostracods) increased
as the ice-free season progressed, but densities of insect larvae
remained relatively constant throughout this time. Total daily

drift of both groups, however, decreased as a function of water
volume. Drift densities, total daily drift and species diversity
were very low in the winter. In contrast, drift rates in a
Florida stream were greater in winter and early spring, and
minimal in the summer (Cowell and Carew, 1976). In the latter
study, total drift showed no significant correlation with
temperature, dissolved oxygen or mean benthic abundance, and only
slight correlation with current velocity. The migration study of
Williams (1980c), in a more temperate location, indicated that
water temperature controlled the seasonal mobility seen in many
species.

     Temporal distribution of benthos on the bottom and in the
water column is very much influenced by the life histories of the
various species. Stoneburner and Smock (1979) showed that there
was a relationship between drift and seasonal growth and emergence
patterns of insects in a South Carolina stream. While Lindegaard
and Thorup (1975) found that species with slow larval development
dominated the benthos of a Danish spring. Hynes (1970a) built a
generalized pattern for seasonal change due to features of the
life histories of temperate lotic species. It may be briefly
summarized as follows: fall is a period of increasing numbers and
biomass due to egg hatching of species which grow in cold winter
water temperatures, and growth of non-insect species hatched during
the summer. Winter is characterized by a slowing down of both the
growth of many species and the rate of egg hatching. Numbers may
decline but biomass need not. In early spring, the emergence of
some species causes a loss in biomass and numbers but this may be
compensated for by hatching of eggs of spring and early summer
species. Late spring is characterized by a drop in biomass and
numbers due to the main emergence period. Towards the end of the
spring, numbers increase as the eggs of summer species hatch and
the Malacostraca breed. During the summer, biomass increases due
to the growth of summer species and benthic densities vary
according to the exact composition of the fauna. Late summer is a
second period of low biomass, due to emergence of summer species,
and increasing numbers due to hatching of eggs of overwintering
species.

     Several workers have shown that the seasonal abundance of
food may strongly influence the life cycles of members of the stream
community. Ross (1963) showed that the life cycles of species in
the "shredder" trophic category, for example, were coincident with
the major surge of CPOM introduced through the autumn leaf fall.
Their eggs are laid and hatch in the fall and growth is completed
during the spring (Cummins, 1977). Changes in the distribution
and abundance of FPOM may be the reason why the net-spinning
caddisfly *Chimarra aterrima* migrates to deeper, slower water in
winter (Williams and Hynes, 1973). Similarly, the seasonal
succession of benthic algae (Neel, 1968; Moore, 1977) may control

the life cycle and temporal occurrence of herbivorous species on
the streambed. Seasonal foraging strategies may also exist in
predatory species (Townsend and Hildrew, 1979).

More marked seasonal microdistributional changes are seen in
temporary streams, and again these are closely tied in with
different life history strategies. Williams and Hynes (1977) noted
that certain major taxa dominated the substrate as a temporary
stream in southern Ontario went from its fall-winter, flowing-
water phase, through its spring, standing-water phase to its
summer, dry phase. The predominant taxa characteristic of each
phase were the Chironomidae, Coleoptera and Sphaeroceridae
(Diptera), respectively.

APPLICATION POTENTIAL OF RESEARCH INTO MIGRATION AND DISTRIBUTION OF STREAM BENTHOS

Although the study of running water animals and their habitat
is not as old as that of its lentic counterpart (Hynes, 1975)
considerable advances have been made in the last twenty years.
Recent stream research has, like most branches of science, posed
more questions than it has answered. Nevertheless, there has now
been accumulated a substantial body of data and theories on the
workings of these particular habitats, sufficient, no doubt, to
enable workers to make at least some simple predictive models that
may have practical application. Unfortunately, such applications
appear to be slow in coming.

Most of the applications made so far have been associated with
the management of stream fishes, particularly those of economic or
recreative importance and especially the salmonids. As part of
this, some consideration has occasionally been given to the benthic
invertebrates, usually in the role as food for the fishes. Many
techniques for the improvement of trout habitat in the wild have
been developed. These include increasing bank stability and cover,
maintenance of sufficient water depth, creation of submerged
cover, creation of additional spawning beds, current deflectors
and removal of predators (White and Brynildson, 1967). Methods
to increase natural production of invertebrates have included
promoting the growth of periphyton by allowing adequate light to
reach the stream, and the addition of nutrients to the water
(Warren et al., 1964).

Recently, Mundie (1974) proposed an innovative, semi-
natural technique for raising young coho salmon (*Oncorhynchus
kisutch*) to the smolt stage. The approach combines some of the
desirable features of rivers with the productive capacity of
hatcheries and applies knowledge of the distribution and movements
of benthos in reaching this goal. Benthic invertebrates on the
riffles of the semi-natural rearing channel are used as part of

the fry's diet (being conveniently transported from the riffles
to the waiting fishes in the pools through drift), and to recycle
excess supplemental prepared food and fish faeces into the benthic
food chain. Microdistribution, productivity and composition of
the benthic community are controllable through manipulation of
substrate particle size and available food on the riffles
(Mundie et al., 1973; Williams and Mundie, 1978). The results of
the first few years of operation of a large-scale experimental
facility based on this model are encouraging.

Other examples of present knowledge put to practical use
might include prediction of the time and place of emergence of
biting adult insects through knowledge of their life cycles, and
prediction that submerged, smooth, concrete surfaces such as occur
below river dams promote the occurrence of certain nuisance
species (e.g. Simuliidae and Hydropsychidae). The control of
certain human tropical diseases that are spread through intermediate
aquatic hosts, for example, schistosomiasis spread by the snail
Biomphalaria glabrata, and onchocerciasis spread by various species
of blackflies, may be greatly speeded up through thorough
knowledge of host ecology (Laird, 1972; Sturrock, 1973). Finally,
studies on the present-day distribution of the fossil remains of
aquatic invertebrates is leading to an understanding of their role
and distribution in past environments (Frey, 1964; Williams and
Morgan, 1977).

REFERENCES

Aiken, D.E. 1968. The crayfish *Orconectes virilis* survival in a
    region with severe winter conditions. *Can. J. Zool.*
    46: 207-211.
Albrecht, M.-L. 1953. Die Plane und andere Flämingbäche.
    Z. Fisch. 1: 389-476.
Ali, A. and Mulla, M.S. 1978. Spatial distribution and day-
    time drift of chironomids in a southern California river.
    Mosquito News 38: 122-126.
Allan, J.D. 1978. Trout predation and the size composition of
    stream drift. *Limnol. Oceanogr.* 23: 1231-1238.
Allee, W.C. 1914. The ecological importance of the rheotactic
    reaction of stream isopods. *Bull. mar. biol. Lab., Woods
    Hole* 27: 52-66.
Allen, K.R. 1951. The Horokiwi stream. A study of a trout
    population. *Fish. Bull. N.Z.* 10: 231 pp.
Anderson, N.H. 1976. The distribution and biology of the
    Oregon Trichoptera. *Agr. Expt. Statn., Oregon St. Univ.
    Tech. Bull.* 134: 152 pp.
Anderson, N.H. and Lehmkuhl, B.M. 1968. Catastrophic drift of
    insects in a woodland stream. *Ecology* 49: 198-206.
Aubert, J. 1959. Plecoptera. Insecta Helvetia Fauna 1.
    Imprimerie la Concorde Lausanne.

Balfour-Brown, F. 1958. British water beetles. Vol. 3.
    Roy. Soc., London.

Ball, I.R. and Fernando, C.H. 1969. Freshwater triclads
    (Platyhelminthes, Turbellaria) and continental drift.
    Nature, Lond. 221: 1143-1144.

Ball, R.C., Wojtalik, T.A. and Hooper, F.F. 1963. Upstream
    dispersion of radiophosphorus in a Michigan trout stream.
    Pap. Mich. Acad. Sci. 48: 57-64.

Barber, W.E. and Kevern, N.R. 1973. Ecological factors
    influencing macroinvertebrate standing crop distribution.
    Hydrobiologia 43: 53-75.

Barlocher, F. and Kendrick, B. 1973. Fungi and food preferences
    of *Gammarus pseudolimnaeus*. *Arch. Hydrobiol.* 72: 501-516.

Beck, W.M. Jr. 1965. The streams of Florida. *Bull. Fla. St.*
    Mus. 10: 3, 91-126.

Bengtsson, J., Butz, I. and Madsen, B.L. 1972. Upstream flight
    by stream insects. *Flora Fauna* 78: 102-104.

Berthelemy, C. 1966. Recherches ecologiques et biogeographiques
    sur les Plécoptères et Coleoptères d'eau courante
    (Hydraena et Elminthidae) des Pyrénées. *Annls. Limnologie.*
    2: 227-458.

Bishop, J.E. 1969. Light control of aquatic insect activity and
    drift. *Ecology* 50: 371-380.

Bishop, J.E. 1973. Observations on the vertical distribution of
    the benthos in a Malaysian stream. *Freshwat. Biol.* 3:
    147-156.

Bishop, J.E. and Hynes, H.B.N. 1969. Upstream measurements of
    the benthic invertebrates in the Speed River, Ontario.
    J. Fish. Res. Bd. Can. 26: 279-298.

Black, J.B. 1963. Observations on the home range of stream-
    dwelling crawfishes. *Ecology* 44: 592-595.

Blackburn, W.B., Trush, W.J. and Buikema, A.J. Jr. 1977. The
    effect of substrate complexity on macrobenthic colonization.
    Virginia J. Sci. 28: 61.

Blum, J.L. 1960. Algal populations in flowing waters. *Spec.*
    publs. Pymatuning Lab. Fld. Biol. 2: 11-21.

Bohle, H.W. 1978. Relation between food supply, drift and micro-
    distribution of larvae of *Baetis rhodani*. Investigations in
    a stream model. *Arch. Hydrobiol.* 84: 500-525.

Boon, P.J. 1978. The preimpoundment distribution of certain
    Trichoptera larvae in the north Tyne River system (northern
    England) with particular reference to current speed.
    Hydrobiologia 57: 167-174.

Bouvet, Y. 1977. Adaptations physiologiques et compartementales
    des *Stenophylax* (Limnephilidae) aux eaux temporaires. *In:*
    Proc. 2nd Int. Symp. on Trichoptera. Crichton, M.I. ed.
    Junk, The Hague, 117-119.

Bovbjerg, R.V. 1959. Density and dispersal in laboratory crayfish
    populations. *Ecology* 40: 504-506.

Bovbjerg, R.V.   1964.   Dispersal of aquatic animals relative to
     density. *Verh. int. Verein. theor. agnew. Limnol.* 15:
     879–884.
Brinkhurst, R.O., Chua, K.E. and Kaushik, N.K.   1972.
     Interspecific interactions and selective feeding by tubificid
     oligochaetes. *Limnol. Oceanogr.* 17: 122–133.
Brinkhurst, R.O. and Jamieson, B.G.M.   1971.   Aquatic oligo-
     chaetes of the world.   Oliver and Boyd, Edinburgh.
Butcher, R.W.   1933.   Studies on the ecology of rivers I.
     On the distribution of macrophytic vegetation in the rivers
     of Britain. *J. Ecol.* 21: 58–91.
Carlsson, M., Nilsson, L.M., Svensson, B., Ulfstrand, S. and
     Wotton, R.S.   1977.   Lacustrine seston and other factors
     influencing the blackflies (Diptera:Simuliidae) inhabiting
     lake outlets in Swedish Lapland. *Oikos* 29: 229–238.
Carpenter, K.E.   1928.   Life in inland waters.   Sidgwick and
     Jackson, London.
Castenholz, R.W. and Wickstrom, C.E.   1975.   Thermal streams.
     In: River ecology.   Whitton, B.A. ed.   Univ. California
     Press.
Chance, M.M.   1970.   The functional morphology of the mouth parts
     of blackfly larvae (Diptera:Simuliidae). *Quaes Ent.*
     6: 245–284.
Chutter, F.M. and Noble, R.G.   1966.   The reliability of a method
     of sampling stream invertebrates. *Arch. Hydrobiol.* 62:
     95–103.
Ciborowski, J.J.H.   1979.   The effects of extended photoperiods
     on the drift of the mayfly *Ephemerella subvaria* McDunnough
     (Ephemeroptera:Ephemerellidae). *Hydrobiologia* 62:
     209–214.
Ciborowski, J.J.H. and Corkum, L.D.   1980.   Importance of
     behaviour to the re-establishment of drifting Ephemeroptera.
     In: Proc. 3rd Int. Conf. on Ephemeroptera.   Flannagan, J.F.
     and Marshall, K.E. eds.   Plenum Press, N.Y., 321–330.
Ciborowski, J.J.H., Pointing, P.J. and Corkum, L.D.   1977.
     The effect of current velocity and sediment on the drift
     of the mayfly *Ephemerella subvaria* McDunnough. *Freshwat.*
     Biol. 7: 567–572.
Claassen, P.W.   1931.   Plecoptera nymphs of America (north of
     Mexico).   C.C. Thomas Publ., Illinois.
Clifford, H.F.   1966.   The ecology of invertebrates in an
     intermittent stream. *Invest. Indiana Lakes, Streams*
     7: 57–98.
Clifford, H.F.   1972.   Drift of invertebrates in an intermittent
     stream draining marshy terrain of west-central Alberta.
     Can. J. Zool. 50: 985–991.
Coker, R.E.   1968.   Streams, lakes and ponds.   Harper and Row,
     N.Y.
Coleman, M.J. and Hynes, H.B.N.   1970.   The vertical distribution
     of the invertebrate fauna in the bed of a stream. *Limnol.*
     Oceanogr. 15: 31–40.

Corbet, P.S. 1963. A biology of dragonflies. Quadrangle Books, Chicago.

Corkum, L.D. 1978. Is benthic activity of stream invertebrates related to behavioural drift? *Can. J. Zool.* 2457-2459.

Corkum, L.D. and Clifford, H.F. 1980. The importance of species associations and substrate types to behavioural drift. *In:* Proc. 3rd Int. Conf. on Ephemeroptera. Flannagan, J.F. and Marshall, K.E. eds. Plenum Press, N.Y. 331-341.

Corkum, L.D. and Pointing, P.J. 1979. Nymphal development of *Baetis vagans* McDunnough (Ephemeroptera:Baetidae) and drift habits of large nymphs. *Can. J. Zool.* 57: 2348-2354.

Cowell, B.C. and Carew, W.C. 1976. Seasonal diel periodicity in the drift of aquatic insects in a subtropical Florida stream. *Freshwat. Biol.* 6: 587-594.

Culver, D.C. 1971. Analysis of simple cave communities III. Control of abundance. *Amer. Midl. Nat.* 85: 173-188.

Cummins, H. 1921. Spring migration of the crayfish *Cambarus argillicola* Faxon. *Trans. Amer. Microscop. Soc.* 40: 28-30.

Cummins, K.W. 1964. Factors limiting the microdistribution of larvae of the caddisflies *Pycnopsyche lepida* (Hagen) and *P. guttifer* (Walker) in a Michigan stream. *Ecol. Monogr.* 34: 271-295.

Cummins, K.W. 1977. From headwater streams to rivers. *Amer. Biol. Teacher.* (May) 305-312.

Cummins, K.W. and Lauff, G.H. 1969. The influence of substrate particle size on the microdistribution of stream macrobenthos. *Hydrobiologia* 34: 145-181.

Daborn, G.R. 1976. Colonization of isolated aquatic habitats. *Can. Fld.-Nat.* 90: 56-57.

Danielopol, D.L. 1976. The distribution of the fauna in the interstitial habitats of riverine sediments of the Danube and the Piesting (Austria). *Int. J. Speleol.* 8: 23-51.

Dansereau, P. 1957. Biogeography: an ecological perspective. Ronald Press, N.Y.

Davies, B.R. 1976. The dispersal of Chironomidae: A review. *J. ent. Soc. Sth Afr.* 39: 39-62.

Davies, L. 1961. Ecology of two *Prosimulium* species with reference to their ovarian cycles. *Can. Ent.* 93: 1113-1140.

DeMarch, B.G.E. 1976. Spatial and temporal patterns in macrobenthic stream diversity. *J. Fish. Res. Bd. Can.* 33: 1261-1270.

Dendy, J.S. 1944. The fate of animals in stream drift when carried into lakes. *Ecol. Monogr.* 14: 333-357.

Dennert, H.G., Dennert, A.L. Kant, P., Pinkster, S. and Stock,J.H. 1969. Upstream and downstream migrations in relation to the reproductive cycle and to environmental factors in the amphipod, *Gammarus zaddachi*. *Bijdr. Dierk.* 39: 11-43.

Dickson, G.W. 1977. Behavioural adaptation of the troglobitic
    amphipod crustacean *Crangonyx antennatus* to stream habitats.
    Hydrobiologia 56: 17-20.
Dimond, J.B. 1967. Evidence that drift of stream benthos is
    density related. *Ecology* 48: 855-857.
Dittmar, H. 1953. Die Bedentung des Ca- und Mg- Behaltes für
    die Fauna fliessender Gewässer. *Ber. limnol. Flussstn
    Freudenthal.* 4: 20-23.
Edington, J.M. 1968. Habitat preferences in net-spinning caddis
    larvae with special reference to the influence of water
    velocity. *J. Anim. Ecol.* 37: 675-692.
Edmunds, G.F. jr., Jensen, S.L. and Berner, L. 1976. The
    mayflies of north and central America. Univ. Minnesota
    Press, Minneapolis.
Egglishaw, H.J. 1964. The distributional relationship between
    the bottom fauna and plant detritus in streams. *J. Anim.
    Ecol.* 33: 463-476.
Egglishaw, H.J. 1968. The quantitative relationship between
    fauna and plant detritus in streams of different
    concentrations. *J. appl. Ecol.* 5: 731-740.
Egglishaw, H.J. 1969. The distribution of benthic invertebrates
    on substrata in fast flowing streams. *J. Anim. Ecol.* 38:
    19-33.
Elliott, J.M. 1967. The life histories and drifting of the
    Plecoptera and Ephemeroptera in a Dartmoor stream.
    J. Anim. Ecol. 36: 343-362.
Elliott, J.M. 1968. The daily activity patterns of mayfly
    nymphs (Ephemeroptera). *J. Zool., London* 155: 201-221.
Elliott, J.M. 1971. Upstream movements of benthic invertebrates
    in a Lake District stream. *J. Anim. Ecol.* 40: 235-252.
Elliott, J.M. 1972. Effect of temperature on the time of
    hatching in *Baetis rhodani* (Ephemeroptera:Baetidae).
    Oecologia 9: 47-51.
Elliott, J.M. 1973. The food of brown and rainbow trout *(Salmo
    trutta* and *S. gairdneri)* in relation to the abundance of
    drifting invertebrates in a mountain stream. *Oecologia*
    12: 329-347.
Eriksen, C.H. 1968. Ecological significance of respiration and
    substrate for burrowing Ephemeroptera. *Can. J. Zool.*
    46: 93-103.
Fast, A.W. and Momot, W.T. 1973. The effects of artificial
    aeration on the depth distribution of the crayfish
    Orconectes virilis (Hagen) in two Michigan lakes. *Am.
    Midl. Nat.* 89: 89-102.
Fernando, C.H. and Galbraith, D. 1970. A heavy infestation of
    gerrids (Hemiptera:Heteroptera) by water mites (Acarina:
    Limnocharidae). *Can. J. Zool.* 48: 592-594.
Fernando, C.H. and Galbraith, D. 1973. Seasonality and dynamics
    of aquatic insects colonizing small habitats. *Verh. int.
    Verein. Theor. Angew. Limnol.* 18: 1564-1575.

Flannagan, J.F. 1977. Emergence of caddisflies from the Roseau
    River, Manitoba. *In:* Proc. 2nd Int. Conf. on Ephemeroptera.
    Pasternak, K. and Sowa, R. eds. Junk, The Hague. 183-197.
Flannagan, J.F. 1978. The winter stoneflies *Allocapnia granulata*
    (Taeniopterygidae), *Taeniopteryx nivalis* and *T. parvula*
    (Capniidae) in southern Manitoba. *Can. Ent.* 110: 111-112.
Ford, J.B. 1962. The vertical distribution of larval
    Chironomidae (Diptera) in the mud of a stream. *Hydrobiologia*
    19: 262-272.
Fox, L.R. 1977. Species richness in streams - an alternative
    mechanism. *Amer. Nat.* 111: 1017-1020.
Freeman, J.A. 1945. Studies on the distribution of insects by
    aerial currents. *J. Anim. Ecol.* 14: 128-154.
Frey, D.G. 1964. Remains of animals in Quaternary lake and bog
    sediments and their interpretation. *Ergeb. Limnol.* 2:
    1-114.
Fryer, G. 1972. Observations on the ephippia of certain
    macrothrieid cladocerans. *Zool. J. Linn. Soc.* 51: 79-96.
Fryer, G. 1974. Attachment of bivalve molluscs to corixid bugs.
    Naturalist, Hull 928: 18.
Gallepp, G.W. 1977. Response of caddisfly larvae (*Brachycentrus*
    spp.) to temperature, food availability and current velocity.
    Am. Midl. Nat. 98: 59-84.
Geijskes, D.C. 1935. Faunistisch-ökologische Untersuchungen am
    Röserenbach bei Leistal im Basler Tafeljura. *Tijdschr. Ent.*
    78: 249-382.
Gislen, T. 1948. Aerial plankton and its conditions of life.
    Camb. Phil. Soc. Biol. Rev. 23: 109-126.
Glass, L.W. and Bovbjerg, R.V. 1969. Density and dispersion in
    laboratory populations of caddisfly larvae(*Cheumatopsyche:*
    Hydropsychidae). *Ecology* 50: 1082-1084.
Gledhill, T. 1971. The genera *Azugofeltria*, *Vietsaxona*,
    Neoacarus and *Hungarohydracarus* (Hydrachnellae:Acari) from
    the interstitial habitat in Britain. *Freshwat. Biol.*
    1: 61-82.
Hall, R.O. and Edwards, A.J. 1978. Observations on the settling
    of *Simulium damnosum* larvae on artificial substrates, in
    the Ivory Coast. *Hydrobiologia* 57: 81-84.
Harker, J.E. 1953. The diurnal rhythm of activity of mayfly
    nymphs. *J. exp. Biol.* 30: 523-533.
Harper, P.P. and Hynes, H.B.N. 1970. Diapause in the nymphs of
    Canadian winter stoneflies. *Ecology* 51: 425-427.
Harris, D.A. and Harrison, A.D. 1974. Life cycles and larval
    behaviour of two species of *Hydrachna* (Acari:Hydrachnidae),
    parasitic upon Corixidae (Hemptera:Heteroptera). *Can. J.
    Zool.* 52: 1155-1165.
Harrison, A.D. 1965. Geographical distribution of riverine
    invertebrates in Southern Africa. *Arch. Hydrobiol.* 61:
    387-394.

Harrison, A.D. 1966. Recolonization of a Rhodesian stream after
    drought. *Arch. Hydrobiol.* 62: 405-421.
Hart, D.D. 1978. Diversity in stream insects: regulation by
    rock size and microspatial complexity. *Verh. int. Verein.*
    theor. angew. Limnol. 20: 1376-1381.
Hart, C.W. Jr. and Fuller, S.L.H. 1974. Pollution ecology of
    freshwater invertebrates. Academic Press, N.Y.
Hastings, E., Kittams, W.H. and Pepper, J.H. 1961. Re-
    population by aquatic insects in streams sprayed with D.D.T.
    Ann. ent. Soc. Amer. 54: 436-437.
Hawkes, H.A. 1975. River zonation and classification. *In:*
    River ecology. Whitton, B.A. ed. Univ. California Press,
    312-374.
Hayden, W. and Clifford, H.F. 1974. Seasonal movements of the
    mayfly *Leptophlebia cupida* (Say) in a brown-water stream
    of Alberta, Canada. *Am. Midl. Nat.* 91: 90-103.
Higler, L.W.G. 1975. Reactions of some caddis larvae
    (Trichoptera) to different types of substrate in an
    experimental stream. *Freshwat. Biol.* 5: 151-159.
Hildebrand, S.G. 1974. The relation of drift to benthos
    density and food level in an artificial stream. *Limnol.*
    Oceanogr. 19: 951-958.
Hildrew, A.G. and Townsend, C.R. 1977. The influence of
    substrate on the functional response of *Plectrocnemia*
    conspersa (Curtis) larvae (Trichoptera:Polycentropodidae).
    Oecologia 31: 21-26.
Hildrew, A.G., Townsend, C.R. and Henderson, J. 1980. Inter-
    actions between larval size, microdistribution and substrate
    in the stoneflies of an iron-rich stream. *Oikos* 35:
    387-396.
Holt, C.S. and Waters, T.F. 1967. Effect of light intensity
    on the drift of stream invertebrates. *Ecology* 48:
    225-234.
Horton, R.E. 1945. Erosional development of streams and their
    drainage basins. *Bull. Geol. Soc. Amer.* 56: 275-370.
Huet, M. 1954. Biologie, profils en long et an travers des
    eaux courantes. *Bull. fr. Piscic.* 175: 41-53.
Hughes, D.A. 1966. The role of responses to light in the
    selection and maintenance of microhabitat by the nymphs
    of two species of mayfly. *Anim. Behav.* 14: 17-33.
Hughes, D.A. 1970. Some factors affecting drift and upstream
    movements of *Gammarus pulex*. *Ecology* 51: 301-305.
Hultin, L. 1971. Upstream movements of *Gammarus p. pulex*
    (Amphipoda) in a south Swedish stream. *Oikos* 22:
    329-347.
Hultin, L., Svensson, B. and Ulfstrand, S. 1969. Upstream
    movements of insects in a south Swedish small stream.
    Oikos 20: 553-557.

Husmann, S. 1971. Ecological studies on freshwater meiobenthon
    in layers of sand and gravel. *In:* Proc. 1st Int. Conf.
    on Meiofauna. Hulings, N.C. ed. *Smithsonian Contr.*
    Zool. 76: 161-169.

Hynes, H.B.N. 1941. The taxonomy and ecology of the nymphs
    of British Plecoptera with notes on the adults and eggs.
    Trans. R. Ent. Soc. Lond. 91: 459-557.

Hynes, H.B.N. 1953. The Plecoptera of some small streams near
    Silkeborg, Jutland. *Ent. Medd.* 26: 489-494.

Hynes, H.B.N. 1955. The reproductive cycle of some British
    freshwater Gammaridae. *J. Anim. Ecol.* 24: 352-387.

Hynes, H.B.N. 1960. The biology of polluted waters. Liverpool
    Univ. Press.

Hynes, H.B.N. 1964. Some Australian plecopteran nymphs.
    Gewäss. Abwäss. 34/5: 17-22.

Hynes, H.B.N. 1970a. The ecology of running waters. Liverpool
    Univ. Press.

Hynes, H.B.N. 1970b. The ecology of steam insects. *Ann. Rev.*
    Ent. 15: 25-42.

Hynes, H.B.N. 1971. Zonation of the invertebrate fauna in a
    West Indian stream. *Hydrobiologia* 38: 1-8.

Hynes, H.B.N. 1974. Further studies on the distribution of
    stream animals within the substratum. *Limnol. Oceanogr.*
    19: 92-99.

Hynes, H.B.N. 1975. The stream and its valley. *Verh. int.*
    Verein. theor. angew. Limnol. 19: 1-15.

Hynes, H.B.N. 1976. Biology of Plecoptera. *Ann. Rev. Ent.*
    21: 135-153.

Hynes, H.B.N. and Williams, T.R. 1962. The effect of DDT on
    the fauna of a Central African stream. *Ann. trop. Med.*
    Parasit. 56: 78-91.

Hynes, H.B.N., Williams, D.D. and Williams, N.E. 1976.
    Distribution of the benthos within the substratum of a
    Welsh mountain stream. *Oikos* 27: 307-310.

Hynes, J.D. 1975. Downstream drift of invertebrates in a
    river in southern Ghana. *Freshwat. Biol.* 5: 515-533.

Ide, F.P. 1935. The effect of temperature on the distribution
    of the mayfly fauna of a stream. *Publs. Ont. Fish. Res.*
    Lab. 50: 1-76.

Illies, J. 1961. Versuch einer allgemeinen biozönatischen
    Gliederung der Fliessgewässer. *Int. Rev. ges. Hydrobiol.*
    Hydrogr. 46: 205-213.

Illies, J. and Botosaneanu, L. 1963. Problèmes et méthodes
    de la classification et de la zonation écologiques des
    eaux courantes, considerées surtout du point de vue
    faunistique. *Mitt. int. Verein. theor. angew. Limnol.*
    12: 1-57.

Jaag, O. and Ambühl, H. 1964. The effect of the current
    on the composition of biocaenoses in flowing water streams.
    Int. Conf. Wat. Pollut. Res. Lond., Pergamon Press, Oxford,
    31-49.

Jackson, D.J. 1956. The capacity for flight of certain water beetles and its bearing on their origin in the Western Scottish Isles. *Proc. Linn. Soc. Lond.* 167: 76-96.

Johnstone, G.W. 1964. Stridulation by larval Hydropsychidae *Proc. R. ent. Soc. Lond. A.* 39: 146-150.

Kaushik, N.K. and Hynes, H.B.N. 1971. The fate of the dead leaves that fall into streams. *Arch. Hydrobiol.* 68: 465-515.

Keetch, D.P. and Moran, V.C. 1966. Observations on the biology of nymphs of *Paragomphus coenatus* (Rambur) (Odonata: Gomphidae) I. Habitat selection in relation to particle size. *Proc. R. Ent. Soc. Lond. A.* 41: 116-122.

Keller, A. 1975. The drift and its ecological significance. Experimental investigations on *Ecdyonurus venosus* (Fabr.) in a stream model. *Schweiz. Z. Hydrol.* 37: 294-331.

Kovalak, W.P. 1978. Diel changes in stream benthos density on stones and artificial substrates. *Hydrobiologia* 58: 7-16.

Kühtreiber, J. 1934. Die Plekopterenfauna Nordtirols. *Ber. naturw.-med. Ver. Innsbruck* 44: 1-219.

Kureck, A. 1967. Über die tagesperiodische Ausdrift von *Niphargus aquilex schellenbergi* Karaman aus Ouellen. *Z. Morph. Ökol. Tiere* 58: 247-262.

Kureck, A. 1969. Tagesrhythmen lapplandische Simuliiden (Diptera). *Oecologia* 2: 385-410.

Laird, M. 1972. A novel attempt to control biting flies with their own diseases. *Science Forum* 5: 12-14.

Lansbury, I. 1955. Some notes on invertebrates other than insects found attached to water bugs (Hemiptera:Heteroptera). *Entomologist* 88: 139-140.

Larimore, R.W. 1972. Daily and seasonal drift of organisms in a warm-water stream. *Illinois Nat. Hist. Surv. (Wat. Res. Centre) Res. Rep.* 55: 1-105.

Lehmann, U. 1967. Drift and Populations dynamik von *Gammarus pulex fossarum* Koch. *Z. Morph. Ökol. Tiere* 60: 227-274.

Lehmann, U. 1972. Tagesperiodisches Verhalten und Habitat wechsel der Larven von *Potamophylax luctuosus* (Trichoptera). *Oecologia.* 9: 265-278.

Lehmkuhl, D.M. 1972. *Baetisca* (Ephemeroptera:Baetiscidae) from the western interior of Canada with notes on the life cycle. *Can. J. Zool.* 50: 1015-1017.

Leonard, J.W. and Leonard, F.A. 1962. Mayflies of Michigan trout streams. *Bull. Cranbrook Inst. Sci.* 43: 1-139.

Lewis, T. and Taylor, L.R. 1967. Introduction to experimental ecology. Academic Press, London.

Lindegaard, C., Thorup, J. and Bahn, M. 1975. The invertebrate fauna of the moss carpet in the Danish spring Ravnkilde and its seasonal, vertical and horizontal distribution. *Arch. Hydrobiol.* 75: 109-139.

Linduska, J.P. 1942. Bottom type as a factor influencing the
    local distribution of mayfly nymphs. *Can. Ent.* 74: 26–30.

Lock, M.A. and Hynes, H.B.N. 1976. The fate of "dissolved"
    organic carbon derived from autumn-shed maple leaves (*Acer
    saccharum*) in a temperate hard-water stream. *Limnol. Oceanogr.*
    21: 436–443.

Logan, S.W. 1963. Winter observations on bottom organisms and
    trout in Badger Creek, Montana. *Trans. Am. Fish. Soc.*
    92: 140–145.

Lowe, R.L. 1974. Environmental requirements and pollution
    tolerance of freshwater diatoms. *U.S. Env. Prot. Agency
    Rep.* 670/4-74-005, 1-333.

Lush, D.L. and Hynes, H.B.N. 1973. The formation of particles
    in freshwater leachates of dead leaves. *Limnol. Oceanogr.*
    18: 968–977.

Macan, T.T. 1939. Notes on the migration of some aquatic
    insects. *J. Soc. Brit. Ent.* 2: 1–6.

Macan, T.T. 1961. Factors that limit the range of freshwater
    animals. *Camb. Phil. Soc. Biol. Rev.* 36: 151–198.

Macan, T.T. 1974. Running water. *Mitt. int. Verein. theor.
    angew. Limnol.* 20: 301–321.

Mackay, R.J. 1977. Behaviour of *Pycnopsyche* (Trichoptera:
    Limnephilidae) on mineral substrates in laboratory streams.
    Ecology 58: 191–195.

Mackay, R.J. and Kalff, J. 1973. Ecology of two related species
    of caddisfly larvae in the organic substrates of a woodland
    stream. *Ecology* 54: 499–511.

Madsen, B.L. 1968. The distribution of nymphs of *Brachyptera
    risi* (Mort.) and *Nemoura flexuosa* Aub. (Plecoptera) in
    relation to oxygen. *Oikos* 19: 304–310.

Madsen, B.L. 1969. Reactions of *Brachyptera risi* (Morton)
    (Plecoptera) nymphs to water currents. *Oikos* 20: 95–100.

Madsen, B.L. 1974. A note on the food of *Amphinemoura
    sulcicollis*. *Hydrobiologia* 45: 169–175.

Madsen, B.L., Bengtson, J. and Butz, I. 1973. Observations on
    upstream migration by imagines of some Plecoptera and
    Ephemeroptera. *Limnol. Oceanogr.* 18: 678–681.

Maquire, B. Jr. 1963. The passive dispersal of small aquatic
    organisms and their colonization of isolated bodies of
    water. *Ecol. Monogr.* 33: 161–185.

Malmqvist, B., Nilsson, L.M. and Svensson, B.S. 1978. Dynamics
    of detritus in a small stream in southern Sweden and its
    influence on the distribution of the bottom animal
    communities. *Oikos* 31: 3–16.

Mann, K.H. 1961. The life history of the leech *Erpobdella
    testacea* Sav. and its adaptive significance. *Oikos*
    12: 164–169.

Mason, W.T., Anderson, J.B. and Morrison, G.E. 1967. A
    limestone-filled artificial substrate sampler-float unit
    for collecting macroinvertebrates in large streams.
    Progr. Fish. Cult. 29: 74.

McCormack, J.C. 1962. The food of young trout (*Salmo trutta*) in two different becks. *J. Anim. Ecol.* 31: 305-316.

McLachlan, A.J. 1969. Substrate preferences and invasion behaviour exhibited by larvae of *Nilodorum brevibucca* Freeman (Chironomidae) under experimental conditions. *Hydrobiologia* 33: 237-249.

McLachlan, A.J. and Cantrell, M.A. 1976. Sediment development and its influence on the distribution and tube structure of *Chironomus plumosus* L. (Chironomidae:Diptera) in a new impoundment. *Freshwat. Biol.* 6: 437-445.

Meadows, P.S. and Campbell, J.I. 1972. Habitat selection by aquatic invertebrates. *Adv. mar. Biol.* 10: 271-382.

Meijering, M.P.D. 1972. Experimental studies on drift and upstream and downstream movements of gammarids in running water. *Arch. Hydrobiol.* 70: 133-205.

Mestrov, M., Lattinger-Penko, R. and Tavcar, V. 1976. Dynamics of population in the isopod *Proasellus slavus* sp.n. and the larvae of chironomids in the hyporheic water of the River Drava with regard to pollution. *Int. J. Speleol.* 8: 157-166.

Mestrov, M. and Tavcar, V. 1972. Hyporheic as a selective biotope for some kinds of larvae of Chironomidae. *Bull. Sci. Conseil des Acad. des Sci. et Arts de la R.S.F. de Yougoslavie* 17: 7-8.

Miller, J.D. and Buikema, A.L. 1977. The effect of substrate on the distribution of the spring form (Form III) of *Gammarus minus* Say, 1818. *Crustaceana Suppl.* 4: 153-163.

Milliger, L.E. and Schlichting, H.E. Jr. 1968. The passive dispersal of viable algae and Protozoa by an aquatic beetle. *Trans. Amer. Microsc. Soc.* 87: 443-448.

Minckley, W.L. 1964. Upstream movements of *Gammarus* (Amphipoda) in Doe Run, Meade County, Kentucky. *Ecology* 45: 195-197.

Minshall, G.W. 1978. Autotrophy in stream ecosystems. *BioScience* 28: 766-771.

Minshall, G.W. and Minshall, J.N. 1978. Further evidence on the role of chemical factors in determining the distribution of benthic invertebrates in the River Duddon. *Arch. Hydrobiol.* 83: 324-355.

Momot, W.T. 1966. Upstream movement of crayfish in an intermittent Oklahoma stream. *Amer. Midl. Nat.* 75: 150-159.

Moore, J.W. 1977. Seasonal succession of algae in rivers II. Examples from Highland Water, a small woodland stream. *Arch. Hydrobiol.* 80: 160-171.

Muller, K. 1954a. Investigations on the organic drift in North Swedish streams. *Rep. Inst. Freshw. Res. Drottningholm* 35: 133-148.

Muller, K. 1954b. Die Drift in fliessenden Gewässern. *Arch. Hydrobiol.* 49: 539-545.

Muller, K. 1966. Die Tagesperiodik von Fliesswasserorganismen.
    Z. *Morph. Okol. Tiere*. 56: 93-142.
Muller, K. 1974. Stream drift as a chronobiological
    phenomenon in running water ecosystems. *Ann. Rev. Ecol.
    Syst*. 5: 309-323.
Mundie, J.H. 1974. Optimization of the salmonid nursery
    stream. *J. Fish. Res. Bd. Can*. 31: 1827-1837.
Mundie, J.H., Mounce, D.E. and Smith, L.S. 1973. Observations
    on the response of zoobenthos to additions of hay, willow
    leaves, alder leaves and cereal grain to stream substrates.
    Fish. Res. Bd. Can. Tech. Rep. 387: 1-123.
Neave, F. 1930. Migratory habits of the mayfly, *Blasturus
    cupidus* Say. *Ecology* 11: 568-576.
Needham, J.G., Travers, J.R. and Hsu, Y-C. 1935. The biology
    of mayflies. Comstock Pub. Co. Inc., N.Y.
Neel, J.K. 1968. Seasonal succession of benthic algae and
    their macroinvertebrate residents in a head-water
    limestone stream. *J. Wat. Polln. Contr. Fed*. 40: 10-30.
Nelson, D.J. and Scott, D.C. 1962. Role of detritus in the
    productivity of a rock-outcrop community in a Piedmont
    stream. *Limnol. Oceanogr*. 7: 396-413.
Noel, M.S. 1954. Animal ecology of a New Mexico spring brook.
    Hydrobiologia 6: 120-135.
Odum, E.P. 1971. Fundamentals of ecology. Saunders, Toronto.
Olsson, T. and Soderström, O. 1978. Springtime migration and
    growth of *Parameletus chelifer* (Ephemeroptera) in a
    temporary stream in northern Sweden. *Oikos* 31: 284-289.
Orghidan, T. 1953. Un nou demeniu de viata acvatica subterana
    "Biotopul hiporeic". *Bull. Sti. sect. Biologie si sti
    Agronau si sect. geologie si geogr. Acad. R.P.R.T*. 7: 3.
Oswood, M.W. 1976. Comparative life histories of the
    Hydropsychidae (Trichoptera) in a Montana lake outlet.
    Amer. Midl. Nat. 96: 493-497.
Patrick, R. 1964. A discussion of the results of the
    Catherwood Expedition to the Peruvian headwaters of the
    Amazon. *Verh. int. Verein. theor. angew. Limnol*.
    15: 1084-1090.
Pearson, W.D. and Franklin, D.R. 1968. Some factors affecting
    drift rates of *Baetis* and Simuliidae in a large river.
    Ecology 49: 75-81.
Peckarsky, B.L. 1979. Biological interactions as determinants
    of distributions of benthic invertebrates within the
    substrate of stony streams. *Limnol. Oceanogr*. 24: 59-68.
Peckarsky, B.L. 1980. Predator-prey interactions between
    stoneflies and mayflies: behavioural observations.
    Ecology 61: 932-943.
Pennak, R.W. 1953. Freshwater invertebrates of the United
    States. Ronald Press, N.Y.

Pennak, R.W.   1958.   Some problems of freshwater invertebrates
    in the western States. *In:* Zoogeography. Hubbs, C.L. ed.
    Amer. Assn. Adv. Sci.   223-230.

Philipson, G.N.   1954.   The effect of water flow and oxygen
    concentrations on six species of caddisfly (Trichoptera).
    Proc. Zool. Soc. Lond.   124: 547-564.

Poole, W.C. and Stewart, K.W.   1976.   The vertical distribution
    of macrobenthos within the Brazos River, Texas.
    Hydrobiologia   50: 151-160.

Proctor, V.W.   1964.   Viability of crustacean eggs recovered from
    ducks. *Ecology*   45: 656-658.

Proctor, V.W., Malone, C.R. and Devlaming, V.L.   1967.
    Dispersal of aquatic organisms: viability of disseminules
    recovered from the intestinal tracts of captive killdeer.
    Ecology   48: 672-676.

Provost, N.W.   1952.   The dispersal of *Aedes taeniorhynchus*
    I. Preliminary studies. *Mosquito News*   12: 174-190.

Rabeni, C.F. and Minshall, G.W.   1977.   Factors affecting
    microdistribution of stream benthic insects. *Oikos*   29:
    33-43.

Rauser, J.   1962.   Zur Verbreitungsgeschichte einer Insekten-
    dauergruppe (Plecoptera) in Europe. *Pr. brn. Zakl. csl.*
    Akad. Ved.   34: 281-383.

Rees, W.J.   1952.   The role of Amphibia in the dispersal of
    bivalve molluscs. *Brit. J. Herpet.*   1: 7.

Reisen, W.K. and Prins, R.   1972.   Some ecological relationships
    of the invertebrate drift in Praters Creek, Picken County,
    South Carolina. *Ecology*   53: 876-885.

Revill, D.L., Stewark, K.W. and Schlichting, H.E. Jr.   1967.
    Passive dispersal of viable algae and Protozoa by certain
    craneflies and midges. *Ecology*   48: 1023-1027.

Ricker, W.E.   1934.   An ecological classification of certain
    Ontario streams. *Univ. Toronto Stud. Biol. Ser.*   37:
    1-114.

Riley, C.F.C.   1920.   Migratory responses of water striders
    during severe drought. *Bull. Brooklyn Ent. Soc. N.Y.*
    15: 1-10.

Robertson, I.C. and Blakeslee, C.L.   1948.   The Mollusca of the
    Niagara frontier region. *Bull. Buffalo Soc. Nat. Sci.*
    19: 1-191.

Roos, T.   1957.   Studies on upstream migration in adult stream-
    dwelling insects I. *Rep. Inst. Freshw. Res. Drottningholm*
    38: 167-192.

Rosine, W.   1956.   On the transport of the common amphipod
    Hyallela azteca in South Dakota by the mallard duck.
    Proc. South Dakota Acad. Sci.   35: 203.

Ross, H.H.   1956.   Evolution and classification of the mountain
    caddisflies. Univ. Illinois Press, Urbana.

Ross, H.H.   1963.   Stream communities and terrestrial biomes.
    Arch. Hydrobiol.   59: 235-242.

Saether, O.A. 1969. Some nearctic Podonominae, Diamesinae and
    Orthocladiinae (Diptera:Chironomidae). *Fish. Res. Bd. Can.*
    Bull. 170: 1-154.
Schneller, M.V. 1955. Oxygen depletion in Salt Creek, Indiana.
    Invest. Indiana Lakes Streams 4: 163-175.
Schwarz, P. 1970. Autecological investigations on the life cycle
    of Setipalpia (Plecoptera) II. Experimental work. *Arch.*
    Hydrobiol. 67: 141-172.
Schwoerbel, J. 1961. Über die Lebensbedingungen und die
    Besiedlung des Hyporheischen Lebensraumes. *Arch. Hydrobiol.*
    Suppl. 25: 182-214.
Schwoerbel, J. 1967. Das hyporheische Interstitial als
    Grenzbiotop zwischen oberirdischem und subterranen
    Ökosystem und seine Bedeutung fur die Primar-Evolution
    von Kleinsthohlen-bewohnern. *Arch. Hydrobiol.* 33: 1-62.
Shelford, V.E. 1913. Animal communities in temperate America.
    Univ. Chicago Press.
Shelly, T.E. 1979. The effect of rock size upon the
    distribution of species of Orthocladiinae (Chironomidae:
    Diptera) and *Baetis intercalaris* McDunnough (Baetidae:
    Ephemeroptera). *Ecol. Ent.* 4: 95-100.
Silver, P.A. 1977. Comparison of attached diatom communities
    on natural and artificial substrates. *J. Phycol.* 13: 402-
    406.
Sprules, W.M. 1947. An ecological investigation of stream
    insects in Algonquin Park, Ontario. *Univ. Toronto Stud. Biol.*
    Ser. 56: 1-81.
Stanford, J.A. and Gaufin, A.R. 1974. Hyporheic communities of
    two Montana rivers. *Science* 185: 700-702.
Steele, V.J. 1967. Resting stage in the reproductive cycles of
    Gammarus. Nature, Lond. 214: 1034.
Steine, I. 1972. The number and size of drifting nymphs of
    Ephemeroptera, Chironomidae and Simuliidae by day and
    night in the River Strenda, western Norway. *Norsk. ent.*
    Tidsskr. 19: 127-131.
Stewart, K.W., Milliger, L.E. and Solon, B.M. 1970. Dispersal
    of algae, protozoans and fungi by aquatic Hemiptera,
    Trichoptera and other aquatic insects. *Ann. Ent. Soc. Amer.*
    63: 139-144.
Stoneburner, D.L. and Smock, L.A. 1979. Seasonal fluctuations
    of macroinvertebrate drift in a South Carolina piedmont
    stream. *Hydrobiologia* 63: 49-56.
Stout, J. and Vandermeer, J. 1975. Comparison of species richness
    for stream-inhabiting insects in tropical and mid-latitude
    streams. *Amer. Nat.* 109: 263-280.
Sturrock, R.F. 1973. Field studies on the population dynamics
    of *Biomphalaria glabrata*, intermediate host of *Schistosoma*
    mansoni on the West Indian island of St. Lucia. *Int. J.*
    Parasit. 3: 165-174.

Sutcliffe, D.W. and Carrick, T.R. 1973. Studies on mountain streams in the English Lake District. III. Aspects of water chemistry in Brownrigg Well, Whelpside Ghyll. *Freshwat. Biol.* 3: 561-568.

Svensson, B.W. 1974. Population movements of adult Trichoptera at a south Swedish stream. *Oikos* 25: 157-175.

Thienemann, A. 1950. The transport of aquatic animals by birds. *Binnengewässer* 18: 156-159.

Thomas, E. 1966. Orientierung der Imagines von *Capnia atra* Morton (Plecoptera). *Oikos* 17: 278-280.

Thorne, M.J. 1969. Behaviour of the caddisfly larva *Potamophylax stellatus* (Curtis) (Trichoptera). *Proc. R. ent. Soc. Lond. A.* 44: 91-110.

Thorp, J.H. 1978. Agonistic behaviour in crayfish in relation to temperature and reproductive period. *Oecologia* 36: 273-280.

Thorup, J. 1966. Substrate type and its value as a basis for the delimitation of bottom fauna communities in running waters. *Pymatuning Lab. of Ecol. Spec. Pubn.* 4, Univ. Pittsburg. 59-74.

Thorup, J. 1974. Occurrence and size-distribution of Simuliidae (Diptera) in a Danish spring. *Arch. Hydrobiol.* 74: 316-335.

Tolkamp, H.H. and Both, J.C. 1978. Organism-substrate relationship in a small Dutch lowland stream. Preliminary results. *Verh. int. Verein. theor. angew. Limnol.* 20: 1509-1515.

Towns, D.R. 1979. Composition and zonation of benthic invertebrate communities in a New Zealand Kauri forest stream. *Freshwat. Biol.* 9: 251-262.

Townsend, C.R. and Hildrew, A.G. 1976. Field experiments on the drifting, colonization and continuous redistribution of stream benthos. *J. Anim. Ecol.* 45: 759-773.

Townsend, C.R. and Hildrew, A.G. 1979. Foraging strategies and coexistence in a seasonal environment. *Oecologia* 38: 231-234.

Travers, S.R. 1925. Observation on the ecology of the mayfly *Blasturus cupidus*. *Can. Ent.* 57: 211-218.

Trivellato, D. and Decamps, H. 1968. L'influence de quelques obstacles simple sur l'ecoulement dans un ruisseau experimental. *Annls. Limn.* 4: 357-386.

Ulfstrand, S., Nilsson, L.M. and Stergar, A. 1974. Composition and diversity of benthic species collectives colonizing implanted substrates in a south Swedish stream. *Ent. Scand.* 5: 115-122.

Verrier, M.-L. 1956. Biologie des Ephémères. Collection Armand, Colin, Paris.

Wallace, R.L. 1978. Substrate selection by larvae of the sessile rotifer *Ptygura beauchampi*. *Ecology* 59: 221-227.

Walton, O.E., Reice, S.R. and Andrews, R.W. 1977. The effects
    of density, sediment particle size and velocity on drift
    of *Acroneuria abnormis* (Plecoptera). *Oikos*   28: 291-298.

Ward, J.V. and Berner, L. 1980. Abundance and altitudinal
    distribution of Ephemeroptera in a Rocky Mountain stream.
    In: Proc. 3rd Int. Conf. on Ephemeroptera. Flannagan, J.F.
    and Marshall, K.E. eds. Plenum Press, N.Y. 169-177.

Ward, G.M. and Cummins, K.W. 1979. Effects of food quality on
    growth of a stream detritivore, *Paratendipes albimanus*
    (Meigen) (Diptera:Chironomidae). *Ecology*   60: 57-64.

Warren, C.E., Wales, J.H., Davis, G.E. and Doudoroff, P.
    1964. Trout production in an experimental stream
    enriched with sucrose. *J. Wildlf. Manag.* 28: 617-660.

Waters, T.F. 1964. Recolonization of a denuded stream bottom
    area by drift. *Trans. Amer. Fish. Soc.* 93: 311-325.

Waters, T.F. 1965. Interpretation of invertebrate drift in
    streams. *Ecology*   46: 327-334.

Waters, T.F. 1972. The drift of stream insects. *Ann. Rev. Ent.*
    17: 253-272.

Westlake, D.F. 1975. Macrophytes. *In:* River ecology. Whitton,
    B.A. ed. Studies in Ecol. 2, Univ. California Press.
    106-128.

White, R.J. and Brynildson, O.M. 1967. Guidelines for
    management of trout stream habitat in Wisconsin. *Dept. Nat.*
    Res. Madison, Wisconsin, Div. Conserv. Tech. Bull.
    39: 1-65.

Whitton, B.A. 1975. Algae. *In:* River ecology. Whitton, B.A.
    ed. Studies in Ecol. 2, Univ. California Press. 81-105.

Wiggins, G.V. 1977. Larvae of the North American caddisfly
    genera (Trichoptera). Univ. Toronto Press.

Wiggins, G.B. and Mackay, R.J. 1978. Some relationships
    between systematics and trophic ecology in Nearctic aquatic
    insects, with special reference to Trichoptera. *Ecology*
    59: 1211-1220.

Williams, D.D. 1977. Movements of benthos during the
    recolonization of temporary streams. *Oikos*   29: 306-312.

Williams, D.D. 1980a. Invertebrate drift lost to the sea
    during low flow conditions in a small coastal stream in
    Western Canada. *Hydrobiologia*   75: 251-254.

Williams, D.D. 1980b. Some relationships between stream benthos
    and substrate heterogeneity. *Limnol. Oceanogr.*   25:
    166-172.

Williams, D.D. 1980c. Temporal patterns in recolonization of
    stream benthos. *Arch. Hydrobiol.* 90: 56-74.

Williams, D.D. 1981. Emergence pathways of adult insects in the
    upper reaches of a stream. *Int. Revue ges. Hydrobiol.*
    (in press)

Williams, D.D. and Hynes, H.B.N. 1974. The occurrence of benthos deep in the substratum of a stream. *Freshwat. Biol.* 4: 233–256.

Williams, D.D. and Hynes, H.B.N. 1976a. The recolonization mechanisms of stream benthos. *Oikos* 27: 265–273.

Williams, D.D. and Hynes, H.B.N. 1976b. The ecology of temporary streams: I. The faunas of two Canadian streams. *Int. Revue ges. Hydrobiol.* 61: 761–787.

Williams, D.D. and Hynes, H.B.N. 1977. Benthic community development in a new stream. *Can. J. Zool.* 55: 1071–1077.

Williams, D.D. and Moore, K.A. 1981. Activity of *Gammarus pseudolimnaeus* (Amphipoda) as affected by its environment (in press).

Williams, D.D. and Mundie, J.H. 1978. Substrate size selection by stream invertebrates, and the influence of sand. *Limnol. Oceanogr.* 23: 1030–1033.

Williams, D.D., Mundie, J.H. and Mounce, D.E. 1977. Some aspects of benthic production in a salmonid rearing channel. *J. Fish. Res. Bd. Can.* 34: 2133–2141.

Williams, N.E. and Hynes, H.B.N. 1973. Microdistribution and feeding of the net-spinning caddisflies (Trichoptera) of a Canadian stream. *Oikos* 24: 73–84.

Williams, N.E. and Morgan, A.V. 1977. Fossil caddisflies (Insecta:Trichoptera) from the Don Formation, Toronto, Ontario, and their use in paleoecology. *Can. J. Zool.* 55: 519–527.

Wilson, D.P. 1970. The larvae of *Sabellaria spinulosa* and their settlement behaviour. *J. mar. biol. Ass. U.K.* 50: 33–52.

Wise, D.H. and Molles, M.C. 1979. Colonization of artificial substrates by stream insects: influence of substrate size and diversity. *Hydrobiologia* 65: 69–74.

Yongue, W.H. Jr. and Cairns, J. Jr. 1978. The role of flagellates in pioneer protozoan colonization of artificial substrates. *Polsk. Arch. Hydrobiol.* 25: 787–801.

TEMPORAL HETEROGENEITY AND THE ECOLOGY OF LOTIC CILIATES

William D. Taylor

National Water Research Institute
Canada Centre for Inland Waters
Burlington, Ontario, Canada

INTRODUCTION

Ciliated protozoa are among the smallest consumers in aquatic ecosystems. Although they are most often studied as predators of bacteria, other foods such as algae, dissolved organic material, other protozoa, and even small invertebrates or the tissue of large animals sustain various free-living species. Almost all of these foods consumed by ciliates are also consumed by larger organisms. In streams, protozoa no doubt compete with, and are eaten by, an array of filter feeding and deposit feeding collectors (Cummins and Klug, 1979; Wallace and Merritt, 1980). To understand the ciliates in the context of the total stream fauna, one must focus upon the two obvious and important attributes in which they differ from other consumers; size and rate of multiplication. One type of habitat where small, rapidly growing organisms may be favored is where there is intense temporal heterogeneity.

Temporal heterogeneity is an important feature of the environment of most organisms, and the way in which they cope with and exploit it is an important part of their interaction with the environment. To discuss the impact of temporal heterogeneity, it is essential to distinguish between predictable and unpredictable events. Predictable events, such as annual or diurnal cycles, often correspond to rhythms in the physiology, behavior and life history of organisms. Unpredictable events are different in that they require a spontaneous response. In the following discussion I will describe the response of ciliates, first as individuals and populations, to the stochastic or unpredictable component of environmental variation. I will then deal with the relationship of environmental uncertainty to the structure of ciliate

communities and the role of ciliates among the rest of the stream
fauna. My emphasis upon ciliates reflects both my own interests
and the relatively abundant information on this part of the micro-
fauna. The discussion should be relevant to other small, heterotrophic organisms, such as zooflagellates and rotifers.

Before proceeding, it is necessary to further describe and
classify temporal heterogeneity. Environmental changes may be
negative, causing mortality, or positive, as in enrichment.
Although these types of events are in a sense opposites, they are
alike in a broader context. A sudden enrichment or an episode of
mortality, if severe enough, will result both in a period of high
resource (e.g., bacteria) levels and density-independent growth of
consumers. A community receiving stochastic enrichments can also
be considered as enduring stochastic episodes of low resource levels
and starvation. For many purposes, these two types of environmental
change can be considered together as disturbances. For the purpose
of this discussion, disturbance can be defined as environmental
change which occurs suddenly enough to alleviate resource
limitation or density-dependence.

A more important distinction is between local and global events.
When disturbance occurs on spatial scales which are large relative
to the organisms involved, then they must be able to endure, either
in an active or domant (hypobiotic) state, the full intensity and
duration of unfavorable periods. When temporal heterogeneity is
patchy on a sufficiently small scale, and asynchronous among patches,
organisms have the alternate option of emmigrating. For example,
for attached protozoans living in the riffles of a stream, episodes
of mortality due to spating are global phenomena; all patches will
be affected simultaneously, and therefore some individuals must
persist through the spate. But for protozoans living on the sediment
surface in a slower reach, local enrichments due to the deposition
of dead organisms from the seston will certainly be patchy and out
of phase. Some species may feed only in these high density patches,
and persist by using their resultant high rate of multiplication to
produce enough progeny to ensure that one will find a new patch.
Nevertheless, some way of persisting through periods of low resource
levels and other temporary stresses will be an important feature of
the biology of most species.

MECHANISMS OF PERSISTENCE

Many ciliates have obvious mechanisms for persisting through
unfavorable periods. Encystment allows some species to resist
physical and chemical stresses, and also to reduce metabolic losses
and the risk of predation. Terrestrial protozoa have in common the
ability to encyst to avoid desiccation. In continuously moist
habitats, the ability to encyst is less common. (but note the
importance of terrestrial protozoa in some streams (Gray, 1952)).

Nevertheless, many aquatic species form cysts in response to exhaustion of their food supply, rather than searching for a new area. As in the case of *Didinium nasatum*, these cysts may not be resistant to desiccation. Other ciliates, such as species of *Ophryoglena*, *Prorodon*, or *Colpoda*, may encyst prior to digestion or cell division. This presumably reduces the risk of mortality while the animal is incapable of feeding. *Tetrahymena corlissi* divides in a cyst when its food is depleted, but not otherwise (Lynn, 1975).

Those ciliates which live in continuously moist habitats, but do not form cysts, may rely on reserves and metabolic conservatism to last until food is again present (see Poljanskij and Chejsin (1965) for a review of cysts and reserve materials). The optimal use of reserves in microbes has been discussed by Parnas and Cohen (1976) and Calow and Jennings (1977). Obviously, it may be advantageous for an organism in a temporally uncertain but spatially homogeneous environment to accumulate food reserves even at the expense of reproduction. Luckinbill and Fenton (1978) found that in *Paramecium* populations, larger individuals survive population crashes while smaller, presumably recently divided cells, may not. *Paramecium* took longer to starve than the smaller *Colpidium campylum*. Exclusive of encysting forms, starvation time probably increases with body size among ciliates as it does among crustacean zooplankton (Threlkeld, 1976). This may contribute to a conflict between selective pressures for two attributes usually considered together, dispersibility and reproductive rate (Gill, 1978), especially as swimming speed also generally increases with size.

Mechanisms of persistence other than encystment, food reserves, and large size may include loricas to reduce predation and withstand mechanical stresses, rigid stalks to withstand strong currents among peritrichs (Taylor and Barton, in preparation) and defensive trichocysts. Some ciliates discourage predators by being less palatable (Taylor, 1980; Berger, 1980).

EXPLOITATION OF FAVORABLE PERIODS

The response of an organism to the sudden availability of food may be separated into two components: the functional (feeding) response and the numerical (reproductive) response. The functional response of a protozoan may begin with a change in behavior and morphology. Some species, e.g., the histiophagous *Tetrahymena corlissi* (Lynn, 1975) and *Ophryoglena* spp. (Canella and Canella, 1976) undergo a pronounced morphological transformation from the more streamlined theront or hunting morphology to the feeding trophont morphology. This is coincident with a change in behavior from a relatively straight, searching locomotion to an apparently frantic stopping, turning and starting until the exact location of the food is found. In bactivorous hymenostomes, the immediate morphological transformation is absent, but the behavioral

transformation from rapid, directional locomotion to local search
and feeding is apparent in some species.

The initiation of the functional response has been studied in
Tetrahymena pyriformis. The importance of both chemical and
mechanical cues has been demonstrated (Ricketts, 1972a, b). In
peritrichs, Vielleux (unpublished manuscript) has shown that
species differ in the relative degree to which they respond to
chemical and mechanical stimuli. The necessity for a stimulus
for phagocytosis in bactivorous forms implies a critical concentration, which in turn reflects that functional response has a
threshold; there is a minimum food concentration at which
phagocytosis and growth will begin (see Taylor, 1978a). On
encountering an abundance of food, initial uptake by phagocytosis
may proceed immediately (Fig. 1). After a period of very rapid
uptake, functional response will decline to the maximum rate at
which the collected material can be processed (Ricketts, 1971).

The numerical response of a ciliate to a sudden abundance of
food can be studied in batch culture. On inoculation into fresh
culture medium, previously starving individuals will undergo a lag
phase. This may be considered a period of purely functional

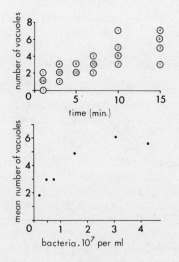

Figure 1. The response of starving *Colpidium campylum* to a sudden
           introduction of prey bacteria. The upper graph shows
           the observed numbers of food vacuoles per cell after
           various times. The lower graph shows the mean number
           of food vacuoles formed in 15 minutes at various
           bacterial densities. (Taylor and Berger, unpublished.)

response. In *Colpidium campylum* (Taylor and Berger, 1976) biomass
growth begins very quickly; the size of the individual ciliates
increases dramatically. The lag phase is only evident when cell
numbers rather than total cell volume are considered. It appears
that cell division (which requires a cessation of feeding) is
postponed as long as possible. This response pattern is
illustrated in Figure 2. Although mechanistic interpretations of
this phenomenon can be put forward (e.g., Williams, 1971) it may
also serve to maximize exploitation of an ephemeral food resource;
if the food is exhausted during the lag or division delay period,
the ciliate will have increased its share relative to competitors
by not dividing. The histiophagous ciliates, which exploit perhaps
the most patchy and ephemeral food resource among free-living
aquatic species, undergo a prolonged feeding period on encountering
food. There is a great increase in cell size, followed by
multiple cell divisions within a reproductive cyst. The
terrestrial colpodids, which are active when soil or vegetation
are moistened by rain, display similar, although perhaps less
dramatic, reproductive patterns. In other forms, maximum cell
size is typically seen at the end of lag phase. This phenomenon is
reminiscent of the observations of Healy (1979). 'When nutrient
limited algae are suddenly supplied with nutrients, carbon fixation
is depressed rather than enhanced, as the cells apparently
sacrifice photosynthesis to maximize their immediate rate of
nutrient uptake and storage. A lag occurs before growth is enhanced.

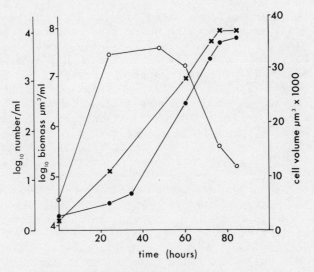

Figure 2. The growth of *Colpidium campylum* in batch culture.
           Solid circles represent cell numbers: crosses – total
           biomass: open circles – mean cell volume. (Redrawn from
           Taylor et al. 1976.)

The other interesting component of numerical response
evident in batch cultures is the response to exhaustion of food
supply. Many ciliates respond by continuing reproduction at the
expense of cell size, e.g., *Tetrahymena pyriformis* (see Cameron,
1973; Lynn 1975), *Colpidium campylum* (Taylor and Berger, 1976) and
Uronema sp. (Hamilton and Preslan, 1969). This presumably
maximizes the number of dispersing forms which might find a new
patch. Cameron and Bols (1975) noted that in populations of
T. pyriformis arrested by starvation, most cells were in the G_1
phase of the cell cycle, as one might expect if the number of
progeny produced was maximized. However, at higher population
densities, a larger proportion of cells was arrested in G_2. These
cells could divide sooner on refeeding. Again, both mechanistic
and adaptive interpretations are possible, and should not be
considered mutually exclusive.

The numerical response of ciliates, of course, has an upper
boundary. It is not limited by functional response; at high food
densities the process of ingestion in *T. pyriformis* becomes
periodic (Ricketts, 1971) and the size of the buccal overture
(mouth) may even decrease (Taylor et al. 1976). Rather, upper
limit to numerical response is determined extrinsically by
temperature and intrinsically by body size (Fenchel, 1968; Finlay,
1977) and genome size (Taylor and Shuter, in press).

PERSISTENCE VERSUS OPPORTUNISM

At the beginning of this essay, I suggested that ciliates
differ from other consumers with which they compete by virtue of
their small size and high potential rate of reproduction. Even for
their size, ciliates are relatively fast-growing, probably because
of nuclear dualism (Taylor and Shuter, in press). But even taking
size into consideration, some ciliates can multiply much faster
than others. Although batch culture studies indicate that ciliates
overshoot the carrying capacity of their environment, and continue
to multiply even at the expense of their size (see above), these
studies are usually conducted on a limited number of easily-
cultured species. They may not be representative. Luckinbill and
Fenton (1978) carefully compared the very fast-growing *Colpidium
campylum* with the larger, slower-growing *Paramecium primaurelia*.
The latter does not reach carrying capacity so quickly, arrests
cell division sooner in depleted culture medium, and assumes more
even densities in a periodic environment. *C. campylum* consistently
overshoots equilibrium, undergoes wild fluctuations in a periodic
environment, and therefore tends to crash to extinction. Although
the latter strategy is no doubt inferior for survival in a test tube,
in an open and spatially heterogeneous system it is a viable
alternative. The large numbers of individuals at the peaks of
population fluctuations would be dispersers in an open system.

Paramecium primaurelia can be considered more adept at persisting
in a spatially uniform but temporally varying habitat, but
Colpidium campylum may be more likely to survive in a habitat of
ephemeral patches. The relatively conservative strategy of
P. primaurelia is particulary interesting in that some of its
predators, notably *Didinium nasutum* and *Woodruffia metabolica*, are
among those species which form resting cysts and are therefore
committed to the waiting rather than searching strategy.

The conservative nature of *Paramecium* relative to *Colpidium*
leads to another issue. Although I have been discussing how
ciliates cope with and exploit temporal heterogeneity, I should
also mention that not all the organisms in a community, not even
all the protozoa, may be dependent upon temporal heterogeneity.
In real ecosystems, although perhaps not in batch cultures, blooms
and periods of exponential growth subside and are replaced by
conditions of relatively low resource levels by constant recycling.
Although opportunistic species may have encysted, be slowly
starving, or have emmigrated, a few species will be able to grow
sufficiently to offset population losses. These equilibrium
species will control resource levels during relatively stable
periods, thereby excluding opportunistic species which may require
higher levels. The number of such equilibrium species which can
be supported by a given number of discrete resources, or which can
be packed along a resource continuum, has received much attention
(Pianka, 1976).

Another topic which has received a great deal of attention is
the characteristics of opportunistic versus equilibrium species, or
r- and K-strategists as they are frequently called. McArthur and
Wilson (1967) introduced these terms, suggesting that organisms
relying on colonization of new habitats would undergo selection
for high r (intrinsic rate of natural increase), while organisms
living in stable communities would be selected for large K
(carrying capacity). K was taken as a measure of an organism's
ability to compete for resources. These parameters were expected
to be negatively correlated, creating a continuum between highly
r-selected and K-selected species.

Luckinbill (1979) used ciliates to test the validity of the
r- K selection hypothesis. He found that although r is negatively
related to competitive ability, as the hypothesis requires, so is
K. As one might expect, K was negatively related to body size
and therefore positively related to r. Earlier studies by Vandermeer
(1969) and Gill (1972) found no consistent relationship. The
lack of relationship between K and competitive ability is not
surprising; K is determined largely by body size and also by the
efficiency with which resources are converted into biomass.
Neither is directly related to competitive ability. There is a

lack of theoretical (Stearns, 1977) and empirical (Luckinbill, 1978) grounds for the belief that K is the compliment of r. The maximum rate of population growth, r, is no doubt related to a ciliate's fitness as an opportunist, but again is complicated by a correlation with body size (Fenchel, 1968, 1974) as are other characteristics related to reproduction and reproductive strategy (Kaplan and Salthe, 1979). Taylor (1978b) considered the residual variation in r not accounted for by body size as an index of a species' position along the continuum from opportunists to equilibrium species. Among the bactivorous ciliates in a small pond, relatively slow-growing species were common, while species relatively fast-growing for their size were relatively rare (Fig. 3). Rare species were thought to be opportunistic, and common species prevalent during relatively stable conditions. The four slow-growing and common species (*Cyclidium glaucoma, Cinetochilum margaritaceum, Halteria grandinella,* and *Paramecium bursaria*), although all bactivorous, appear to be quite different in their mode of feeding and nutrition, as one might expect of equilibrium species. Taylor and Shuter (in press) have found that this residual variation in r not attributable to body size is related to the size of the micronucleus. Ciliates which are relatively fast-growing for their size have relatively small micronuclei (genomes) for their size, and *vice versa*. This relationship promises a useful alternative measure of the degree to which a species has been "r-selected" for high growth rate. It is interesting to note that laboratory data on feeding and growth are available mostly for relatively fast-growing, presumably opportunistic ciliates, such as *Tetrahymena, Paramecium, Colpidium,* and *Uronema*.

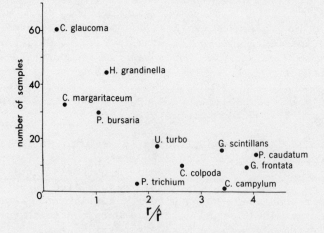

Figure 3.   The commonness of several ciliate taxa versus the growth index r/r̂ (observed maximum rate of population growth divided by the rate expected for a ciliate of that size). (Redrawn from Taylor 1978b.)

REPRODUCTIVE STRATEGIES AND COMMUNITY ECOLOGY

The study of life history strategies and the r-K continuum has largely been concerned with the attributes of species in constant versus variable environments. However, most natural communities comprise both equilibrium species and species which are opportunistic, although the relative abundance or commonness of these kinds of species will vary. It is not adequate to assume that all species from harsh environments will be opportunistic weeds, while species from benign environments will be competitively superior. Environmental variability and disturbance should be taken as the independent variables, not a subjective assessment of severity. It is variance that is of interest, not the mean. Chronic stress and episodes of disturbance should have very different effects on the structure of communities.

The recent theoretical work by Levins (1979) serves as a foundation for understanding the relationship of environmental variation to coexistence of species with different response times. He demonstrated that just as different discreet resources can each allow the existence of a species under equilibrium conditions, variance in each of those resources can support additional species. To expand further, higher moments of the mean and covariances, if distinct, could all support additional species, as if they were discreet resources. A stream ecologist might rather consider that environmental variation, expressed in terms of frequency as in a power spectrum, represents a resource continuum which can be divided among species. For each resource, or combination of resources, one might envision continuous spectra of variance in both temporal and spatial scales, which are the resources of species with different capacities for migration and functional and numerical response. This view has produced new insights into the ecology of phytoplankton (Harris, 1980).

To return to assemblages of lotic ciliates, we would expect to find species which persist at equilibrium (in some relative sense) and species which exploit higher moments or frequencies of environmental variation.

In the pond study mentioned above (Taylor, 1978b; Fig. 3), relatively slow-growing species were common, and species designated as relatively fast-growing on the basis of their r_m and size were relatively rare. This relationship contains implicit predictions about the response of this community to perturbations; following an enrichment or sudden episode of mortality, species from the right side of Fig. 3 might appear. Under stable conditions, species from the left would be expected alone. If one could manipulate this community to increase or decrease the frequency or severity of disturbance, the relative commonness of the two

types of species should shift accordingly. These phenomena
could lead to a relationship between species richness and disturbance as outlined by Connell (1978), and illustrated in Fig. 4.

Communities which are not subject to disturbances should
contain a relatively low number of equilibrium species, each
limited by a different resource or using a particular part of the
resource spectrum. Some disturbance will cause an increase in
species richness by giving other species opportunities to grow.
As disturbance increases, more resources will be captured by
opportunistic species, and slower, equilibrium species may not be
able to recover between disturbances, despite the intervening
periods of high (out of control) resource levels. Eventually,
species richness may decline as only a few very fast-growing
species can persist in the face of frequent episodes of
mortality. This relationship between disturbance frequency and
species richness can also be understood in terms of species
richness of communities of different average ages. New or well
established communities might not be as rich in species as ones
intermediate in age (see Sousa (1980) and references therein).

There are some data suggesting ciliate communities might
follow this model. Eddison and Ollason (1978) found that
laboratory communities of ciliates maintained with a fluctuating
temperature regime had increased species diversity compared to
communities kept at a constant temperature. Cairns et al. (1971)
illustrated that protozoan communities in new artifical substrates
(sponges) increased in species richness until an equilibrium

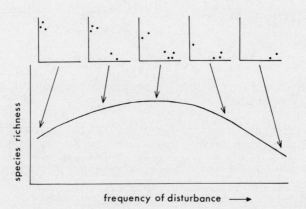

Figure 4.   A diagram to illustrate the possible relationship
            between Figure 3 and the intermediate disturbance
            hypothesis. The upper graphs illustrate the commonness
            of species with different values of $r/\hat{r}$ as in Figure 3.

number of species was reached. If these communities were disturbed
by squeezing the sponges, a short-term increase in species richness
occurred, followed by a return to the former level. A recent
comparative study on the sessile, filter-feeding ciliates of several
small streams (Taylor and Barton, in preparation) has found that
stable rural streams and streams below reservoirs had fewer species
than more spate-prone and intermittent streams. In extremely spate-
prone urban streams, species richness was also low. Highest
richness occurred in streams intermediate in their frequency and
severity of disturbance from spating. Similar relations between
richness and habitat stability were observed in macroinvertebrates
and fish. Macroinvertebrates appeared to be more severely affected,
perhaps because their generation times are closest to the frequency
of the disturbance.

The succession of protozoan species which occurs when a
container of pond or stream water is collected and enriched with
organic material illustrates many of these points. The disturbance
of enrichment (or frequently collection alone is sufficient)
causes a temporary increase in the numbers of individuals and
species (Legner 1973, 1975), although severe enrichments may reduce
richness (Legner, 1975). Different degrees of enrichment cause
different species to appear. Legner (1975), in his studies on
enrichments of water from rivers and impoundments, found that lesser
enrichments favored *Cyclidium glaucoma*, which is slow-growing for
its size, while severe enrichments favored the relatively fast-
growing *Colpidium campylum*. The order of appearance of the species
in the various succession experiments is predicted by their highest
observed specific growth rates.

Enriched field collections, if allowed to proceed through
succession long enough, will be taken over by relatively slow-
growing ciliates, especially hypotrichs, then eventually by the
faster growing metazoa; rotifers, gastrotrichs, naidids,
cladocerans, etc. This succession of species, which presumably
leads to the replacement of faster species by increasingly better
competitors, emphasizes the continuum of response times evident
among consumers and the difficulty of dealing simply with
equilibrium and perturbation. Systems might be in a state of
transition on coarse time scales (e.g., years) but essentially
constant at fine scales (e.g., days) or, conversely, a system
with intense fine-scale temporal and spatial heterogeneity might
be relatively constant year to year. The spectrum of frequencies
characteristic of a given environment might fruitfully be compared
to the response times and amplitudes of its biota. Particular
frequencies and amplitudes may be more common in some habitats
than others, and this may account for variations in the relative
abundance of certain taxa.

Figure 5 is an attempt to map, very approximately, the potential functional and numerical responses of various taxonomic groups onto a graph of frequency and amplitude. Of course, taxa should be represented by area, not lines, and my vertical axis is still unlabelled for lack of suitable scale. Additional axes would be required for other components of variation in resource concentration, e.g., size of the food particles and spatial frequencies for patch size.

Frequencies low enough to be below the range of biological response, that is, to the right of Fig. 5, will not be directly observed. Complete damping by the consumers will erase such variation, and it will only be reflected as changes in consumer biomass. High frequency, high amplitude variation (upper left in Fig. 5) will swamp the biological response and will be observable as high resource levels. These conditions will permit the growth of relatively fast-growing taxa, such as ciliates.

CONCLUDING DISCUSSION

I started this essay with the contention that the important differences between ciliated protozoa and other consumers were their small size and high rates of reproduction, and have argued that temporal heterogeneity is important to their place in natural assemblages of organisms. In discussing temporal heterogeneity at length, I have probably over-emphasized reproductive

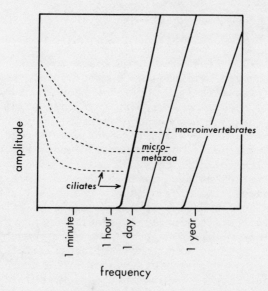

Figure 5. The potential functional responses (broken lines) and numerical responses (solid lines) of different groups of invertebrates on a map of frequency and amplitude.

rate. Small size in itself may be an important reason why ciliates can persist in some aquatic communities.

In plankton communities, ciliates must compete with, and are preyed upon, by larger zooplankton (Tezuka, 1974; Porter et al., 1979). Size selective predation on crustacean zooplankton may relieve these pressures on ciliates and help account for their commonness and abundance in lake plankton. In the absence of planktivorous fish, they might be replaced in a shift towards larger and competitively superior metazoan zooplankters. Streams and other benthic habitats for ciliates are probably quite different from plankton in this respect; predation on the metazoan predators and competitors of ciliates is probably less intense, permitting the existence of relatively long-lived species. Small particles of high food value (i.e., those which are not repeatedly collected and rejected) may be rarer in the seston of a stable stream than in a lake into which it flows, reflecting more regulation of consumers by availability of resources at lower trophic levels in streams and by predation from higher trophic levels in lake plankton. In summary, I suspect that small size is very important in understanding the ecology of ciliates versus crustacean zooplankton in lakes, but it may be much less important in streams, where ciliates likely depend on exploiting disturbance via their high rates of multiplication.

Small size might also be of direct importance to ciliates when resources occur in small patches. Samples always integrate resource levels to a mean over an area or volume. To an organism searching for food, concentration might be more meaningfully expressed as frequency of patches of various sizes and densities. As the mean concentration of a resource declines, patches of a critical concentration required for a positive energy budget become rarer and smaller. Common, slow-growing ciliates, such as *Cyclidium* and *Cinetochilum*, are often small, while ephemeral taxa with higher growth rates, such as *Colpidium* and *Glaucoma*, are often larger. Small size may allow species to feed at higher effective resource concentrations when resource levels are low. To use the same analogy that I used above for temporal variation, when mean levels are low, high amplitude variation may still occur at high frequencies. The interaction between body size and patch size is of doubtful relevance to ciliates filter-feeding from the seston of streams, but it may be important to grazing species.

In general, it is probable that the short generation time of ciliates and other small organisms allows them to exploit high frequency temporal heterogeneity at scales of hours to weeks, especially when the amplitude is sufficient to saturate the functional response of larger organisms. This environmental variation may largely determine the relative importance of ciliates and larger metazoan consumers, the species composition of ciliate community, and the characteristics of the species present.

REFERENCES

Berger, J. 1980. Feeding behavior of *Didinium nastum* on
     Paramecium bursaria with normal and apochlorotic zoochlorellae.
     J. Gen. Microbiol. 118: 397-404.
Cairns, J. Jr., Dickson, K.L. and Yongue, W.H. Jr. 1971. The
     consequences of nonselective periodic removal of portions of
     freshwater protozoan communities. *Trans. Am. Microscop. Soc.*
     90: 71-80.
Calow, P. and Jennings, J.B. 1977. Optimal strategies for the
     use of reserve materials in microbes and metazoa. *J. Theor.*
     Biol. 65: 601-604.
Cameron, I.L. 1973. Growth characteristics of *Tetrahymena*. *In*:
     Biology of *Tetrahymena*. A.M. Elliott, ed. Dowden,
     Hutchinson and Ross, Stroudsberg, Pennsylvania.
Cameron, I.L. and Bols, N.C. 1975. Effect of cell population
     density on G_2 arrest in *Tetrahymena*. *J. Cell Biol.* 67:
     518-522.
Canella, M.F. and Rocchi-Canella, I. 1976. Biologie des
     Ophryoglenina. *Annali Dell'Università di Ferrara*, Sezione 3,
     Vol. 3. *Suppl.* 2: 1-510.
Connell, J.H. 1978. Diversity in tropical rain forests and
     tropical reefs. *Science* 199: 1302-1310.
Cummins, K.W. and Klug, M.J. 1979. Feeding ecology of stream
     invertebrates. *Ann. Rev. Ecol. System.* 10: 147-172.
Eddison, J.C. and Ollason, J.G. 1978. Diversity in constant and
     fluctuating environments. *Nature* 275: 309-310.
Fenchel, T. 1968. The ecology of the marine microbenthos III.
     The reproductive potential of ciliates. *Ophelia* 5: 123-136.
Fenchel, T. 1974. Intrinsic rate of natural increase; the
     relationship with body size. *Oecologia* 14: 317-326.
Finlay, B.J. 1977. The dependence of reproductive rate on cell
     size and temperature in freshwater ciliated protozoa.
     Oecologia 30: 75-81.
Gill, D.E. 1972. Intrinsic rates of increase, saturation
     densities, and competitive ability. I. An experiment with
     Paramecium. *Amer. Nat.* 106: 461-471.
Gill, D.E. 1978. On selection at high population density.
     Ecology 59: 1289-1291.
Gray, E. 1952. The ecology of the ciliate fauna of Hobson's
     Brook, a Cambridgeshire chalk stream. *J. Gen. Microbiol.*
     6: 108-122.
Hamilton, R.E. and Preland, J.E. 1969. Cultural characteristics
     of a pelagic marine hymenostrome ciliate, *Uronema* sp.
     J. Exp. Mar. Biol. Ecol. 4: 90-99.
Harris, G.P. 1980. Temporal and spatial scales in phytoplankton
     ecology. Mechanisms, methods, models, and management.
     Can. J. Fish. Aquat. Sci. 37: 877-900.

Healy, F.P. 1979. Short-term responses of nutrient deficient
    algae to nutrient addition. *J. Phycol.* 15: 289-299.
Kaplan, R.H. and Salthe, S.N. 1979. The allometry of reproduction:
    an empirical view in salamanders. *Amer. Nat.* 113: 671-689.
Legner, M. 1973. Experimental approach to the role of protozoa in
    aquatic ecosystems. *Amer. Zool.* 13: 177-192.
Legner, M. 1975. Concentration of organic substances in water as
    a factor controlling the occurrence of some ciliate species.
    Int. Revue ges. Hydrobiol. 60: 639-654.
Levins, R. 1979. Coexistence in a variable environment. *Amer.
    Nat.* 114: 765-783.
Luckinbill, L.S. 1978. *r* and K selection in experimental
    populations of *Escherichia coli*. *Science* 202: 1201-1203.
Luckinbill, L.S. 1979. Selection and the r/k continuum in
    experimental populations of protozoa. *Amer. Nat.* 113:
    427-437.
Luckinbill, L.S. and Fenton, M.M. 1978. Regulation and
    environmental variability in experimental populations of
    protozoa. *Ecology* 59: 1272-1276.
Lynn, D.H. 1975. The life cycle of the histophagous ciliate
    Tetrahymena collissi Thompson, 1955. *J. Protozool.* 22:
    188-195.
McArthur, R.H. and Wilson, E.O. 1967. The theory of island
    biogeography. Princeton University Press, Princeton,
    New Jersey.
Parnas, H. and Cohen, D. 1976. The optimal strategy for the
    metabolism of reserve materials in micro-organisms.
    J. Theor. Biol. 56: 19-55.
Pianka, E.R. 1970. On "*r*" and "K" selection. *Amer. Nat.* 104:
    592-597.
Pianka, E.R. 1976. Competition and niche theory. *In*:
    Theoretical ecology: principals and applications. R.M. May,
    ed. W.B. Saunders, Toronto.
Poljanskij, J.I. and Chejsin, E.M. 1965. General Protozoology.
    (second revised edition) Clarendon Press, England.
Porter, K.G., Pace, M.L. and Battey, J.F. 1979. Ciliate
    protozoans as links in freshwater planktonic food chains.
    Nature 277: 583-565.
Ricketts, T.R. 1971. Periodicity of endocytosis in *Tetrahymena
    pyriformis*. *Protoplasma* 73: 387-396.
Ricketts, T.R. 1972a. The induction of endocytosis in starved
    Tetrahymena pyriformis. *J. Protozool.* 19: 373-375.
Ricketts, T.R. 1972b. The interaction of particulate material
    and dissolved foodstuffs in food uptake by *Tetrahymena
    pyriformis*. *Arch. Mikrobiol.* 81: 344-349.
Sousa, W.P. 1980. The response of a community to disturbance:
    the importance of successional age and species' life
    histories. *Oecologia* 45: 72-81.

Stearns, S.C. 1977. The evolution of life history traits. *Ann. Rev. Ecol. System.* 8: 145-171.

Taylor, W.D. 1978a. Growth responses of ciliate protozoa to the abundance of their bacterial prey. *Microbial Ecol.* 4: 207-214.

Taylor, W.D. 1978b. Maximum growth rate, size and commonness in a community of bactivorous ciliates. *Oecologia* 36: 263-272.

Taylor, W.D. 1980. Observations on the feeding and growth of the predacious oligochaete *Chaetogaster longi* on ciliated protozoa. *Trans. Am. Microscop. Soc.* 99: (in press).

Taylor, W.D. and Berger, J. 1976. Growth of *Colpidiom campylum* in monoxenic batch culture. *Can. J. Zool.* 54: 392-398.

Taylor, W.D., Gates, M.A. and Berger, J. 1976. Morphological changes during the growth cycle of axomic and monoxenic *Tetrahymena pyriformis*. *Can. J. Zool.* 54: 2011-2018.

Taylor, W.D. and Shuter, B.J. Body size, genome size and intrinsic rate of increase in ciliated protozoa. *Amer. Nat.* (in press).

Tezuka, Y. 1974. An experimental study of the food chain among bacteria, *Paramecium* and *Daphnia*. *Int. Revue ges. Hydrobiol.* 59: 31-37.

Threlkeld, S.T. 1976. Starvation and the size structure of zooplankton communities. *Freshwat. Biol.* 6: 489-496.

Vandermeer, J.H. 1969. The competitive structure of communities: an experimental approach with protozoa. *Ecology* 50: 362-371.

Veilleux, B.G. The functional response of five bactivorous ciliates isolated from an activated-sludge ecosystem. (Unpublished manuscript.)

Wallace, J.B. and Merritt, R.W. 1980. Filter-feeding ecology of aquatic insects. *Ann. Rev. Ent.* 25: 103-132.

Williams, F.M. 1971. Dynamics of microbial populations. *In*: Systems analysis and simulation in ecology. B.C. Patten, ed. Academic Press, New York.

THE ECOLOGY OF *GAMMARUS* IN RUNNING WATER

R. Marchant

National Museum of Victoria
Victoria
South Australia

INTRODUCTION

Species of the amphipod *Gammarus* are widespread and often abundant in the streams and rivers of the northern hemisphere. They can usually be sampled quantitatively, are well known as consumers of allochthonous organic detritus (a major source of food in running water), are eaten by a variety of riverine fish and are commonly found drifting or moving upstream. As a typical benthic invertebrate from flowing water, *Gammarus* appears more convenient than many for quantitative study in the field, because of its conspicuousness and general ease of handling. In this chapter I will review data, largely from field studies, on its life history, movements, density, population dynamics, growth, production and energy flow.

Hynes (1955) summarised the sequence of events in the reproduction of *Gammarus*; and his study and others since then have concentrated on interpreting field data to understand the topics mentioned above. The following species from running water have been studied: *G. pseudolimnaeus* Bousfield, *G. lacustris limnaeus* Smith and *G. minus* Say from North America; and *G. pulex* (L.) and *G. fossarum* Koch from Europe. Details of the distribution of these species are given by Bousfield (1958), Meijering (1972a), Gledhill, et al., (1976) and Pennak (1978). Unfortunately, it is only *G. pulex* and *G. pseudolimnaeus* for which much quantitative data from the field exists. I intend to omit species which have been found in running water, but usually occur in brackish water, e.g. *G. duebeni* Lilljeborg and *G. tigrinus* Sexton, or still water, e.g. *G. fasciatus* Say.

225

LIFE HISTORY

   The life history of *Gammarus* has been well studied in running
water: Table 1 gives details for the various species. At all the
localities mentioned, egg production of *Gammarus* is largely confined
to the spring and summer when water temperatures are highest.
Temperature appears to have a major affect on life spans, the time
taken until first breeding begins and the winter resting period.

   In the populations of *G. pulex* studied by Hynes (1955) and
G. minus studied by Kostalos (1979), the earliest born individuals
could begin breeding in the late summer and early autumn, i.e.
within 4-5 months of birth, before the onset of the winter resting
period. Those that survived the winter bred again the following
spring, although one population (Richmondhill) continued breeding,
but at a reduced rate, during the winter. Later born individuals
grew over the autumn and winter to mature and breed the following
spring after about 8 months. The periods of high temperature
(>15°C) during the summer could have enabled the first born to
reach maturity quickly because populations that did not experience
these temperatures (Rold Kilde, Tadnoll Brook) took 1 yr before at
least some individuals were mature and 2 yr before all were. *G. l.*
limnaeus experienced low temperatures (<5°C) for more of the year
than *G. pulex*, never experienced >15°C, and took about 22 months
before any individuals matured. *G. pseudolimnaeus* appears
intermediate experiencing <5°C for as long as *G. l. limnaeus*, but
>15°C for as long as some *G. pulex* populations, with individuals
taking 12 months to mature. These differences in life span are
reflected in the duration of the resting period which is longer for
G. pseudolimnaeus than for *G. pulex* and *G. minus* and shorter than
that for *G. l. limnaeus*. *G. fossarum* is an exception with a life
cycle similar to *G. pulex* at Shotwick, but in a stream whose
temperature does not exceed 14°C and is <5°C for about 4 months of
the year.

   The length of the resting period seems to depend directly on
the duration of low minimum temperatures as those species which have
the longest resting periods (*G. l. limnaeus* and *G. pseudolimnaeus)*
also spend the longest time at < 5°C. However, it is not so
straightforward because when minimum temperatures are a little higher,
e.g. 6 or 7°C, some populations of *G. pulex* (Richmondhill, Tadnoll
Brook) do not have a definite resting period, although breeding is
certainly diminished over winter; this suggests either that only
some of the females become non-reproductive or that they all do, but
not simultaneously. Hynes (1955) suggested that rapidly falling
temperatures in autumn initiate a definite resting period: at
Shotwick and the River Terrig temperature fell by about 6°C in a
month (to 7°C) immediately before the resting period while at
Richmondhill and Tadnoll Brook temperature never fell at this rate.
In the Credit River, Ontario temperature fell by about 7°C in a

Table 1. Details of the life history of various species of *Gammarus*. Temperature ranges are based on monthly spot readings and are thus approximate except where indicated by an asterisk (*) when ranges are based on mean monthly temperatures. Life span indicates the maximum age of an average adult. The resting period occurs during winter and refers to the period when the mature female is non-ovigerous; in this case it represents any month when < 3% of specimens (of both sexes) > 6 mm in length were ovigerous. Periods above or below certain temperatures are only approximate when based on spot readings of temperature, and could not be determined in some cases because monthly reading were not given.

Species	Locality	Annual temperature range (°C)	Period >15°C (months)	Period temperature <5°C (months)	Life span (months)	Resting period	Authority
G. pseudolimnaeus	Credit River, Ontario	0–21*	4	5	12	4	Hynes & Harper, 1972; Marchant & Hynes, 1981b
G. l. limnaeus	Lutteral Creek, Ontario	4–11	0	5	24	8	Hynes & Harper, 1972
	Alton Stream, Ontario	2–13	0	5	24	8	
G. fossarum	Breitenbach, West Germany	5–14	0	4	12	3	Lehmann, 1967
G. pulex	Rold Kilde, Denmark	5–14	0	–	24	2	Iversen & Jessen, 1977
	Richmondhill stream, Isle of Man	6–16	4	0	12	0	
	River Terrig, Scotland	3–16	2	2	12	3	Hynes, 1955
	Shotwick Stream, England	4–15	5	1	12	3	
	Tadnoll Brook, England	7–14*	0	0	24	0	Welton, 1979
G. minus	Falls Ravine Creek, Pennsylvania	0–22	–	–	12	<2	Kostalos, 1979

month (to 11°C) before the resting period began. However, such
large decreases in water temperature did not occur at Lutterall
Creek, Alton or Breitenbach and it is not clear what induces the
resting period at these localities. Possibly decreasing day length
combined with low mean temperatures (< 5°C) is important, as it may
be with all populations in maintaining the resting period; in most
populations the resting period ends after the shortest day. It
should be noted that G. fossarum and G. l. limnaeus generally inhabit
cold streams and the apparently irregular features of their life
cycles may indicate adaptation to lower temperatures.

Temperature is not the only factor which can affect life
history; food quantity and quality may be important because these
can alter growth rate in the laboratory (see following section on
Growth). Thus, one reason why some G. pseudolimnaeus born early in
spring cannot reach maturity at the high summer temperatures (as do
some G. pulex) may be that sufficient high quality food is not
available. To determine the comparative influence of temperature
and nutrition in the field required detailed data on mean temperatures, which are lacking for most of the studies quoted in Table 1.
Such data can be readily obtained (at least for shallow rivers) with
continuous temperature recorders or maximum-minimum thermometers,
e.g. Marchant and Hynes (1981b) or Macan (1959).

Hobrough (1973) has shown that juvenile G. pulex survive
organic pollution in the summer better than the adults. Thus no
individuals from the generation that hatches in spring mature before
the end of summer; they take until late autumn. In this case
pollution causes an artificial slowing in development rate by killing
any individuals that mature in summer.

DRIFT AND UPSTREAM MIGRATION

Gammarus is well known to move up and downstream, and its
movements have been studied in the field: G. pulex (Elliott, 1969;
Hultin, 1971; Iversen and Jessen, 1977), G. fossarum (Lehmann, 1967;
Meijering, 1977), G. pseudolimnaeus (Marchant and Hynes, 1981b;
Waters, 1962, 1965, 1976); and in artificial streams in the labor-
atory: G. pulex (Hughes, 1970; Meijering, 1972b), G. fossarum
(Meijering, 1972b), G. pseudolimnaeus (Wallace et al. 1975).

Both the field and laboratory studies show that the highest
numbers of Gammarus drift and migrate upstream generally at night
and when the current is slow, but probably above some minimal value;
the rates of both movements are generally lowest in winter or at low
temperatures. The only exception to this pattern was found in the
laboratory by Meijering (1972) who showed that drift, but not
upstream movement, decreased as temperature decreased. However, his
subsequent field study in which the movements (both up and downstream)
of G. fossarum were monitored almost continuously, revealed that both

decreased in winter. Presumably it is the behaviour in the
artificial stream which is anomalous. In the river most *Gammarus*
move upstream at the side where current is lowest (Elliott, 1971a;
Hultin, 1971).

The drift density measured in most studies (Table 2) is
10^4-10^5 *Gammarus* $24h^{-1}$ per m^3 s^{-1}. None of the studies showed that
drift was compensated by upstream movement, although Marchant and
Hynes (1981b) found no significant statistical difference between
the numbers of *G. pseudolimnaeus* moving in either direction.
However, drift of *G. pseudolimnaeus* was unusually low in the Credit
River, whereas it was much higher in Valley Creek where, by
denuding areas, Waters (1965) demonstrated that most re-colonising
G. pseudolimnaeus entered from upstream, implying that drift
exceeded upstream movement.

The degree of compensation also varies seasonally. Elliott
(1971a) in a study of the upstream movements of many taxa including
G. pulex, in Wilfin Beck, found such movements compensated for 30%
of the drift in winter but only 7-10% in spring and summer.
Meijering (1977), who sampled *G. fossarum* continuously during his
study, found most compensation in summer when, occasionally, numbers
migrating upstream exceeded those drifting; in winter few or none
moved upstream, but drift continued.

The distance drifted by *Gammarus* has been estimated by Waters
(1965) and Elliott (1971b). Waters showed that average numbers
drifting $24h^{-1}$ were restored 38 m downstream of a blockage. McLay
(1970) calculated from Waters' data a mean drift distance of 28.5 m
by assuming an exponential decline with distance of numbers drifting.
Elliott (1971b) confirmed that the exponential model fitted such
data by releasing known numbers of *G. pulex* and other stream
invertebrates at various distances upstream of blocking nets. He
found the mean distance drifted by *G. pulex* varied from 1.01 m at
a current of 10 cm s^{-1} to 5.66 m at 60 cm s^{-1}. In his other study
(Elliott, 1971a) of upstream movements he determined, with marked
individuals, that *G. pulex* could move about 4 m $24h^{-1}$ upstream and
drift 3-5 m $24h^{-1}$ downstream at currents of 12-45 cm s^{-1}. These
results suggest that more *Gammarus* drift past a point than move
upstream not because they move faster by drift, but because they do
so in greater numbers.

Once the distance a species of *Gammarus* will drift is known,
then any study to determine whether drift causes significant drops
in population density must be made on a length of a river (from
which there is a net loss and not just replacement by drift from
upstream of the reach) that is longer than the average distance
drifted daily; thus irregular fluctuations in density (measured
over a day) arising from up and downstream movements will be less

Table 2. Drift density and degree of upstream movement for various species of *Gammarus*.

Species	Locality	Mean drift density (numbers 24h^{-1} per m^3 s^{-1})	Upstream movement as a percentage of drift (mean %)	Authority
G. pulex	Rold Kilde, Denmark	4.1×10^5	n.m.	Iversen & Jessen, 1977
	Wilfin Beck, England	1.2×10^4	n.m.	Elliott, 1969
G. fossarum	Breitenbach, West Germany	4.1×10^5	56	Lehmann, 1967
	Breitenbach, West Germany	†	42	Meijering, 1977
	Rohrweisenbach, West Germany	†	54	
G. pseudolimnaeus	Valley Creek Minnesota	1.7×10^5	n.m.	Waters, 1962
	Credit River, Ontario	2×10^2	69	Marchant & Hynes, 1981b

n.m. = not measured, † = could not be calculated because no value for current given; but mean drift 24h^{-1} was 10^2 in both streams whereas Lehmann (1967) gave a rate of 1.7×10^3 24h^{-1}.

likely to occur. This is perhaps why Waters (1965) could show no
significant decline in the population density of G. pseudolimnaeus
immediately upstream of his drift nets, although the density of
drift was high (Table 2). He should not have expected any decline
until about 28.5 m upstream of his nets.

MICRODISTRIBUTION AND DENSITY

Gammarus is generally readily sampled quantitatively as it does
not burrow far (< 10 cm) into the stony substrate of streams (Waters,
1976; Marchant and Hynes, 1981b) as do some invertebrates (Williams
and Hynes, 1974); and compared with many stream taxa it is more
catchable (Carle, 1977) because it does not stick to rocks or sink
rapidly. Marchant and Hynes (1981b) calculated the probability of
catching G. pseudolimnaeus in one sweep of a hand net over an area
of 0.15 m^2 as 0.65 and in three sweeps 0.95.

Generally, Gammarus occurs at highest densities in macrophytes
(G. pulex, Welton, 1979) or in other areas with low current
(< 25 cm s^{-1}) such as the shallows near the bank (G. pseudolimnaeus,
Marchant and Hynes, 1981b), probably because it is least active in
strong currents (see above) and thus avoids them. Rees (1972) showed
that G. pseudolimnaeus was not found at currents above about 55 cm
s^{-1} in the field and preferred substrates with a particle size of
1.6-3.2 cm; Marchant and Hynes (1981b) found the highest densities
of this species on a substrate of sand and stones < 20 cm in diameter.
G. pulex and G. fossarum have been found on similar substrates
ranging from sand and small stones (Lehmann, 1967) to flints (< 7 cm
in diameter) in a chalk stream (Welton, 1979) and fist-sized stones
(Iversen and Jessen, 1977); and in currents up to 70 cm s^{-1}.
Maitland (1966) found G. pulex most commonly in the weeds and riffles
of the River Endrick and least commonly in pools, whereas Kostalos
(1979) found G. minus equally abundant in pools and riffles; but his
creek was small (< 1 m across) and current was slow (3 cm s^{-1}), while
the River Endrick was 20-35 m wide and current varied from 5-50 cm
s^{-1}.

Table 3 gives the annual range of densities of various species
of Gammarus whose populations have been censused regularly for at
least a year. The highest densities generally occur in the smallest
rivers, although it should be noted that most studies have been done
only in creeks and streams. Macan and Mackereth (1957), in a survey
of some of the rivers of the Lake District of England, also recorded
the lowest densities of G. pulex in rivers 10-30 m wide; Maitland
(1966) found lower densities of this species in the main channel than
in the tributaries of the River Endrick.

The 95% confidence limits for some of the estimates of density
given in Table 3 are: ±50% for G. pseudolimnaeus in the Credit
River; ±10% and ±63% for G. pulex in Tadnoll Brook and Rold Kilde

Table 3. The annual range of density of various species of *Gammarus* and the size of the streams and rivers in which populations were studied. Densities are mean values for a number of stations in each river, except in the case of Ford Wood Beck where density at only one station is given.

Species	Locality	Range of density (numbers m^{-2})	Period of density $>10^3$ m^{-2} (months)	Width of river (m)	Authority
G. pulex	Rold Kilde, Denmark	$1.3 \times 10^3 - 5.5 \times 10^3$	12	0.5-0.7	Iversen & Jessen, 1977
	Tadnoll Brook, England	$8.4 \times 10^2 - 9.7 \times 10^3$	10	4-5	Welton, 1979
	Ford Wood Beck, England	$6.0 \times 10^1 - 5.4 \times 10^2$	0	2	Macan & Mackereth, 1957
G. fossarum	Breitenbach, West Germany	$9.3 \times 10^2 - 1.8 \times 10^4$	11	0.7-1.3	Lehmann, 1967
G. pseudolimnaeus	Valley Creek, Minnesota	$3.5 \times 10^3 - 2.6 \times 10^{4+}$	-	5	Waters, 1965
	Credit River, Ontario	$1.3 \times 10^1 - 8.2 \times 10^2$	0	12-15	Marchant & Hynes, 1968b
	Cone Spring, Iowa	$8 \times 10^1 - 1.2 \times 10^3$	2	-	Tilley, 1968

+ does not represent annual range, but variation between habitats.

respectively. The sizes of the variances on which these confidence
limits are based indicate a contagious or clumped distribution of
individuals within the streams, as usually occurs with benthic
invertebrates in this habitat (Hynes, 1970). Welton (1979) showed
that to obtain confidence limits of 10% he had to take 120 samples
on each visit; to obtain the same confidence limits for the other
two populations would have required 404 samples of *G. pseudolimmaeus*
and 352 samples of *G. pulex* on each visit. One possible cause of
such clumped distributions is that potential sources of food such as
leaves are usually not evenly dispersed in streams. Both Kostalos
(1979) and Haeckel et al. (1973) showed that higher densities of
Gammarus occurred in accumulations of leaves than in bare areas.
Such heterogeneity in the distribution of *Gammarus* within a small
area requires stratification of samples to make sampling efficient
and to reduce confidence limits (Welton, 1979; Marchant and Hynes,
1981b). It should also be noted that if the logarithms of the
population densities are considered (the best way to display
relative change) then 95% confidence limits of 50% on an arithmetic
scale reduce to about 10% (Marchant and Hynes, 1981b).

It is apparent, if Tables 2 and 3 are compared, that those
populations of *Gammarus* with high population densities also have
high drift densities, suggesting competition for space occurs with
this genus, at least in small streams. Nilsson and Sjöström (1977)
showed that the number of *G. pulex* colonising substrate depended
on the carrying capacity of the substrate rather than the number in
the drift. Macan and Mackereth (1957) explained the reasonably
constant density of *G. pulex* in Ford Wood Beck by proposing that
overcrowding results in drift; the ratio of the maximum to minimum
density recorded by them in a year was only 2.1-2.7 (Table 3 only
gives densities at one out of five stations sampled), whereas such
ratios in the other studies were always > 4 and usually about 10.
Marchant and Hynes (1981b) suggested the absence of overcrowding of
G. pseudolimmaeus in the Credit River, where it occurs at comparatively low densities (Table 3), explains its low drift density.
Neither they nor Waters (1972, 1976), however, were able to show a
correlation for this species between drift and population density,
and if there is a relation it appears to be between populations, not
within them, and is probably not linear (Waters, 1972). Waters
(1976) has shown, though, that drift is correlated with the rate of
production of *G. pseudolimmaeus* (see below), which supports his
suggestion that excess production drifts.

To settle the type of relation between drift and population
density, more populations with densities intermediate between those
shown in Table 3 need to be studied. To settle the associated
problem of how much loss drift (which is generally greater than
upstream movement, Table 2) causes to a population of *Gammarus*, a
simultaneous study of drift, upstream movement, life cycle and
population density must be made along a reach of suitable length.

Then, the net loss by migration from the study area can be compared
with the rate of mortality of the generation present. Marchant and
Hynes (1981b) made such a comparison and showed that drift caused
negligible loss of individuals from their study area. However, drift
density was atypically low in their river (Table 2). In the other
localities it seems quite likely that drift causes significant loss
as, indeed, Iversen and Jessen (1977) claimed in their study. They
also showed that the size-frequency of individuals in the drift
occasionally differed from that of the benthic population. Such
differences will only have significance when the amount of loss
caused by drift can be gauged.

MORTALITY

     The mortality of *Gammarus* along a stretch of river, which
includes all other forms of loss to the population except drift,
has seldom been measured. This is surprising considering the number
of studies of its density (Table 3) and the fact that cohorts of
individuals can be traced in most populations (Table 1).

     The causes of mortality of a population of *Gammarus* in running
water could be predation, starvation or death through old age. Fish
are major predators, e.g. brown trout (*Salmo trutta*) (Welton, 1979).
Benthic invertebrates may also eat *Gammarus*, e.g. some perlodid and
perlid stoneflies (Macan and Mackereth, 1957), and Hynes (1954) notes
that cannibalism can occur. Parasitism of *Gammarus* by acanthocephalans
e.g. *G. pulex* by *Polymorphus minutus* in the U.K., is common (Hynes
and Nicholas, 1963), but only heavy infections, which are rare, are
likely to kill the amphipod.

     The patterns of relative mortality for two species of *Gammarus*
are given in Table 4. Loss due to drift was negligible for *G.
pseudolimnaeus* (see above) and for the population (from a spring) of
G. pulex quoted in the table; the other, downstream, populations of
G. pulex studied by Iversen and Jessen, however, did suffer loss by
drift. The first period of mortality in *G. pseudolimnaeus* resulted
from the death of eggs (for unknown reasons), predation on the
smallest individuals by fish, and possibly cannibalism of young by
adults. The lack of mortality of *G. pulex* over the same range of
sizes was due apparently to there being no fish in the spring;
Iversen and Jessen did not estimate potential recruitment and thus it
is impossible to determine whether there was substantial death of
eggs at the beginning of the cohort. At downstream stations, *G.
pulex* suffered mortality averaging 0.66 for the size range 2-5 mm,
apparently due to predation by fish, although some of this loss was
perhaps due to drift.

     The next two periods of mortality of *G. pseudolimnaeus* in
September and over winter were due partly to predation and, it was
suggested, partly to starvation. There was also mortality of

Table 4. Relative mortality (M = $\log_{10} N_2 - \log_{10} N_1$) of two species of *Gammarus* throughout their life spans. The range of length classes indicates the mean length at the beginning and end of the periods of mortality.

G. pseudolimnaeus (Marchant & Hynes, 1981b)			*G. pulex* (Iversen & Jessen, 1977)		
Range of length classes (mm)	Months	$\underline{M}$	Range of length classes (mm)	Months	$\underline{M}$
egg*–5	June–August	0.87	2–5	September–May	0
5–6	August–September	0	5–6	May–November	0
6–8	September–October	0.54	6–7	November–January	0.46
8–11	October–December	0	7–7	January–March	0
11–13	December–April	0.77	7–9	March–September	2.7
13	April–May	0			
13	May–July	0.54			

* length class of newly hatched individuals was 2 mm.

G. pulex over its second winter though not at the beginning of its
second autumn. Iversen and Jessen state that the bottom of the
spring was covered, even in summer, with decaying leaves which would
provide more potential food, than in the Credit River, before the
input of leaves and thus enable *G. pulex* to survive this period
better than *G. pseudolimnaeus*. Marchant and Hynes (1981b) suggested
G. pseudolimnaeus may have starved during winter because it became
difficult to gather food under the ice. There was apparently no ice
on Rold Kilde, but *G. pulex* was larger during its second winter than
its first and thus required more energy (Nilsson, 1974), which makes
it possible that starvation was again the cause of the observed
mortality. Welton (1979) recorded mortality in autumn and early
winter (September to February) of all length classes of *G. pulex* in
Tadnoll Brook, where also no ice formed during winter.

The final periods of mortality of both species were undoubtedly
the result of ageing at the end of the life cycle. Welton (1979)
and Hynes (1955) both recorded such mortality among mature
individuals after breeding. More studies of survivorship over the
whole life span are, or course, necessary before definite causes of
mortality can be recognized at all stages of the life cycle. It is
particularly important to understand winter mortality of *Gammarus*,
and to determine whether it is a general feature of their life cycles,
because it is those that survive that breed in spring. In such
future studies it will be essential to distinguish loss due to drift
from loss due to mortality.

RECRUITMENT

As shown above, to enable mortality to be assessed at the
beginning of the life cycle the potential recruitment of *Gammarus*
must be known. It is also important to estimate this value because
it constitutes part of the production of a generation (see below).
Recruitment can be calculated from data on clutch size and
development time of the eggs provided data on mean temperature in
the stream are available, as development times appear to depend only
on temperature. Marchant and Hynes (1981b) calculated recruitment by
estimating the number of egg days from a graph of eggs m^{-2} against
time and dividing this figure by the mean development time.

Table 5 gives the clutch sizes of various species of *Gammarus*.
In all cases there is a linear increase of clutch size with length
of female, although the rate of this increase varies between species
being greatest for *G. pseudolimnaeus* and least for *G. fossarum*.
There is no evidence that clutch size at a specific length varies
other than randomly during the recruitment season (Hynes, 1955;
Hynes and Harper, 1972; personal observation of *G. pseudolimnaeus*),
but clutch size can vary between seasons, e.g. *G. pseudolimnaeus*,
perhaps due to variations in food supply, or can be remarkably

Table 5. The clutch size of various species of *Gammarus*, according to the length of the female. The range of lengths represents those lengths at which females were usually found to be ovigerous.

length (mm)	G. pseudolimnaeus* (Marchant & Hynes, 1981b)	G. pulex (Welton, 1979)	G. pulex (Hynes, 1955)	G. fossarum (Lehmann, 1967)	G. l. limnaeus (Hynes & Harper, 1972)
	Number of eggs				
6		3.3	2.3	4.8	
7		7.7	7.0	5.5	5.0
8		12.1	11.7	6.4	10.5
9	15.0, 22.9	16.5	16.5	7.3	13.7
10	22.1, 30.2	20.8	21.2		17.1
11	30.8, 39.6	25.2	25.9		20.8
12	37.0, 45.6	29.6	30.6		24.3
13	45.0, 54.3				
14	52.5, 62.0				31.1

* left column: 1977 clutches; right: 1978 clutches

constant between different sites, e.g. *G. pulex*. According to
Marchant and Hynes (1981b) *G. pseudolimnaeus* suffered substantial
egg mortality; the cause was unknown, but the eggs died during
incubation.

Table 6 indicates how development time of the eggs varies with
temperature for two species of *Gammarus*. In neither case do the
eggs develop for a constant number of degree days at all
temperatures; below 5-8°C the number of degree days required
decreases. The longer development time required by *G.
pseudolimnaeus* at these low temperatures (which prevail when
breeding begins) helps to account for the fact that none of its
young reach maturity and begin breeding before the onset of the
winter resting period as do some of the young of *G. pulex*. The
eggs of *G. l. limnaeus* take 45 days to develop at 12°C which is
about twice as long as taken by the eggs of either of the previous
two species and thus it has a longer life span of 2 years.

Welton (1979) measured a constant value of 370 degree days for
the development of the eggs of *G. pulex*. This appears to be an
approximation compared with the data in Table 6, which clearly
indicate that the number of degree days is not constant. However,
the data in Table 6 are probably themselves approximate because

Table 6.   Development times at constant temperatures of the eggs
           of two species of *Gammarus*. The figures in brackets
           give the number of degree days at each temperature
           (above 0°C).

Temperature (0°C)	Mean development time	
	G. pseudolimnaeus (Marchant & Hynes, 1981b) (days)	*G. pulex* (Nilsson, 1977) (days)
1	176.9 (176.9)	110.5 (110.5)
5	70.9 (354.5)	59.3 (296.5)
8	44.2 (353.6)	41.3 (330.4)
10	33.9 (339.0)	33.6 (336.0)
12	26.5 (318.0)	27.2 (326.4)
15	19.8 (297.0)	20.6 (309.0)

they were obtained at constant temperatures while temperatures that
fluctuate to the same extent as in the stream are known to decrease
the development time of *G. pulex* and *G. fossarum* eggs by about 10%
(Roux, 1975; Koch-Kallnbach and Meijering, 1977). Thus Marchant
and Hynes (1981b) and Welton (1979) slightly underestimated recruitment because they overestimated development times by basing
them on constant temperatures.

Marchant and Hynes (1981b) showed that variation in potential
recruitment resulted largely from variation in the density of
ovigerous females rather than from variation in clutch size. Thus
comparisons of potential recruitment between species can only be
made if data on density and mortality of ovigerous females are
considered as well as, of course, information on frequency of
breeding. Comparisons based only on potential number of broods of
a certain size raised by individual females (Hynes, 1955) do not
allow for differences in mortality and are thus not reliable. From
Tables 3 and 5 it would be reasonable to conclude that a generation
of *G. pulex* from Tadnoll Brook produces more eggs than a generation
of *G. pseudolimnaeus* from the Credit River, not only because the
densities of *G. pulex* are higher, but also because they breed for
a greater percentage of their life span.

GROWTH

As many populations of *Gammarus* have distinguishable cohorts of
individuals (Table 1), growth can be calculated from increases in
the mean size of these individuals. This assumes that losses to
the population are not selective of particular sizes; neither Iversen
and Jessen (1977) nor Marchant and Hynes (1981b) could show that
losses were selective.

Growth in length of *G. pulex* and *G. pseudolimnaeus* is linear
with time, but growth in weight ($\underline{B}$) is exponential with time
(because the weight of *Gammarus* is approximately proportional to
its length cubed) and consequently Table 7 gives specific growth
rates ($\underline{G}$). Specific growth rate is higher at the beginning of the
life cycle than at the end when it decreases abruptly once the
animals reach a size (11 mm for *G. pseudolimnaeus* in the Credit
River; 6 mm for *G. pulex* in the Rold Kilde) close to their maximum
mean length. Such a pattern indicates *Gammarus* does not grow
beyond fixed upper limits, which vary between species: a mean of
13 mm or 10.7 mg dry weight for *G. pseudolimnaeus* and 9 mm or 4.5 mg
dry weight for *G. pulex*; and which possibly vary between localities
for one species.

Specific growth rate also varies between species and within a
species at similar stages in the life cycle. Part of this
variability is due to the species experiencing different

Table 7. Specific growth rate ($G = 100$ (ln B_2 − ln B_1) days^{-1}, i.e. % day^{-1}) and specific growth per degree day ($S = 100$ (ln B_2 − ln B_1) degree days^{-1}, i.e. % degree day^{-1}) for two species of *Gammarus*. Where the degree days are marked with an asterisk (*) they are only approximate because they were calculated from spot rather than mean temperatures.

Species	$\underline{G}$	Period (months)	Degree days	$\dfrac{S}{(\times 10^{-1})}$	Degree days to grow 1 mm in length	Authority
				In the field		
G. pseudolimnaeus	1.86	5a	1997	1.4	298 }	Marchant & Hynes, 1981b
	0.25	6b	503	0.9	252 }	
G. pulex	0.96	11a	2400*	1.3	600 }	Iversen & Jessen, 1977
	0.28	12b	2700*	0.4	900 }	
	0.75	9a	2300*	0.9	852	Hynes, 1955
	3.03	5a	2000*	2.3	308	
	1.23	12b	3541	1.3	494c }	Welton, 1979
	0.44	12b	3541	0.5	494c }	
				In the laboratory		
G. pulex	0.76–1.18 (♂)d	5	1184–1765	1.0–1.2	-- }	Nilsson, 1974
	0.66–0.91 (♀)d	5	1184–1765	0.8–0.9	-- }	
	0.37–1.69	1–2	480–975	0.2–1.1	--	Willoughby & Sutcliffe, 1976
G. pseudolimnaeus	0.12–0.85	2	1054	0.1–0.5	--	Bärlocher & Kendrick, 1973

a = beginning of life cycle
b = end of life cycle
c = measured in laboratory and used to calculate $\underline{G}$ and $\underline{S}$ from degree days accumulated in the field
d = measured in laboratory, but in stream water under natural diurnal fluctuations of light and temperature

temperatures while they grow. Low temperatures slow the specific
growth rate as demonstrated by Nilsson (1974) in the laboratory with
G. pulex. In the field the specific growth rates of both *G.
pseudolimnaeus* and *G. pulex* decreased noticeably in winter (Iversen
and Jessen, 1977; Marchant and Hynes, 1981b). If temperature is
excluded as a factor by calculating specific growth per degree day,
then patterns of growth for *Gammarus* in the field are quite
consistent both between and within species: in the field, specific
growth at the beginning of the life cycle is 0.9-2.3×10^{-1} per
degree day (but usually 1.3-1.4×10^{-1}) and at the end is
0.4-0.9×10^{-1}. It must be realised that some of these figures
are approximate because in some cases the number of degree days
are only approximate. A similar range of values is found in the
laboratory, which gives confidence in extrapolating such values for
growth to the field. The fact that males grow faster than females
in the laboratory agrees with the field studies which always show
that males reach a larger maximum size than females.

Thus temperature seems to account for many of the differences
in growth rate and therefore in life span, as already discussed.
The ability, however, of some individuals in some populations of
G. pulex (River Shotwick; Hynes, 1955) to reach maturity and breed
during their first summer is not completely accounted for by
differences in temperature. Specific growth per degree day of
these individuals (2.3×10^{-1}) is higher than usually found at the
beginning of the life cycle, and probably results from a larger or
more nutritious supply of food.

The effect of diet on the growth of *Gammarus* has been clearly
shown in the laboratory and there is no reason to think that this
does not also apply in the field. Cultures of fungal hyphomycetes
from **streams** generally promoted faster growth of *G. pseudolimnaeus*
($\underline{G} = 0.85$) than decaying elm or maple leaves ($\underline{G} = 0.19$-0.35)
(Bärlocher and Kendrick, 1973); decaying oak and elm leaves
promoted better growth of *G. pulex* ($\underline{G} = 1.69$) than hyphomycetes
($\underline{G} = 0.97$) or algae ($\underline{G} = 0.37$) (Willoughby and Sutcliffe, 1976).
It will only be possible to separate the influences of temperature
and food on growth measured in the field when detailed data on mean
temperature are routinely collected, as suggested before in the
section on life history.

PRODUCTION

Data on growth and density can be combined to estimate the
rate at which the biomass of *Gammarus* increases by growth, regardless
whether this biomass survives to the end of the life span. This
rate is known as production and can be estimated by a number of
methods which Waters (1977) has reviewed. As *Gammarus* is a
conspicuous consumer of the major source of food in flowing water,

allochthonous detritus, and as it in turn is a frequent item in the diet of riverine fish, it could be an important link in the flow of organic matter or energy in streams and rivers and values for its production are thus valuable data.

Table 8 presents estimates of production for *G. pulex* and *G. pseudolimnaeus*. These fall within the range of values given by Waters (1977) for fresh water amphipods of various genera from both still and running water. The estimate of Marchant and Hynes also includes production due to the eggs, although this represents only 2% of total production; none of the estimates include losses due to moulting, but this is probably less than 10% of the total (Marchant and Hynes, 1981b). If Tables 3 and 7 are compared with Table 8 it is clear that changes in density are a much greater influence than changes in growth on production and as densities are lowest in the biggest rivers so is production (the Thames is 73 m wide where *G. pulex* was sampled). Consequently, as mortality determines change in density during the life cycle so it must also control production; a decrease in the mortality of small *Gammarus* would increase production probably more than any other factor because in most of the studies in Table 8 it is the growth of small individuals that die before maturity that contributes most to production. Such a decrease would also increase the percentage of the total due to egg production because the percentage of females surviving to breed would increase

Table 8. Production (*P*) and *P/B* ratio (*B* is mean biomass) for two species of *Gammarus* in various localities. All values are in kg dry weight; the value for the River Thames was converted from wet weight assuming dry weight is 22% of wet weight (Nilsson, 1974).

Species	Locality	$\frac{P}{(\text{kg ha}^{-1}\text{yr}^{-1})}$	$\frac{P}{B}$	Authority
G. pulex	Rold Kilde, Denmark	38–70	2.0	Iversen & Jessen, 1977
	Tadnoll Brook, England	129	2.8	Welton, 1979
	River Thames, England	0.2	6.7	Mann, 1971
G. pseudolimnaeus	Credit River, Ontario	15	4.7	Marchant & Hynes, 1981b
	Valley Creek, Minnesota	68–316	6.0	Waters, 1976
	Cone Spring, Iowa	366	–	Tilley, 1968

and, as discussed above, it is the density of ovigerous females that most influences recruitment.

These conclusions are supported by the annual P/B ratios. As Waters (1977) has already noted, these ratios are primarily affected by life span which results in those from populations with a 2 year life span being about half those from populations with a 1 year life span. Waters found univoltine populations had a mean value of 4.5; this is close to the value for the univoltine *G. pseudolimnaeus* in the Credit River. In other words, the cohort P/B ratio for *Gammarus* is more or less constant, indicating that changes in biomass are strongly correlated with changes in production. Biomass itself is influenced by changes in density more than by variation in the weight of individuals, e.g. Welton (1979), thus supporting the previous idea that production is most influenced by the density of *Gammarus*. Therefore, although growth rates of *Gammarus* measured in the laboratory are comparable with those measured in the field, inaccuracies caused by extrapolating them from the laboratory to the field are not likely to produce major errors when estimating production. This is a significant advantage if cohorts are not distinguishable and growth cannot be estimated in the field.

As a corollary of the influence of density of production, Marchant and Hynes (1981b) showed that the 95% confidence limits of their estimate of production were dominated by errors associated with measuring density: the average confidence limits for density were ±50% of the mean whereas those for production only slightly exceeded ±50%. Thus the 95% confidence limits for production in Rold Kilde and Tadnoll Brook are probably about equal to the confidence limits for density, i.e. ±60% and ±10% respectively.

ENERGY FLOW

A complete energy budget for a species of *Gammarus* based on field data has not been produced yet. However, it is possible to give a composite picture.

Allochthonous detritus constitutes the majority of the diet of stream dwelling *Gammarus* and consists mainly of decaying leaves, although small amounts of algae and animal remains are also eaten (Hynes, 1954; Moore, 1975; Marchant and Hynes, 1981a). Hyphomycete fungi and bacteria which colonise the decaying leaves considerably improve the leaves' nutritional value by raising their protein content (Kaushik and Hynes, 1971; Bärlocher and Kendrick, 1976). The input of leaves varies seasonally and to streams flowing through temperate deciduous forests it peaks in autumn. Anderson and Sedell (1979) give values for the total input of detritus to such streams of $3.5 \times 10^3 - 6.7 \times 10^3$ kg ha^{-1} yr^{-1}, with an average of 5.1×10^3 kg ha^{-1} yr^{-1}. Marchant and Hynes (1981a, 1981b)

measured the feeding rate of *G. pseudolimnaeus* in the field and
calculated that the population consumed 1.5×10^3 kg ha^{-1} yr^{-1} of
organic matter in the half of the river occupied by the amphipod.
Thus in the river as a whole *G. pseudolimnaeus* consumed 0.8×10^3 kg
ha^{-1} yr^{-1} which is about 16% of the average input to such a river
given above.

This percentage seems a substantial amount of the total input
to be consumed by one species. However, the assimilation
efficiency of *G. pseudolimnaeus* feeding on detritus in the
laboratory is low, being about 10% (Marchant and Hynes, 1981a).
This efficiency varies slightly between species of leaves and
increases to 43-76% if pure cultures of fungi are fed (Bärlocher
and Kendrick, 1975). Marchant and Hynes concluded that
G. pseudolimnaeus probably only digests the microbial component of
the detritus as Hargrave (1970) has shown for the aquatic talitrid
amphipod *Hyalella azteca*. Nilsson (1974) measured an assimilation
efficiency of 30-40% for *G. pulex* feeding on decaying leaves in the
laboratory, but it seems likely that this is somewhat of an overestimate (Marchant and Hynes, 1981a).

Production of *G. pseudolimnaeus* (Table 8) is about 2% of
consumption and thus about 20% of assimilation, if it is valid to
extrapolate the value of 10% to the field as Marchant and Hynes
(1981a) conclude. Nilsson (1974) estimated production of *G. pulex*
in the laboratory indirectly by measuring respiratory rate and
subtracting this from the rate of assimilation; in this case
production was about 60% of assimilation and 20% of consumption. He
also measured growth (or production) directly under almost natural
field conditions (Table 7) and found it was about a third of the
value calculated by subtraction. If his direct measurements of
growth and his measurements of respiration are summed to give
assimilation rate, then assimilation efficiency of *G. pulex* becomes
about 20%, which agrees much better with previous estimates for
Gammarus and other amphipod genera, e.g. *H. azteca* has an
assimilation efficiency of 7-15% on detritus (Hargrave, 1970); the
ratio of production to consumption is also lowered to 7%. Nilsson
suggested that the difference between the direct and indirect
estimates of production was due to increased expenditure of energy
during the direct measurements, which took longer (160 days) than
the indirect ones (2 days), for locomotion and reproduction.
However, the amphipods appeared to have the same freedom for
movement in both his experiments and I think it more probable that
the differences are due to errors in estimating assimilation
efficiency. Reproduction of *G. pseudolimnaeus* accounts for about
6% of the production of those individuals that survive to breed.
Thus a decrease of this extent might be expected when comparing the
direct with the indirect values of production for *G. pulex*; but a
decrease of 60% seems unlikely.

The large amount of detritus not digested by *Gammarus*,
apparently about 80-90% of the consumed, is expelled as faecal
pellets which themselves become colonised by fungi and bacteria
(Lautenschlager et al., 1978) after a few days. Undoubtedly, these
pellets are eaten by other stream invertebrates and may be re-
ingested by *Gammarus*, particularly small individuals that may not
be as capable as adults at shredding leaves and other large particles
of organic matter. Therefore, although populations of *Gammarus* may
be the first to eat a substantial amount of the detritus that gets
into running water, they egest large amounts of finely chewed leaf
material which decay more readily than whole leaves because of
increased surface area and provide what may be a major source of
food for other benthic invertebrates in flowing water.

Of the small amount of ingested detritus that is transformed
into *Gammarus* tissue (about 5% from the two studies discussed
above) a large percentage may be eaten by predators. Welton (1979)
calculated that fish ate at least 60% of the annual production of
G. pulex in Tadnoll Brook; and it is possible to calculate from
the data of Marchant and Hynes (1981b) that if the first period of
mortality during the life cycle was due solely to predation then at
least 20% of the annual production of *G. pseudolimnaeus* was lost to
predators. Generally, predation is probably heaviest on the
smallest individuals whose growth contributes most to production.

The above estimates and efficiencies of energy transfer are
based on short term studies. However, only long term field studies
of *Gammarus* will provide sufficient data for statistical analysis
of variation in production and feeding rate, which is necessary
before energy flow can be understood. Much of such variation could
be due to fluctuations in the annual temperature regime as temperature
influences so many aspects of the performace of *Gammarus* in the
field.

Two relevant papers have appeared recently by Waters and
Hokenstrom (1980) and Welton and Clarke (1980). The first paper
is an expanded version of Waters (1976). During five years of
study Waters and Hokenstrom measured a mean density of
G. pseudolimnaeus of $1.6 \times 10^3 - 7.2 \times 10^3 \ m^{-2}$, a mean drift of
$10^4 - 10^5$ amphipods day^{-1} per m^3 s^{-1} and an annual production of
$64 - 271$ kg ha^{-1}. Drift and production were positively correlated
which the authors suggest is the result of drift removing biomass
that exceeds the stream's carrying capacity. Annual drift past a
point was equivalent to production in $20 - 25$ m of stream implying
no upstream movement was necessary to compensate for such a comparatively small loss. Thus the drift densities of other species
of *Gammarus* (Table 2) are probably within the productive capacity
of the amphipods. Waters and Hokenstrom confirmed that negligible
numbers of *G. pseudolimnaeus* occurred > 10 cm deep in the substrate.

Welton and Clarke (1980) showed that the degree days required
for egg development by *G. pulex* varied from 349 to 384 between 5
and 20°C in contrast to the constant value of 370 degree days
quoted earlier by Welton (1979). Otherwise their data on
development are similar to those obtained by Nilsson (1977; Table 6).
Specific growth rate of *G. pulex* in length (mm) varied from 0.28
to 0.96% day^{-1} between 5 and 20°C; growth of 0.69 mm % day^{-1} (at
15°C) was equivalent to 1.01 µg % day^{-1} and comparable to data
(based on weight) in Table 7. Although Welton and Clarke
considered growth in length to be exponential, it seems from their
graphs that it could equally well be interpreted as linear.

REFERENCES

Anderson, N.H. and Sedell, J.R. 1979. Detritus processing by macro-
    invertebrates in stream ecosystems. *Ann. Rev. Ent.* 24:
    351–377.
Bärlocher, F. and Kendrick, B. 1973. Fungi in the diet of
    Gammarus pseudolimnaeus (Amphipoda). *Oikos* 24: 295–300.
Bärlocher, F. and Kendrick, B. 1975. Assimilation efficiency of
    Gammarus pseudolimnaeus (Amphipoda) feeding of fungal mycelium
    or autumn-shed leaves. *Oikos* 26: 55–59.
Bärlocher, F. and Kendrick, B. 1976. Hyphomycetes as intermediaries
    of energy flow in streams. *In*: Recent Advances in Aquatic
    Mycology. E.B. Gareth Jones, ed. Elek, London.
Bousfield, E.D. 1958. Freshwater amphipod crustaceans of glaciated
    North America. *Can. Field Nat.* 72: 55–113.
Carle, F.L. 1977. The removal sample method for estimating benthic
    populations and diversity (abstract only). *N. Amer. Benth.
    Soc.* Twenty-Fifth Annual Meeting.
Elliott, J.M. 1969. Diel periodicity in invertebrate drift and the
    effect of different sampling periods. *Oikos* 20: 524–528.
Elliott, J.M. 1971a. Upstream movements of benthic invertebrates
    in a Lake District stream. *J. Anim. Ecol.* 40: 235–252.
Elliott, J.M. 1971b. The distances travelled by drifting
    invertebrates in a Lake District stream. *Oecologia* 6: 350–
    379.
Gledhill, T., Sutcliffe, D.W. and Williams, W.D. 1976. Key to
    British freshwater Crustacea: Malacostraca. *Sci. Publ.
    Freshwat. Biol. Assoc.* U.K., 32: (2nd edition) 1–72.
Haeckel, J.W., Meijering, M.P.D. and Rusetzki, H. 1973.
    Gammarus fossarum Koch als Fallaubzerseter in Waldbächen.
    Freshwat. Biol. 3: 241–249.
Hargrave, B.T. 1970. The utilization of benthic microflora by
    Hyalella azteca (Amphipoda). *J. Anim. Ecol.* 39: 427–437.
Hobrough, J.E. 1973. The effects of pollution on *Gammarus pulex*
    (L.) subspecies *pulex* (Schellenberg) in the inlet streams of
    Rostherne Mere, Cheshire. *Hydrobiologia* 41: 13–35.

Hughes, D.A.   1970.   Some factors affecting drift and upstream
    movements of *Gammarus pulex*. *Ecology*   51:   301-305.
Hultin, L.   1971.   Upstream movements of *Gammarus pulex pulex*
    (Amphipoda) in a south Swedish stream.   *Oikos*   22:   329-347.
Hynes, H.B.N.   1954.   The ecology of *Gammarus duebeni* Lilljeborg
    and its occurrence in freshwater in western Britain.   *J.
    Anim. Ecol.*   23:   38-84.
Hynes, H.B.N.   1955.   The reproductive cycle of some British
    freshwater Gammaridae.   *J. Anim. Ecol.*   24:   352-387.
Hynes, H.B.N.   1970.   The Ecology of Running Water.   Liverpool
    University Press, Liverpool.
Hynes, H.B.N. and Harper, F.   1972.   The life histories of
    Gammarus lacustris and *Gammarus pseudolimmaeus* in southern
    Ontario.   *Crustaceana Suppl.*   3:   329-341.
Hynes, H.B.N. and Nicholas, W.L.   1963.   The importance of the
    Acanthocephalan *Polymorphus minutus* as a parasite of
    domestic ducks in the United Kingdom.   *J. Helminth.*   37:
    185-198.
Iversen, T.M. and Jessen, J.   1977.   Life cycle, drift and
    production of *Gammarus pulex* L. (Amphipoda) in a Danish
    spring.   *Freshwat. Biol.*   7:   287-296.
Kaushik, N.K. and Hynes, H.B.N.   1971.   The fate of dead leaves
    that fall into streams.   *Arch. Hydrobiol.*   68:   465-515.
Koch-Kallnbach, M.E. and Meijering, M.P.D.   1977.   Duration of
    instars and praecopulae in *Gammarus pulex* (Linnaeus, 1758)
    and *Gammarus roeselii* Gervais 1835 under semi-natural
    conditions.   *Crustaceana Suppl.*   4:   120-127.
Kostalos, M.S.   1979.   Life history and ecology of *Gammarus minus*
    Say (Amphipoda, Gammaridae).   *Crustaceana*   37:   113-122.
Lautenschlager, K.P., Kaushik, N.K. and Robinson, J.B.   1978.
    The peritrophic membrane and faecal pellets of *Gammarus
    lacustris limnaeus* Smith.   *Freshwat. Biol.*   8:   207-211.
Lehmann, U.   1967.   Drift and Populationsdynamik von *Gammarus
    pulex fossarum* Koch.   *Z. Morph. Ökol. Tiere.*   60:   227-274.
Macan, T.T.   1959.   The temperature of a small stony stream.
    Hydrobiologia   12:   89-105.
Macan, T.T. and Mackereth, J.C.   1957.   Notes on *Gammarus pulex*
    in the English Lake District.   *Hydrobiologia*   9:   1-11.
Maitland, P.S.   1966.   Notes on the biology of *Gammarus pulex*
    in the River Endrick.   *Hydrobiologia*   28:   142-151.
Mann, K.H.   1971.   Use of the Allen curve method for calculating
    benthic production.   *In*:   A Manual on Methods for the
    Assessment of Secondary Productivity in Fresh Waters.
    W.T. Edmondson and G.G. Winberg, eds.   I.B.P. Handbook.
    17:   (1st edition) Blackwell, Oxford.
Marchant, R. and Hynes, H.B.N.   1981a.   Field estimates of
    feeding rate for *Gammarus pseudolimmaeus* (Crustacea:
    Amphipoda) in the Credit River, Ontario.   *Freshwat. Biol.*
    11:   (in press).

Marchant, R. and Hynes, H.B.N. 1981b. The distribution and
    production of *Gammarus pseudolimnaeus* (Crustacea: Amphipoda)
    along a reach of the Credit River, Ontario. *Freshwat. Biol.*
    11: (in press).
McLay, C. 1970. A theory concerning the distance travelled by
    animals entering the drift of a stream. *J. Fish. Res. Bd.
    Canada* 27: 359-370.
Meijering, M.P.D. 1972a. Physiologische Beitrage zur Frage der
    Systematischen Stellung von *Gammarus pulex* (L.) und
    Gammarus fossarum Koch (Amphipoda). *Crustaceana Suppl.*
    3: 313-325.
Meijering, M.P.D. 1972b. Experimentelle Untersuchungen zur Drift
    und Aufwanderung von Gammariden in Fliessgewässern. *Arch.
    Hydrobiol.* 70: 133-205.
Meijering, M.P.D. 1977. Quantitative relationships between drift
    and upstream migration of *Gammarus fossarum* Koch 1835.
    Crustaceana Suppl. 4: 128-135.
Moore, J.W. 1975. The role of algae in the diet of *Asellus
    aquaticus* L. and *Gammarus pulex* L. *J. Anim. Ecol.* 44:
    719-730.
Nilsson, L.M. 1974. Energy budget of a laboratory population of
    Gammarus pulex (Amphipoda). *Oikos* 25: 35-42.
Nilsson, L.M. 1977. Incubation time, growth and mortality of the
    amphipod *Gammarus pulex* under laboratory conditions. *Oikos*
    29: 93-98.
Nilsson, L.M. and Sjöstrom, P. 1977. Colonisation of implanted
    substrates by differently sized *Gammarus pulex* (Amphipoda).
    Oikos 28: 43-48.
Pennak, R.W. 1978. Freshwater invertebrates of the United States.
    (2nd edition). John Wiley & Sons, New York.
Rees, C.P. 1972. The distribution of the amphipod *Gammarus
    pseudolimnaeus* Bousfield as influenced by oxygen concentration,
    substratum, and current velocity. *Trans. Amer. Microscop. Soc.*
    91: 514-529.
Roux, A.L. 1975. Température stable et température fluctuante.
    II Étude comparative de leurs effets sur la durée d'intermue
    de Gammaridae femelles. *Verh. int. Verein. theor. angew.
    Limnol.* 19: 3014-3021.
Tilly, L.J. 1968. The structure and dynamics of Cone Spring.
    Ecol. Monogr. 38: 169-197.
Wallace, R.R., Hynes, H.B.N. and Kaushik, N.K. 1975. Laboratory
    experiments on factors affecting the activity of *Gammarus
    pseudolimnaeus* Bousfield. *Freshwat. Biol.* 5: 533-546.
Waters, T.F. 1962. Diurnal periodicity in the drift of stream
    invertebrates. *Ecology* 43: 316-320.
Waters, T.F. 1965. Interpretation of invertebrate drift in
    streams. *Ecology* 46: 327-334.
Waters, T.F. 1972. The drift of stream insects. *Ann. Rev. Ent.*
    17: 253-272.

Waters, T.F.   1976.   Annual production and drift of the amphipod
    Gammarus pseudolimmaeus in Valley Creek, Minnesota (abstract
    only).   *N. Amer. Benth. Soc.* Twenty-Fourth Annual Meeting.
Waters, T.F.   1977.   Secondary production in inland waters.
    Adv. Ecol. Res.   10:   91-164.
Waters, T.F. and Hokenstrom, J.C.   1980.   Annual production and
    drift of the stream amphipod *Gammarus pseudolimmaeus* in Valley
    Creek, Minnesota.   *Limnol. Oceanogr.*   25:   700-710.
Welton, J.S.   1979.   Life-history and production of the amphipod
    Gammarus pulex in a Dorset chalk stream.   *Freshwat. Biol.*
    9:   263-275.
Welton, J.S. and Clarke, R.T.   1980.   Laboratory studies on the
    reproduction and growth of the amphipod *Gammarus pulex* (L.).
    J. Anim. Ecol.   49:   581-592.
Williams, D.D. and Hynes, H.B.N.   1974.   The occurrence of benthos
    deep in the substratum of a stream.   *Freshwat. Biol.*   4:
    233-256.
Willoughby, L.G. and Sutcliffe, D.W.   1976.   Experiments on feeding
    and growth of the amphipod *Gammarus pulex* (L.) related to its
    distribution in the River Duddon.   *Freshwat. Biol.*   6:
    577-586.

EFFECTS OF HYDRODYNAMICS ON THE DISTRIBUTION OF LAKE BENTHOS

David R. Barton

Department of Biology
University of Waterloo
Waterloo, Ontario, Canada

Most studies of the benthos of lakes have been aimed toward evaluating productivity, either in terms of the lake's capacity for fish production or its general trophic status, including the degree of man-made change. Both qualitative and quantitative approaches have been used depending on the specific problem being studied, but in either case conclusions are based on samples of the bottom fauna. Interpretation of these samples usually assumes that they represent a biological expression of environmental conditions at a given point or area over a period of time. That is, benthic animals have relatively long life cycles and live, by definition, on the bottom, so are exposed to conditions at a given site for a relatively long period of time - an argument often put forward to justify biological monitoring of water quality.

I have no doubt that this is a valid justification; benthic animals generally are more persistent in time than the water above them. However, as shown by most year-round or longer term studies, the animals within a lake often exhibit considerable temporal and spatial heterogeneity. These fluctuations in the abundance of organisms in one portion of the lake may be related to their annual cycle of recruitment and mortality, to unpredictable events such as storms or prolonged stratification, or to the normal behavior of the animals in response to predictable environmental changes. As I hope to show in this essay, all of these factors are related, at least in part, to movements of water, the lake's hydrodynamics.

Four classes of motion seem to be of greatest importance: surface waves, seiches (both surface and internal), wind-induced currents and convection currents. In the following I will briefly discuss some of the ways in which benthic invertebrates are

affected by these motions. The physics of these phenomena is
beyond the scope of this essay, as well as being beyond the
sophistication of the biological examples, so for these aspects the
interested reader is referred to any standard limnological text
(e.g. Hutchinson, 1957).

Surface waves

Since the size of surface waves is determined by the strength
and duration of the wind in combination with the fetch and depth of
the lake, larger lakes can have larger waves. It is possibly of
greater biological significance, however, that the shores of larger
lakes are affected by wave action more of the time. Waves generated
by strong local winds can travel long distances (in deep water)
with little loss of energy, and lighter winds perpendicular to the
shore are generated by the lake itself (Bruschin and Schneiter,
1979). This more or less continuous wave action prevents the
accumulation of fine sediment and creates conditions along the shore
which are more lotic than lentic. The benthic invertebrates inhabiting this wave- or surf-zone are those characteristic of running water
and may include members of the Heptageniidae, Baetidae
(Ephemeroptera), Plecoptera, Hydropsychidae (Trichoptera), Psephenidae
Elmidae (Coleoptera), and certain Chironomidae (Diptera) (Wesenberg-
Lund, 1908; Krecker and Lancaster, 1933; Fryer, 1959; Barton and
Hynes, 1978a). The presence of these groups, especially the
Hydropsychidae and Psephenidae, has been used to define the lower
limit of the wave-zone, e.g. 1 m in Lake Mendota, Wisconsin
(Muttkowski, 1918), 2 m in Lake Simcoe, Ontario (Rawson, 1930) and
5 m in western Lake Erie (Shelford and Boesel, 1942).

The relationship between fetch and the maximum potential height
of surface waves, estimated as $h = 0.105\sqrt{fetch(cm)}$ (Wetzel, 1975),
seems to account very well for the relative depths of the wave-zones
of smaller lakes such as Lake Simcoe (f = 24 km) and western Lake
Erie (f = 80 km). In larger lakes other factors become important.
For example, in eastern Lake Superior the maximum fetch is about
480 km yet I have never observed hydropsychids or psephenids deeper
than 5 m even on the most exposed shores. Perhaps the depth of the
thermocline limits the size of waves except during very high winds.

On a local scale the relationship between fetch and the depth
of wave action is clearly reflected in the vertical zonation of the
benthic fauna. In a recently completed study (Barton and Carter,
in prep.), we compared benthic communities on the north, south,
east and west sides of several bedrock islands in eastern Georgian
Bay, Lake Huron. The prevailing winds are from the west with a
fetch of about 100 km. Easterly fetches are in the range of 8-15
km while north and south shores are somewhat protected by adjacent
islands. Quantitative samples were collected down to a depth of
4 m in May and July. Ordination of these samples revealed clusters

of sites ranked along the predicted gradient of decreasing exposure
to wave action: 1) depths of 1 m on west shores and 0.2 m on east
shores; 2) 2 m west; 3) 1 m east; 4) 4 m west, 2 m east and 1-2 m
north and south; and, 5) 3 and 4 m on north, south and east shores.
There was a tendency for total invertebrate abundance to increase
with decreasing exposure in May but this was less apparent in July
when large numbers of *Alona costata* and *A. setulosa* (Cladocera)
were found at the most exposed sites. Other species of *Alona*,
Chydorus gibbus, Naididae and Chironomidae dominated more sheltered
sites. Species richness was inversely related to exposure in both
months. Shannon-Wiener diversity tended to be greatest at the
intermediate sites.

Earlier studies have revealed the importance of wave action in
the distribution of certain molluscs. In western Lake Erie, Dennis
(1938) observed that *Goniobasis* required wave action or at least a
solid, unsilted substratum. *Physa*, *Lymnea* and *Pleurocera* also
preferred moving water but could not inhabit areas of extreme
exposure. *Campeloma*, *Planorbis*, *Amnicola* and eight other genera
completely avoided exposed areas. Krecker's (1924) detailed
observations on the distribution of *Goniobasis livescens* in the
same region indicated that this snail could withstand fairly heavy
surf but was more abundant in slightly sheltered areas. With
increasing exposure they attached to larger objects and, when waves
broke over them, moved to the sheltered side of the stone. Krecker
found that *Goniobasis* maintained its position despite having its
shell experimentally rotated through an arc of about 30° to
simulate the wash of a passing wave. If the stone to which it was
attached was struck with a hammer but did not move, the snail
stayed in place; if the stone moved, the snail dropped off
immediately.

Neither of these authors mentioned morphological differences
among populations inhabiting areas of differing exposure, a subject
which has received considerable attention in marine littoral
situations (e.g. Harger, 1970). Evidence of such differences was
provided by Brown et al. (1938) who compared populations of clams
(Unionidae) from a sheltered bay, a moderately exposed bay and a
completely exposed shoal, all in western Lake Erie. For all species
collected in sufficient numbers, size (length, weight and obesity)
was inversely proportional to exposure. Though noting that plankton
was less rich on the shoal, the authors did not conclude whether
this stunting was due to the relative abundance of food, the extra
energy required by the continual effort to stay in place, or,
perhaps, selective dislodgement of larger specimens.

Prolonged high winds during storms produce unusually large
waves which may have severe effects on benthic organisms.
Wesenberg-Lund (1908) and Dendy (1944) found large numbers of
various invertebrates, especially caddisfly larvae and molluscs,

stranded on the beach following such storms. Strahlendorf (1980)
suggested that amphipods and other invertebrates beached among
detached mats of *Cladophora* might be important forage for shore
birds in eastern Lake Ontario. Barton and Hynes (1978b) attributed
a sudden reduction in the abundance of benthic organisms at depths
of 0-2 m in north central Lake Erie to scouring by storm waves in
autumn. Dendy (1944) found that stream animals drifting into a
lake could survive in the sandy littoral zone if they could find
stable cover from wave action. Even vertebrates are not immune to
such events: in Oneida Lake, New York, Clady and Hutchinson (1975)
found egg masses of *Perca flavescens* on the beach after one day of
strong winds. These eggs had been deposited at depths of 1.5 to
3.7 m. Total mortality was conservatively estimated at nearly 1%
of all eggs spawned.

     Wind-induced wave action may also have more prolonged effects.
The disappearance of the mayfly *Hexagenia* from western Lake Erie has
been frequently cited as an example of the effects of eutrophication
(e.g. Britt, 1955). These mayflies do not seem to have returned
despite marked improvements in water quality (IJC, 1980). As
pointed out by Taft and Kishler (1973), the western basin was
originally covered by extensive beds of macrophytes which stabilized
the bottom allowing rapid clearing of the water even after severe
storms. Heavy siltation following agricultural development of the
watershed eliminated the macrophytes leaving an unstable mud bottom
which is frequently resuspended by wave action. Since the macrophytes disappeared well before the mayflies, perhaps the activities
of the mayflies themselves (e.g. production of fecal pellets) were
essential to stabilize the mud enough to allow larvae of succeeding
generations to become established. If so, re-establishment of the
population would be severely hampered by the loss of most or all of
the individuals present at any one time. Instability of the
substratum has also been used to explain the absence of *Asellus
racovitzai racovitzai* (Isopoda) from most of the western basin
(Kerr, 1978). This species seems to be thriving on enriched mud in
somewhat deeper or more sheltered parts of the lake.

Seiches

     In addition to producing surface waves, prolonged wind often
sets up long standing waves, seiches, which may be apparent at the
surface or, more frequently, at the thermocline. Surface seiches
are seldom of sufficient magnitude or frequency to be of great
biological importance although temporary lowering of the water
level does subject animals to desiccation and may substantially
increase the molar action of waves on gently sloping beaches
(Krecker, 1931). Increased sedimentation in the nearshore zone
from erosion of the shoreline during the high cycle may also prove
detrimental to the fauna.

More important, and much more frequent, are internal waves
set up by wind stress during thermal stratification. The piling up
of warm epilimnetic water depresses the thermocline at the downwind
side of the lake with a corresponding elevation at the upwind side.
When the wind stops, the epilimnion and hypolimnion oscillate relative to one another resulting in drastic changes in temperature
(and often various chemical parameters) at the substratum contacted
by the thermocline as it moves on- and offshore. Since these standing
waves appear to rotate around the basin due to the Coriolis effect,
the entire perimeter of the lake can be subject to sudden temperature
fluctuations. This may be the cause of the 'sublittoral-minimum'
in the abundance of benthic animals observed in many large, deep
lakes (Brinkhurst, 1974).

The critical field test of this hypothesis has not yet been
done but there is at least one published report which describes
severe biological consequences associated with an internal seiche
(Emery, 1970). In mid August 1969, there was little difference
in temperature between the surface (20°C) and bottom (18.7°C) in
Little Dunks Bay (northern Georgian Bay) where observations were
being made from an underwater chamber. During a 4 hr period cold,
murky, hypolimnetic water moved to within 4 m of the surface lowering
the temperature to 7°C and dropping visibility from 10 to 2 m.
Mobile fish left the area. Benthic sculpins and crayfish did not,
and many dead ones were observed after the seiche. No mention was
made of the fate of other invertebrates but it seems likely that
non-burrowing forms, such as Trichoptera and most Ephemeroptera,
would also have suffered extreme thermal stress.

Seiches are not necessarily biologically detrimental. As the
epilimnion and hypolimnion oscillate relative to one another, strong
currents are generated, particularly near the nodes. Boltt (1969)
felt that planktonic larvae of benthic species utilized these
seiche currents to reach and recolonize areas of St. Lucia Lake
(South Africa) which had been denuded by a previous saltwater
intrusion.

Wind-induced currents

It is becoming increasingly evident that currents are of
critical importance in the distribution of many benthic animals in
lakes. In addition to the motions associated with seiches,
horizontal currents are generated directly by wind stress, or by
variations in atmospheric pressure, horizontal density gradients
(temperature or solution) and inflows from streams. Wind-induced
currents are by far the most common and powerful. Current velocity
tends to increase with wind speed up to some critical level, then
decreases, probably because of turbulence from waves. Hamblin
(1971) observed sustained surface currents of up to 18 cm sec^{-1}
in the open waters of Lake Erie. These may have been augmented

by the permanent circulation of the lake since they represented 7%
of wind speed, a very high value - 2% is more common (Wetzel, 1975).

Currents are, of course, responsible for the distribution of
substances which influence the composition and abundance of the
benthic fauna. This is especially apparent near point sources of
allochthonous organic matter such as polluted or enriched rivers,
whose inflows may induce more or less permanent currents due to
density differences. In large bodies of water these currents tend
to be deflected to the right (in the northern hemisphere) by the
Coriolis effect and this may be reflected in the distribution of
organisms which thrive in enriched situations. An illustration of
this is seen in the distributions of tubificids in Saginaw Bay,
Lake Huron (Schneider et al., 1969) and Green Bay, Lake Michigan
(Howmiller and Beeton, 1970) which show distinct zones of greatest
abundance parallel to the right-hand shores.

Ephemeral surface currents, wind drift, change direction with
the wind and the net result may range from a general distribution
of material and organisms throughout the lake to sudden, massive
shifts in population abundance. For example, Wolfert and Hiltunen
(1968) found heavy concentrations of the snail *Viviparous japonicus*
along the downwind shores, either north or south, in Sandusky Bay,
Lake Erie. Davies (1976b) found that under the influence of a
17-23 km h^{-1} wind egg masses of *Chironomus anthracinus* in Loch Leven
tended to accumulate 30° to the right of the wind direction
regardless of the site of oviposition.

This posed an interesting problem, as: 1) the bottom of Loch
Leven consists of sandy shallows and fine organic mud at depths
greater than 3 m; 2) egg masses of *C. anthracinus* accumulate in the
sandy shallows if strong winds blow during or just after
oviposition, and 3) larvae of *C. anthracinus* live almost exclusively
in the offshore mud. How do the larvae then reach a suitable
habitat? Based on his own earlier work (1974) and that of Lellak
(1968), Davies suggested that the positive phototaxis of first
instar larvae allows them to become planktonic and take advantage
of the same wind-induced currents to locate favorable conditions.

As any student of invertebrate or marine biology knows, planktonic larvae are the rule for marine benthic animals. In freshwater
this is generally not thought to be so, perhaps because so much
work has been done with coarse nets or in rivers where it seems
intuitively obvious that planktonic activity is likely to be
disadvantageous. First instar chironomid larvae, at least of lake
dwelling species, are the best known exception - positive phototaxis and morphological adaptations for planktonic life have been
reported by many authors (reviewed by Lellak, 1968; Oliver, 1971).
Less well known are observations of similar behavior in other
groups such as Neuroptera and Trichoptera (Wesenberg-Lund 1943;

Elliott, 1977).

It is likely that many, if not all, lentic insects are
potentially planktonic in the earliest instars. Weerekoon (1956)
noted the presence of insects typical of the shallow nearshore
zone on a permanently submerged (2-4 m) bank 500-700 m offshore in
Loch Lomond. Deep water with a muddy bottom separated the bank
from the shore. Since these insects leave the lake as adults,
Weerekoon wondered how they got back to the bank. Through a series
of experiments and observations, he successively ruled out wind
drift of eggs, active migration (walking along the bottom) of
larvae, selective oviposition over the bank and indiscriminant
oviposition all over the lake with development only in favorable
habitats. After observing that first instar *Phryganea*
(Trichoptera) are planktonic swimmers and young *Caenis*
(Ephemeroptera) swim much more than older ones, Weerekoon
hypothesized that the bank was repopulated by planktonic larvae of
these and other species utilizing wind-induced currents. Though
such currents were not measured, they undoubtedly occurred during
periods of offshore wind, need not have been very strong, and
would have accounted for the slow increases in population sizes
observed on the bank.

Is such behavior unique to lentic species? It is easy to
imagine that larvae which hatch from eggs deposited in an
unfavorable situation would swim up into the water column and be
carried away, finally settling only when suitable conditions were
encountered. For species whose eggs are laid in masses, crowding
by siblings might prompt the same behavior. Both factors could
apply as easily in streams as in lakes. Unfortunately, drift
nets fine enough to capture very young larvae are seldom used in
streams, but planktonic activity may be largely suppressed in
streams since lotic insects can be more precise in the selection
of oviposition sites than lentic ones and thus relatively few
larvae would hatch in unfavorable situations. For lotic species
hatching from an egg mass, planktonic existence might be very
brief since turbulent flow would quickly bring the larvae into
contact with less populous substrata. One might also hypothesize
that isolated offshore shoals in lakes are less likely to be
inhabited by insects whose eggs are laid singly.

Entry of older larvae and other organisms into the water
column in streams is a well-known phenomenon and has engendered
considerable debate over whether this is an active or a passive
process. The numerous reports of benthic animals in the water
column of lakes in the absence of scouring by waves would tend to
support the active drift theory. Mundie (1959) towed nets near
the surface of Lac La Ronge, Saskatchewan at night over depths of
5 and 13.5 m and caught several instars of chironomid larvae as
well as *Caenis, Hexagenia, Triaenodes, Hyallela, Pontoporeia* and

leeches. Davies (1976a) caught 18 species of chironomids in traps
suspended offshore in Loch Leven, 17 or which were represented by
third or fourth instar larvae. Hydras, *Hyallela azteca*,
Chaetogaster, *Nais simplex* and chironomids were captured in a
plankton net towed near the surface of Lake Superior, probably in
the daytime (Hiltunen, 1969). Hilsenhoff (1966) observed that
fourth (and possibly third) instar *Chironomus plumosus* often became
limnetic in Lake Winnebago, Wisconsin.

The pelagic occurrence of most nearshore benthic species has
recently been documented in both Lake Erie and Lake Michigan.
Barton and Hynes (1978b) caught the same species in nearly the
same proportions in drift traps as were collected in bottom
samples from the wave-zone of north central Lake Erie. More
animals were in the water column at night than in the day despite
calmer conditions at night. They concluded that longshore drift
was the primary mechanism allowing recolonization of an unstable
shoreline which is severely scoured by storms in autumn and spring.
The results of Wiley and Mozley (1978) in Lake Michigan were even
more analogous to stream drift since the proportion of the benthic
community (0.11% by numbers) in the water column was in the same
range reported for streams. Though nearly all of their sampling
was done in calm weather, they concluded that pelagic activity is
important in temporally unstable habitats since it allows rapid
colonization and exploitation of ephemeral deposits of rich organic
matter or periodically denuded substrata.

Temporal instability does not seem to be an absolute
requirement for pelagic activity by benthic animals. In a small
pond not subject to severe wave action, Macan and Kitching (1976)
found nearly every benthic species on plastic sheets or artificial
plants suspended near the surface. Kreis et al. (1971) collected
the same species on artificial substrates suspended 46 m from shore
as on those 366 m out in a southern Oklahoma reservoir. More
animals and more species colonized baskets hung near the surface
than at depths of 1.2 and 2.4 m. Walton (1980) has shown that the
same phenomenon occurs in stream pools.

Convection currents

Pelagic activity by benthic organisms thus appears to be an
active behavior and may not necessarily be linked to temporal
habitat instability. It may have originated as a response to
normal seasonal events. In temperate and northern lakes,
migration into the profundal in autumn and back to the littoral in
spring (Berg, 1938; Eggleton, 1931; Clampitt, 1970) would allow
animals to avoid ice scour, extremely low temperatures and even
freezing during the winter, and to complete their development
rapidly in warmer water in spring. Furthermore, pupation in the
shallows would reduce the distance through which emerging pupae

would be vulnerable to fish predation and pre-emergence aggregation
would ensure the close proximity of mates. These advantages might
be negated by the energy costs of such migrations if they involve
sustained, directed swimming. If suitable currents exist at the
appropriate times, energy costs would be minimized.

One of the necessary currents, offshore in autumn, is a
normal feature of temperate lakes during the breakdown of thermal
stratification. Cooling tends to be more rapid in the shallows
near shore and the cooler, denser water flows downward along the
bottom. Lake Washington provides a striking example of the
potential strength of these convection currents: the lake's
bottom is W-shaped in cross-section due to the annual scour of
newly deposited limnetic peat by autumnal currents flowing down
the lake's steep slopes (Gould and Budinger, 1958). In the English
Lake District, Gorham (1958) found that convection currents during
autumn turnover may mix as much as the top 22 mm of profundal
sediment. Similar mixing did not occur in spring. That benthic
animals respond to these water movements is indicated by Mundie's
(1956) observation of a four-fold increase in the abundance of
pelagic chironomids at overturn.

A return flow should occur in spring, at least in larger
lakes where the water warms much faster in the littoral leading to
the formation of a thermal bar. This would provide a shoreward
current at least from the depth of initial stratification (Bennett,
1971). For animals overwintering in deeper water or in smaller
lakes, the strong winds and isothermal conditions of spring should
provide strong currents which would transport them toward shore,
although perhaps indirectly.

The overall importance of water movements to aquatic insects
can hardly be overstated. By far the majority of Chironomidae
(Davies, 1976c) and Ephemeroptera (Edmunds et al., 1976)
oviposit near shore, not over the general lake surface, so the
establishment of offshore populations of larvae depends on
migration. These migrations seem to be dependent on currents. In
fact, the distributions of larval insects might give an indication
of the strength of the lake's internal circulation. In very large
lakes a few chironomids penetrate even the greatest depths
(Rawson, 1953; Adams and Kregear, 1969) but numbers are substantially reduced below about 150 m. The virtual absence of
chironomids and other insects from the offshore parts of certain
smaller lakes might also be attributable to weak internal
circulation. This could apply in the far north such as Lake
Thingvallavatn, Iceland, which is deep (114 m), does not become
thermally stratified and has very few chironomids in the profundal
region (Lindegaard, 1980). At the other extreme, both of latitude
and size, is a small lake in Nigeria studied by R. Landis Hare
(pers. comm., Univ. of Waterloo), which is sheltered from the wind

by the surrounding forest and exhibits weak daily thermal
stratification. Chironomids and insects other than *Chaoborus* are
found only around the margin.

In summary, movements of the water would seem to be at least
as important to the benthic fauna in lakes as in rivers; perhaps,
because they are less predictable, more so. Under average
conditions surface waves effect a fine depth zonation of the shallow
water fauna, which may include distinctly lotic elements.
Community responses to the degree of exposure to wave action
include differences in total abundance, species richness and
diversity. Individual species may show morphological differences.
Larger storm waves often denude large areas. In combination with
human-related environmental stress, waves may drastically alter
the benthic community over even larger areas. Internal waves may
account for the 'sublittoral-minimum' observed in some stratified
lakes, and in extreme cases cause heavy mortality in the benthic
 fauna. Currents, largely induced by wind or temperature
differences, are critical in the dispersal of allochthonous
material and organisms to various parts of lakes. Most benthic
organisms are capable of actively entering the water column where
they are transported by currents. Such migrations allow animals
to exploit different depths on a seasonal basis as well as ephemeral
resources such as newly scoured substrata. In the absence of
suitable currents, large areas of lakes may be inaccessible to
certain benthic animals, particularly insects.

In view of these factors, I conclude as did Mundie (1959),
that lentic benthic organisms are not as static and dependent on
bottom conditions as has usually been supposed.

REFERENCES

Adams, C.E. and Kregear, R.D. 1969. Sedimentary and faunal
    environments in eastern Lake Superior. *Proc. 12th Conf.
    Great Lakes Res.*: 1-20.
Barton, D.R. and Hynes, H.B.N. 1978a. Wave-zone macrobenthos of
    the exposed Canadian shores of the St. Lawrence Great Lakes.
    J. Great Lakes Res. 4: 27-45.
Barton, D.R. and Hynes, H.B.N. 1978b. Seasonal variations in
    densities of macrobenthic populations in the wave-zone of
    north-central Lake Erie. *J. Great Lakes Res.* 4: 50-56.
Bennett, J.R. 1971. Thermally driven lake currents during the
    spring and fall transition periods. *Proc. 14th Conf. Great
    Lakes Res.*: 535-544.
Berg, K. 1938. Studies on the bottom animals of Esrom Lake.
    K. Danske Videsnk. Selsk. Naturv. Math. Afd. 9: 1-255.
Boltt, R.E. 1969. The benthos of some southern African lakes.
    Park V. The recovery of the benthic fauna of St. Lucia
    Lake following a period of excessively high salinity.
    Trans. R. Soc. S. Afr. 41: 295-323.

Brinkhurst, R.O. 1974. The Benthos of Lakes. MacMillan, London.

Britt, N.W. 1955. Stratification in western Lake Erie in summer of 1953: effects on the *Hexagenia* (Ephemeroptera) population. *Ecology* 36: 239-244.

Brown, C.J.D., Clark, C. and Gleissner, B. 1938. The size of certain naiads from western Lake Erie in relation to shoal exposure. *Amer. Midl. Nat.* 19: 682-701.

Bruschin, J. and Schneiter, L. 1979. Wind-waves in the bay of Geneva. *In* "Hydrodynamics of lakes." Graf, W.H. and Mortimer, C.H., eds, Elsevier, Amsterdam.

Clady, M. and Hutchinson, B. 1975. Effect of high winds on eggs of yellow perch, *Perca flavescens*, in Oneida Lake, New York. *Trans. Am. Fish. Soc.* 104: 524-525.

Clampitt, P.T. 1970. Comparative ecology of the snails *Physa gyrina* and *Physa integra* (Basommatophora:Physidae). *Malacologia* 10: 113-151.

Davies, B.R. 1974. The planktonic activity of larval Chironomidae in Loch Leven, Kinross. *Proc. R. Soc. Edinb.*, B. 74: 275-283.

Davies, B.R. 1976a. A trap for capturing planktonic choronomid larvae. *Freshwat. Biol.* 6: 373-380.

Davies, B.R. 1976b. Wind distribution of the egg masses of *Chironomus anthracinus* (Zetterstedt) (Diptera:Chironomidae) in a shallow, wind-exposed lake (Loch Leven, Kinross). *Freshwat. Biol.* 6: 421-424.

Davies, B.R. 1976c. The dispersal of Chironomidae larvae: a review. *J. ent. Soc. Sth Afr.* 39: 39-62.

Dendy, J.S. 1944. The fate of animals in stream drift when carried into lakes. *Ecol. Monogr.* 14: 334-357.

Dennis, C.A. 1938. Aquatic gastropods of Bass Island region of Lake Erie. *F. Theo. Stone Lab.*, *Contrib.* 8: 1-34.

Edmunds, G.F., Jr., Jensen, S.L. and Berner, L. 1976. The mayflies of North and Central America. Univ. Minnesota Press, Minneapolis.

Eggleton, F.E. 1931. A limnological study of the profundal bottom fauna of certain freshwater lakes. *Ecol. Monogr.* 1: 233-331.

Elliott, J.M. 1977. A key to British freshwater Megaloptera and Neuroptera. *Freshwat. Biol. Assoc. Sci. Publ.* 35.

Emery, A.R. 1970. Fish and crayfish mortalities due to an internal seiche in Georgian Bay, Lake Huron. *J. Fish. Res. Bd. Canada* 27: 1165-1168.

Fryer, G. 1959. The trophic interrelationship and ecology of some littoral communities of Lake Nyasa with special reference to the fishes, and a discussion of the evolution of a group of rock-frequenting Cichlidae. *Proc. Zool. Soc. Lond.* 132: 153-281.

Gorham, E. 1958. Observations on the formation and breakdown of the oxidized microzone at the mud surface in lakes. *Limnol. Oceanogr.* 3: 291-298.

Gould, H.R. and Budinger, T.F. 1958. Control of sedimentation and
    bottom configuration by convection currents, Lake Washington,
    Washington. *J. mar. Res.* 17: 183-197.
Hamblin, P.F. 1971. Circulation and water movement in Lake Erie.
    Dept. Energy, Mines, Resources. Ottawa, Can. Sci. Ser. 7:
    50 pp.
Harger, J.R.E. 1970. The effect of wave impact on some aspects
    of the biology of sea mussels. *Veliger* 12: 401-404.
Hilsenhoff, W.L. 1966. The biology of *Chironomus plumosus* in Lake
    Winnebago, Wisconsin. *Ann. ent. Soc. Amer.* 59: 465-473.
Hiltunen, J.K. 1969. Invertebrate macrobenthos of western Lake
    Superior. *Mich. Acad.* 1: 123-133.
Howmiller, R.P. and Beeton, A.M. 1970. The oligochaete fauna of
    Green Bay, Lake Michigan. *Proc. 13th Conf. Great Lakes Res.*:
    15-46.
Hutchinson, G.E. 1957. A treatise on limnology. *1.* Geography,
    physics, and chemistry. John Wiley & Sons, Inc., N.Y.
International Joint Commission. 1980. Report for years 1978-1979,
    Washington, Ottawa. 64 pp.
Kerr, J.R. 1978. Some aspects of life history and ecology of the
    isopod *Asellus r. racovitzai* in western and central Lake Erie.
    Ohio J. Sci. 78: 298-300.
Krecker, F.H. 1924. Conditions under which *Goniobasis livescens*
    occurs in the Island Region of Lake Erie. *Ohio J. Sci.*
    24: 299-310.
Krecker, F.H. 1931. Vertical oscillations or seiches in lakes as
    a factor in the aquatic environment. *Ecology* 12: 156-163.
Krecher, F.H. and Lancaster, L.Y. 1933. Bottom shore fauna of
    western Lake Erie: a population study to a depth of six
    feet. *Ecology* 14: 79-93.
Kreis, R.D., Smith, R.L. and Moyer, J.E. 1971. The use of
    limestone-filled basket samplers for collecting reservoir
    invertebrates. *Wat. Res.* 5: 1099-1106.
Lellak, J. 1968. Positive phototaxis der Chironomiden larvulae
    als regulierender faktor ihrer verteilung in stehenden
    gewasser. *Ann. Zool. Fenn.* 5: 84-87.
Lindegaard, C. 1980. Bathymetric distribution of Chironomidae
    (Diptera) in the oligotrophic Lake Thingvallavatn, Iceland.
    In "Chironomidae: ecology, systematics, cytology and
    physiology." Murray, D.A., ed. Pergamon, Oxford.
Macan, T.T. and Kitching, A. 1976. The colonization of squares
    of plastic suspended in mid water. *Freshwat. Biol.* 6: 33-40.
Mundie, J.H. 1956. Invertebrate animals (Director's Report).
    Freshwat. Biol. Assoc. 24th Annual Rep. p. 29.
Mundie, J.H. 1959. The diurnal activity of the larger inverte-
    brates at the surface of Lac La Ronge, Saskatchewan.
    Can. J. Zool. 37: 945-956.
Muttkowski, R.A. 1918. The fauna of Lake Mendota. *Trans. Wis.
    Acad. Sci., Arts and Letters* 19: 374-482.

Oliver, D.R. 1971. Life history of the Chironomidae. *Ann. Rev. Entomol.* 16: 211-230.

Rawson, D.S. 1930. The bottom fauna of Lake Simcoe and its role in the ecology of the lake. *Univ. Toronto Stud. biol. Ser.* 40: 1-183.

Rawson, D.S. 1953. The bottom fauna of Great Slave Lake. *J. Fish. Res. Bd. Canada* 10: 486-520.

Schneider, J.C., Hooper, F.F. and Beeton, A.M. 1969. The distribution and abundance of benthic fauna in Saginaw Bay, Lake Huron. *Proc. 12th Conf. Great Lakes Res.*: 80-90.

Shelford, V.E. and Boesel, M.W. 1942. Bottom animal communites of the island area of western Lake Erie in the summer of 1937. *Ohio J. Sci.* 42: 179-190.

Strahlendorf, P.W. 1980. Migrant shorebird ecology, with special reference to shorebird migration along the north-eastern shore of Lake Ontario. *Parks Branch, Ont. Min. Nat. Res., Napanee* 96 pp.

Taft, C.E. and Kishler, W.J. 1973. *Cladophora* as related to pollution and eutrophication in western Lake Erie. *Wat. Res. Center, Ohio State Univ., Proj. Compl. Rep.* 332X, 339X, 103 pp.

Walton, O.E., Jr. 1980. Active entry of stream benthic macroinvertebrates into the water column. *Hydrobiologia* 74: 129-139.

Weerekoon, A.C.J. 1956. Studies on the biology of Loch Lomond: 2. The repopulation of McDougall Bank. *Ceylon J. Sci., Section C. Fisheries* 7: 95-133.

Wesenberg-Lund, C. 1908. Die littoralen Tiergesellschaften unserer grossern Seen. *Int. Revue ges. Hydrobiol.* 1: 574-607.

Wesenberg-Lund, C. 1943. Biologie der Susswasserinsekten. Gyldendalske Boghandel-Nordisk Forlag, Kopenhagen.

Wetzel, R.G. 1975. Limnology. W.B. Saunders Co., Philadelphia.

Wiley, M.J. and Mozley, S.C. 1978. Pelagic occurrence of benthic animals near shore in Lake Michigan. *J. Great Lakes Res.* 4: 201-205.

Wolfert, D.R. and Hiltunen, J.K. 1968. Distribution and abundance of the Japanese snail, *Viviparous japonicus*, and associated macrobenthos in Sandusky Bay, Ohio. *Ohio J. Sci.* 68: 32-39.

SECTION III

ASPECTS OF AQUATIC ENVIRONMENT PERTURBATIONS:

MANAGEMENT AND APPLICATION

MAN'S IMPACT ON TROPICAL RIVERS

Letitia E. Obeng

United Nations Environment Programme
Regional Office for Africa
P.O. Box 30552
Nairobi, Kenya, Africa

> "And the waters hugged their
> earthy beds in harmony to
> serve all creatures – till
> man muddied and confused
> them ......"

INTRODUCTION

The close dependence of man on water for life accounts for
the association which has existed between him and several rivers
over the centuries. His contact with major world rivers –
Tigris, Euphrates, Indus, Nile – encouraged the definition and
practice of various water management technologies such as dam
construction and crop irrigation which, in turn, contributed to
development and the progress of civilization. In the process of
using surface waters for development, man has interfered so much
with streams, rivers and lakes that now, they can hardly be
described as natural.

The interference has taken many forms. Water has been
withdrawn, often excessively, and faunal associations and
population balances have been grossly affected by the
disturbances; where water flow has been checked by a barrier,
ecological systems have been drastically changed downstream.
Upstream of such barriers, impoundments created manifest
characteristics which have little in common with the parent river
ecosystem and only over a period of time do they gradually begin
to acquire some definitive character. Water quality is affected
by dissolved and suspended matter through the passage of silt,
organic matter and other allochthonous materials and the leaching

of excesses of agricultural chemicals into surface waters.
Physico-chemical values are then altered and confusion in the
ordered systems affects growth and production patterns. Furthermore, since surface waters are often used for the disposal of
human waste, municipal sewage and industrial effluents, the organic
and inorganic elements introduced into them in this way affect
delicately balanced functional systems. Acid rain sometimes of pH
range 4.5 - 4.2, as reported from Sweden, has a serious impact
and disturbs surface waters. It is often a direct consequence of
coal combustion which releases oxides of sulphur and nitrogen into
the atmosphere, and of mining and similar operations by man.
Sometimes also, man quite deliberately introduces exotic plants
and animals into unaccustomed waters and thereby causes far-
reaching and, in some instances, irreversible repercussions in
aquatic ecosystems. *Eichhornia crassipes* (water hyacinth) for
example, originally South American, is now a serious plant pest
in many countries. *Lates niloticus* (the Nile perch), previously
alien to the Lake Victoria ecological system undoubtedly disturbed
the balanced faunal associations of the local fisheries when it
first appeared and, its effect on speciation of the indigenous
fishes is still unresolved.

The ecological changes which accompany the many activities
of man on surface waters may be traced roughly to three basic
reasons; namely, the physico-chemical and biological changes
which follow:
  (i)   increase in area, depth and volume of a water body,
  (ii)   decrease in the area, depth and volume, and
  (iii) changes in water quality which may be traced to a
         variety of causes (or a combination of these three
         categories of change).

The effects of these modifications on tropical waters are far
from being fully understood. Only limited studies, generally,
have been carried out on tropical streams and rivers. In Africa
for example, apart from the Nile and a few of the east African lakes
which have been investigated to some extent, until quite recently,
there was very little on record. Studies have been undertaken on
aspects of the Nile by Brook (1954), Brook and Rzoska (1954),
Talling (1957) Rzoska (1976) and others which cover its physical
and chemical characteristics, plankton, fishes, papyrus and
aquatic weeds.

On the other hand, considerable studies have been undertaken,
for many years now, on temperate streams and rivers. They have
been studied sectorally, according to the interests of specialists,
as well as comprehensively. Hynes (1970) has recorded over a
thousand references dating from Steinmann and Theinemann who were
pioneers in this field during the early part of the century to

recent times and reasonable information exists on aspects of the
ecology of streams and rivers.

In tropical regions, most of the significant ecological
changes which have occurred in surface waters have been
associated with the damming of streams and rivers by man. True,
drastic changes have also resulted from the degradation of water
quality by man, but attention is here focussed on the influence
of man on tropical streams and rivers through the construction of
dams.

It may not be an exaggeration to state that almost all rivers
on the African continent which are capable of generating power
have been dammed; those which are not large enough to produce power
have been dammed all the same for flood control, reliable supply of
water, fisheries or for some other reason. Consequently there are
innumerable small impoundments all over the African continent.

Whether it is a large river or a small stream which is dammed,
ecological upsets result as the moving water environment is
changed to that of a stationary water body. Small impoundments
hardly receive any consideration or study. Yet many schistosomiasis
transmission sites are found on hundreds of small impoundments in
endemic areas all over tropical Africa. However, it is on large
dams to which public attention is drawn that studies have been
concentrated.

In the last two decades, the construction of large dams in
Africa has encouraged the intensification of freshwater
investigations and studies, with emphasis on the ecological
changes which accompany them. But more studies and observations
are needed to cover other aspects such as geological, hydrological,
social and economic factors which have relevance to the nature and
severity of the impacts caused by the impoundment of the water
upstream of dams, and also downstream where there is reduction in
water quantity and other serious drawbacks. Because technical
information on tropical streams and rivers is limited, the base
data for comparing and evaluating results of studies on the changes
which occur from the activities of man are also often inadequate.

In the following, the effect of interference by man on four
large African rivers, the Volta, Niger, Zambezi and Nile is
discussed. The limnological changes observed in the Volta River
as it changed into a lake are presented first together with the
effects of the change on the fauna and flora. Later, this will be
compared with similar observations on the other three lakes in an
attempt to identify common features which may contribute to our
further understanding of what to expect when man slows down flow
and accumulates large quantities of water in rivers. Some

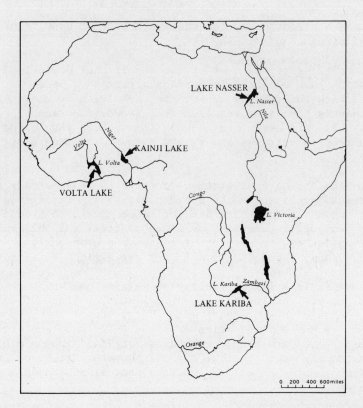

Figure 1. Map of Africa showing four major dammed rivers.

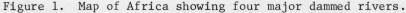

information on the social and human ecology aspects of interference
with these rivers will be given also.

Between 1958 and 1968, four large rivers were dammed and four
large artificial lakes were created in parts of Africa. Lake
Kariba was the first man-made lake to be created in December 1958
on the River Zambezi between Zambia and Zimbabwe. Volta Lake
started to form in May 1964 on the Volta River in Ghana. Lake
Kainji followed the completion of the damming of the River Niger
in Nigeria in December 1968, and the construction of the Aswan
High Dam in 1968 resulted in the formation of Lake Nasser/Lake
Nubia respectively in Egypt and in the Sudan (Fig. 1).

These man-made lakes share a common dendritic form with
extensive irregular shores (Fig. 2) however there are differences
in area, maximum depth, volume, holding capacity and discharge
rates and they lie in different geographical, climatic and
ecological zones. Lake Nasser on the Nile is almost entirely in a

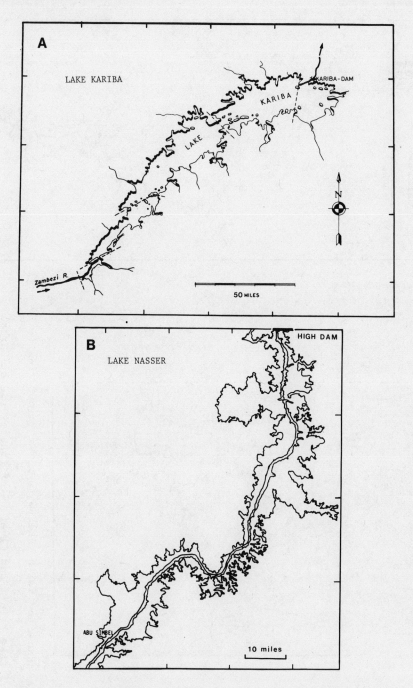

Figure 2a. Lake Kariba on the Zambezi (M. Van der Lingen).

Figure 2b. Lake Nasser and River Channel (P. Raheja).

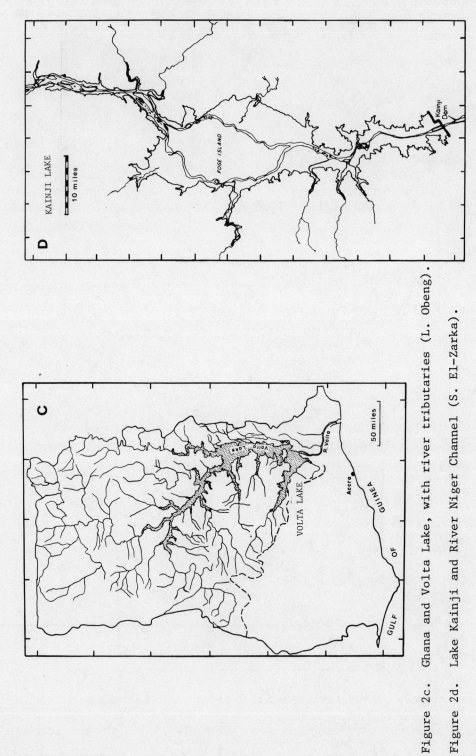

Figure 2c.   Ghana and Volta Lake, with river tributaries (L. Obeng).

Figure 2d.   Lake Kainji and River Niger Channel (S. El-Zarka).

Table 1. Characteristics of the four large African man-made lakes.

	River			
	Volta River	Zambezi	River Niger	The Nile
Name of dam	Volta Lake	Lake Kariba	Lake Kainji	Lake Nasser
Country	Ghana	Zambia/Zimbabwe	Nigeria	Egypt
Dam site	Akosombo	Kariba	Bussa	Aswan
Completion date	May 1964	December 1958	December 1968	May 1968
Location:				
Latitude	$6^{o}15'-9^{o}10'N$	$16^{o}28'-18^{o}04'S$	$10^{o}-10^{o}55'N$	$23^{o}58'-20^{o}27'N$
Longitude	$1^{o}40'W-0^{o}20'E$	$26^{o}42'-29^{o}03'E$	$4^{o}20'-45^{o}50'E$	$4^{o}25'-4^{o}45'E$
Altitude (in m. ASL)	85	485	142	180
Lake:				
Length (km)	402	277	137	482
Maximum width (km)	32	19.4	24	13
Area (km^2)	8,500	5,364	1,280	6,276
Maximum depth (m)	90	75	50	130
Mean depth (m)	18.6	29.2	12.3	25.2
Shoreline (km)	5,000	2,164	720	5,416
Volume (km^3)	165	156	15.8	158
Ecological zone	Forest/Savannah	Forest	Forest/Savannah	Desert
Population displaced	80,000	86,000	50,000	120,000

The figures are approximate. Various authors differ.

desert region and it does not receive any river tributaries or
substantial rainfall; Lakes Kariba, Kainji and Volta lie within
forested and savannah regions. Each has a different morphometry.
The characteristics of the four dams are presented in Table 1.
Some of the observations which have been made on the four lakes
show remarkable similarities and perhaps contribute further to
our knowledge of the ecological changes involved in the formation
of tropical man-made lakes.

Among the early overall impacts of the modification of these
large rivers by man was the innundation of large areas of land and
the restriction of access to subterranean minerals and similar
resources in the river basins. Indirect effects appeared through
changes which occurred in hydrology and levels of water tables
and in increased seismic activity. In the aquatic ecosystem
itself, in addition to changes in the water flow and the upsets in
the substratum, there were variations in temperature distribution
patterns as well as in the density of suspended matter, turbidity
and light penetration, and in the concentration and distribution
of oxygen, other gases, dissolved salts and chemical factors. The
variety and abundance of plankton, invertebrates, fishes and other
members of aquatic plant and animal communities were also greatly
affected through a boom in productivity during the early days.
The range of complex changes which were observed appears to be
characteristic of the early periods of impoundments; they typified
neither river nor lake conditions.

EFFECTS OF DAMMING THE VOLTA RIVER

With the completion of the construction of the dam, on the
Volta, the enlarging river flooded the lands beyond its banks,
wetting and churning up the dry earth, uprooting and toppling
adamant and unresisting trees alike, submerging homes and town,
till, in time, it spread over an area of about 8,500 km^2, creating
the world's largest single man-made lake, 402 km long, dendritic in
form, with an irregular shore line of over 4,800 km and a water
holding capacity of 165 km^3.

The White Volta enters Ghana from Togo through the extreme
north-eastern corner of the country. Shortly after, it receives
the Red Volta from Upper Volta. Further south it meets with the
Black Volta with which it forms the start of Volta River. Eight
rivers, the Daka, Oti, Asukawkaw, Dayi, Pru, Obosom, Sene and Afram
are the main tributaries on the Volta River.

Apart from the immediate changes caused in the ecology of the
river, there were enormous upsets in the lives of the people who
lived in the Volta basin. 80,000 people (1% of the population)
in 700 villages were flooded out of their homes. 69,249 had to
be relocated in 52 new towns specially built for them. They had

the traumatic experience of abandoning familiar and long accustomed lands (man is supposed to have been in the basin for at least half a million years) with ancestral resting places, farms and homes for strange places. They had to learn new skills to exploit the facilities brought by the change. They had to attune themselves to different social conditions, re-establish viable communities while clinging to, protecting, and preserving their cultural identities. Some people had to change from being farmers to fishermen, and they had soon to face the problem of sharing the fishery benefits of the new lake with an estimated 12,500 seasoned fishermen who invaded their basin. They brought with them diseases, including urinary schistosomiasis to which the un-prepared relocated people became exposed.

During the decade following the closure of the dam, ecological studies were undertaken jointly by national institutions and a number of United Nations bodies as part of a comprehensive, multi-disciplinary research programme. Many interesting and useful observations were made.

The bulk of the information used here has been drawn from results (mostly unpublished except in mimeographed Annual Reports) of studies by the staff of Ghana's Institute of Aquatic Biology. The information has been checked against, and complemented with, accounts recorded independently by other workers during the period. To avoid repetition of references from the same work and to make for smooth reading, acknowledgement and credit are given here to contributors to the following major works on African man-made lakes which have been used extensively not only for the account of the early days of the formation of the Volta Lake, but also for information on Lakes Kainji, Kariba and Nasser/Nubia (Lowe-McConnell, 1966; Obeng, 1969; Ackermann et al., 1973; Balon and Coche, 1974; Water Supply and Management, 1980; Proceedings of the International Conference on Kainji Lake and river basins development in Africa, 1980).

Volta Lake - early physical and chemical changes:

An initial and quite obvious change was the colour of the water in the enlarging river as turbid waters flowed in. Although the early measurements made were only qualitative and readings were taken with a Secchi disc, the results have been valuable for comparing turbidity and, possibly, sedimentation patterns along the lake. Readings taken near Yeji, about 140 km north ot the dam in 1968 for example showed light penetration to be as low as 10 cm while, near the dam site, on the other hand, readings of up to 380 cm were recorded at the time. Throughout the study period, in the rainy season the turbidity was generally high and light pene-tration was greatly reduced. Under reversed conditions in the dry season and, during the annual harmattan season when dry winds from

the Sahara blow across the West Coast to the Atlantic and which
last from November to April, the transparency increased. In that
period, around Yeji, the Secchi disc readings showed a range of
117 cm, 130 cm, 190 cm for November, January and April
respectively. Algal blooms also clouded the water in season and
the bodily mixing of the upper to lower waters increased the
turbidity. The overall distribution pattern which emerged was a
definite decrease in turbidity from the northern section southwards towards the dam which would suggest a loss of much of the
suspended matter to the areas northwards from the dam. In
succeeding years, transparency increased as the suspended matter
settled down.

During May - August 1964, in the period of filling, a mass
cooling of the water was reported at a time when there was hardly
any difference in temperature ($27^{o}C/26^{o}C$) between the top and
bottom. Later, there was a rise in surface temperature and a
significant difference between the surface and bottom values.
From 1966 and 1969, a surface to bottom temperature average range
of $31^{o}C$ - $27.3^{o}C$ was recorded in the northern section of the lake.
Further readings taken from 1970 and 1972 showed similar values.
Subsequent observations showed that surface temperatures remained
fairly even with only small fluctuations. During the day,
morning temperatures, as was to be expected, were usually lower
than the afternoon readings. At all stations the upper limit of
the surface temperatures did not vary much and neither did the
bottom temperatures. During eight years of observations, there
were only few surface temperatures above $33^{o}C$, and no bottom
temperatures below $23^{o}C$ were recorded from any station. However,
as with turbidity values, under specific conditions there were
variations. Rainfall for example, had an immediate effect on
lowering surface temperatures. Also, during the harmattan period,
there was a general reduction in surface temperature which was more
pronounced in the northern section. In such a season, the average
surface temperature dropped from 29.6 to $28.2^{o}C$ in the southern
section, whilst further north around Yeji, temperatures taken in
November, January and April averaged $28.8^{o}C$, $26.4^{o}C$ and $32^{o}C$
respectively. Temperature readings on the forming lake did not
show the typical temperature distribution pattern of stratification
as had been observed during the early period of the Kariba. The
absence of any permanent stratification during the period was not
surprising since the water was constantly affected by the mixing
process caused by external factors including waves, winds, storms
and rain, the harmattan, dry and flood seasons, to a depth of
20 - 35 m.

The hydrogen ion concentration (pH) values recorded from
1966 to 1972 were between pH 6.8 - pH 8.5 for most of the surface
samples taken. The total ion budget for the lake was low and it
was reflected in the conductivity readings (40 - 70 µmhos).

There was a definite decrease of hydrogen ion concentration values
with depth and, in some areas, values of pH 6.5 were recorded for
bottom samples.

Dissolved oxygen distribution in the forming lake environment
was interesting. The initial flooding of the area and impoundment
of water, with the subsequent decay of vegetable matter, caused a
shortage of oxygen for a period. Thereafter, the concentration
improved and, by 1968, oxygen could be measured to the bed in most
sections of the lake. As with temperature there was an oxygen
decrease with depth, in the early days. The difference was then
often substantial with readings ranging from 90% saturation at the
surface to 5% at a depth of 45 m. However winds and waves
encouraged a downward movement and bodily mixing of the surface
waters taking oxygen down to the bottom.

Chemical analyses included Ca^{++}, Mg^{++}, K^+, Mn^{++}, Na^+ total
iron, phosphate, nitrate, chloride, ammonia, hydrogen sulphide
and silicate. Iron was generally abundant in the surface waters
and there was a correlation between its concentration and the
turbidity readings. It seemed to be highest in flood waters and
to diminish as turbidity decreased. The values of ammonia, H_2S,
phosphates and managanese were highest in bottom samples. Ca^{++},
Mg^{++}, K^+, and silicate did not seem to have any specific distribution
pattern. Along the length of the lake, in a north to south
direction, there was little variation in the levels of concentration of the chemical factors.

There were limited Volta River data for comparison with the
results from the Lake. However measurement made on streams and
rivers in the basin brought out some differences as shown in
Table 2. There were higher values from the rivers and streams than
from the Lake for Fe^{++}, Na^{++} and Cl^- and lower readings for K^+ and
Mg^{++}. Exceedingly high concentrations were recorded from streams
which received large quantities of waste effluents from factories.

Plankton changes:

Samples taken from rivers and streams in the basin, and on the
Volta downstream of the dam showed a variety of plankton and algae.
They included diatoms, desmids and chlorophytic algae, *Navicula*
sp., *Synedra* sp., *Cocconeis* sp., *Gomphonema* sp., *Diploneis* sp.,
Cosmarium sp., *Closterium* sp., *Euglena* sp., *Phacus* sp., *Pediastrum*
sp., *Peridinium* sp., *Colacium* sp., and others.

Some of the diatoms were collected from algae, rocks, and
the river bed. The blue-green algae *Oscillatoria* sp. and *Anabaena*
sp. were present. Some Cladocera and copepods, especially *Cyclops*
sp., were also recorded.

Table 2. Values of chemical factors in streams and rivers in Volta Basin (mg l^{-1}).

River/Place	T°C	pH	O_2	Cl^-	NO_3^-	NH_3	SO_4^{2-}	PO_4^{3-}
Volta Lake	30.3	6.9	-	1.6	-	0.01	-	-
Dayi	-	5.7	-	19.7	-	0.16	-	-
Kwabenya	23°-28°	6.3-8.7	1.6-8.8	44.60-192.07	0.01-0.25	0-0.75	6.25-37.00	0-0.16
Tinkong	24°-27°	7.4-8.6	1.6-13.6	119.11-302.7	0-0.13	0.01-0.08	3.10-100.00	0.01-0.06
Asuoya	23°-28°	6.2-7.7	2.6-8.0	29.77-63.52	0.03-0.16	0.02-0.45	6.25-50.00	0-0.63
Astuare (Volta)*	25°-30°	4.7-8.2	0-16.0	0-44.67	0-0.28	0.01-2.00	0-25.00	0-1.89
Senchie	29°-31°	6.8-7.9	3.9-13.6	0-14.89	0-0.19	0-0.04	0-3.10	0-0.08
Senchie (Volta)	29°-30°	7.0-7.6	8.0-8.8	4.96-19.85	0.01-0.02	0-0.01	0-0	0-0
Akotex I **	36°-40°	9.0-11.5	0-12.0	0.99-709.70	0.03-1.25	0.04-3.13	12.50-75.00	0.05-10.0
Densu	26°-30°	7.2-8.4	2.4-9.6	9.92-729.56	0.01-0.13	0.02-0.40	6.25-100.00	0-0.05
Ayensu	26°-32°	7.0-8.3	4.8-8.8	29.78-94.29	0.01-0.62	0.01-0.06	3.10-12.50	0-0.03

* Receiving discharge from Sugar factory.
** Receiving discharge from Textile factory.

Table 2. cont'd.

River/Place	SiO$_2$	Na$^+$	K$^+$	Ca^{2+}	Mg^{2+}	Fe	Mn$_2$	Cond. (mho)
Volta Lake	15.7	3.7	5.0	7.8	2.0	0.01	–	–
Dayi	17.6	4.1	2.9	4.4	1.3	7.4	–	–
Kwabenya	0.05-100.00	5.00-75.00	0.40-6.40	12.0-68.0	0.15-24.30	0.01-1.00	0-0.47	500-5,000
Tinkong	10.00-15.00	7.20-76.00	0.40-4.20	28.0-104.0	0.29-0.95	0.01-7.00	0-0.31	610-5,900
Asuoya	5.00-100.00	29.00-47.00	1.90-9.30	16.0-60.0	0.04-26.70	0.18-1.00	0-0.31	70-3,200
Astuare (Volta)*	2.50-150.0	3.00-95.00	1.00-30.00	8.0-64.0	0.24-0.34	0-7.50	0-1.88	520-4,900
Senchie	1.25-5.00	1.20-4.80	0.40-2.20	4.0-28.0	0.01-0.12	0-0.18	0-0.01	23-600
Senchie (Volta)	2.50-3.75	2.60-3.60	1.40-2.60	4.0-20.0	0.02-0.07	0.09-2.00	0-0	520-540
Akotex I **	2.50-50.00	112.5-387.5	26.25-387.50	0-4.8	0.02-0.12	0.37-2.00	0-0.31	8,200-22,000
Densu	5.00-10.00	1.6-2000.0	0.80-76.00	16.0-116.0	0.02-2.70	0.02-1.00	0-0.31	2,100-3,800
Ayensu	2.50-5.00	2.4-40.0	1.20-13.00	10.0-32.00	0.07-0.15	0.02-2.00	0-0.31	200-3,000

In the forming lake, the plankton populations observed were
abundant and quite varied. It would seem that the changes in
temperature and dissolved nutrients provided an adequate stimulant
for explosive growths. Most easily noticed among the phytoplankton
were the blue-green algae *Microcystis, Oscillatoria* and *Anabaena*
which formed extensive blooms and were first prominent in the
northern section and later extended southwards where the
Bacillariophyceae were then dominant. The Chloroccocales were not
generally as widespread but they were abundant in the shallow
shores in 1966. Common Chlorococcales and Bacillariophyceae
included *Pediastrum, Synedra, Cosmarium, Scenedesmus, Eudorina*
and *Volvox*. Observations made by several workers suggested that
the local distribution and density of phytoplankton populations
were influenced by the intensity of turbidity. The vertical
distribution and the depths at which phytoplankton were found
were affected also by the movement of the waters. The plankton
crop was a mixture of river plankton, mostly diatoms and rotifers,
and lacustrine organisms with copepods and cladocerans dominant.
Of the zooplankters, rotifers, copepods and cladocerans were
sampled from various places in the lake. The Cladocera recorded
included *Diaphanosoma, Ceriodaphnia, Bosmina* and *Moina*. Copepods,
particularly *Cyclops* and *Diaptomus* were common. Rotifers were
represented by the genera *Keratella, Filinia, Polyarthra, Trichocerca*
and *Asplanchna* which were fairly well distributed. During these
early days there was so much instability that the populations and
distribution patterns were quite variable.

Changes in the invertebrate fauna:

The larval stages of various terrestrial insects were collected
in samples especially from the littoral zones. There were
chironomid and *Mansonia* mosquito larvae, larval caddisflies and
mayfly (especially *Caenis* and *Baetis*) and dragonfly nymphs as well
as Hemiptera and Coleoptera. There were also aquatic oligochaetes.
Of special interest was the mayfly *Povilla adusta* whose large
nymphs were quite common on the submerged trees while most of the
other invertebrates were associated with aquatic weeds.

Benthic organism distribution followed dissolved oxygen
availability, even though oxygen was not a limiting factor, and
therefore, they were not found in deep waters.

The invasion of the lake by aquatic snails started
imperceptibly, then, suddenly, it seemed as if every rooted,
floating or submerged plant in the shallow littoral zones of the
lake carried *Bulinus (Physopsis) rholfsi* which, in this region,
is the intermediate host for *Schistosoma haematobium*. Other common
molluscs were *Bulinus forskalii* and *Pila africana*, and they all
thrived well among the weeds. Invertebrates characteristic of fast

flowing waters, such as the Simuliidae, and other benthic
organisms which had inhabited the rapids disappeared with the
slowing down of flow and the increased depth of the river. However,
the impoundment did not eliminate the Simuliidae from the basin.
Below the dam, there were many breeding sites and large populations
of *Simulium damnosum* on the Volta River. It was often collected
with *S. adersi*, *S. griseicolle*, *S. alcocki* and *S. unicornutum*.
But since it preferred the bigger turbulent flows to the smaller
streams, which had other species, the downstream Volta River
provided ideal habitats allowing enormous, dense masses of *S.
damnosum* larvae and pupae to develop, mostly on trailing
vegetation.

Changes in Aquatic plants Population:

The establishment of aquatic plants in the lake was quite
impressive. Both the physical and nutrient changes seemed to
have triggered off a burst of growth. *Pistia stratiotes* a common
plant on some surface waters in the country in particular was
abundant, with individual plants attaining enormous sizes, some
roots up to four feet long. The plants formed extensive mats
interspersed with grasses, mostly *Scirpus* sp. and *Cyperus* sp.
Striking also were sheets of small floating plants which covered
large areas. They included the waterfern *Salvinia nymphellula*,
Spirodella polyrhiza, *Azolla africana* and *Lemna* sp. *Ceratophyllum
demersum*, submerged, later became very important because *Bulinus*
snails became associated with it. It had wide distribution in
the shallow shores. The appearance of the grass *Vossia cuspidata*
in the southern section near the dam was specially observed. At
first, there were only few scattered groups of growths along the
shore - but they were all the most noticeable because of the
absence of other plants. Up to two thirds of the individual tall
grasses were submerged in the water. Even at that early time, it
was possible, to find snails and their eggs on the plants which
at that time provided them with suitable micro-habitats. There
was close association between the aquatic weeds and the spread of
snails and other invertebrates on the lake. *Pistia* in particular
in the early days effectively contributed to the spread of *Bulinus*
spp. over the lake as individuals, as well as small groups of
plants which broke from the mats and floated down the lake.
Altogether, aquatic weeds belonging to some fifty genera were
recorded on the lake and in the basin.

The invasion of artifical lakes by plants is one of the
serious impacts of the slowing down of the water flow in rivers.
Pistia was a nuisance not only because of its implication in the
spread of invertebrate intermediate hosts of parasites, but also
because it physically caused obstruction and interfered with ease
of passage for boats. Large mats of weeds at one time prevented

the passage of a large ferry. The classic case of *Salvinia auriculata* on the Kariba and the numerous problems caused by water hyacinth, *Eichhornia crassipes* in Sudan and elsewhere are examples of what headaches can be caused by aquatic weeds provided with suitable ecological conditions for growth.

Changes in fish populations:

An initial effect of the flooding was that the fish species which could neither adapt to the changed environment, or escape, perished. Some were affected because of the drastic physical changes of increased depth and volume of water, slowing down of water flow, high turbidity levels and also because their habitats, feeding and breeding sites became permanently uninhabitable and they were displaced. Then later, as the inundated vegetation decayed, there was an oxygen deficiency which was also a hazard to others. But most of the species survived and the Volta Lake did not have to be re-stocked with exotic fish. The river had some fishes which had been documented (Irvine 1949), but the fishery was limited. *Tilapia* spp., *Clarias* spp. and *Labeo* spp. had been among the common river species. A survey conducted within five months of the closure of the dam in 1964 recorded 60 species from 37 genera of fish, most of them commercially important. After the initial limnophysical and chemical upsets and, as the conditions near the dam became more lacustrine, decaying plants and other organic matter provided abundant food which, with the enlarged medium made conditions conducive to explosive growths of fish. The species which survived the ecological catastrophy moved around and settled in parts of the lake which they found most suitable. As they drew on the resources of the new medium and adapted to it, various species became prominent. A number of them, including the *Tilapia* spp., became abundant and individual size increased. The herbivorous, omnivorous and bottom feeding fish groups, including *Labeo* became abundant later. By 1966, the carnivorous fishes particularly *Lates niloticus* and *Chrysichthys* spp. had become dominant. The latter, originally caught in small numbers became a common species in the southern section.

An interesting account was made of the appearance of the clupeids, *Pellonula afzelinsi* and *Cynothrissa mento* at the time. They were exceedingly abundant, forming migrating shoals as they pursued food (mostly ostracods, copepods, cladocerans and larvae of *Chaoborus* sp.) along the lake. The clupeids in turn formed food for a number of other fish species. Even at that early time, a connection existed between the fish catch and distribution, the rainy season and the inflow of water from the tributary rivers into the lake. Within five years of closure of the dam, fish catches which had been minimal on the river, had risen to a conservative estimate of 60,000 metric tons.

The fish, electricity, the large store of water which could
be used for water supply, irrigation and industry, together with
the opportunities for navigation and recreation and the bonus
benefit of the destruction of the breeding sites of *S. damnosum*
on the lake more than made up for the consequences of the ecological
changes which took place. But the fact remains that people lost
their homes and lands to the floods and, most serious of all, with
the *Bulinus* snails that invaded and became established on the
shallow western shore, and the infected persons who came to the
lake for fishing and other purposes, came schistosomiasis, caused
by *S. haematobium*. This has now become widespread and demands
much resources to free the infected areas for safer exploitation.

CHANGES ON THE OTHER MAN-MADE LAKES

Together, the damming of the four rivers Volta, Niger, Nile
and Zambezi flooded an area of about 20,300 km^2. About 300,000
people were living in the areas which were subsequently flooded.
In the case of the Kariba, more so than the other three lakes,
it also became necessary to launch a special programme to rescue
stranded wildlife.

The records made during the early days of formation of the
three large African Lakes Kariba, Kainji and Nasser/Nubia show a
number of similarities to what had been recorded for Volta, in
spite of basic difference between them. The physical changes that
took place as the Kainji formed for instance, as with the Volta,
were greatly influenced by the annual harmattan season. The
temperature values on the Niger River, which had a range from
29° - 30°C, dropped, considerably during the harmattan to about
22°C, and although lake temperature recordings on occasion went
up to 37°C in the harmattan (and also because of the mixing of
the waters) the average temperature recorded was from 26° - 27.5°C.

On the Kariba and Nasser, the temperature ranges were 17 - 32°C
and 16 - 32°C respectively. In the former, inflowing waters very
much dictated the temperature levels and distribution, while the
latter, the winter winds had a similar effect. Stratification
was definitely present during the early times in the Kariba but
was never permanent in the Nasser.

Turbidity intensity and distribution changed as the dam
ecosystems became more lake-like. In the Kainji, the black
seasonal floods of the River Niger produced interesting and significant turbidity patterns. The white floods brought large
amounts of suspended clay into the lake and at the time, reduced
the transparency to as low as 10 cm. With the inflow of the
clearer waters of the black floods, readings of up to 300 cm were
recorded. In Lake Nasser, inflowing turbid waters reduced
transparency to values of 20 - 25 cm, presumably due largely to

silt and other suspended matter. It has been stated that
sediments in some places in the lake are up to 25 m deep mostly
in the old river bed and in the littoral zones. As in Volta
Lake, the water was much clearer near the dam, where readings
were from 100 - 200 cm suggesting a deposition of the suspended
matter away from the dam site. For the Kariba also, the depth
of visibility recorded increased towards the dam to an average
value of 405 cm giving Kariba, the clearest water of the four
lakes.

With the close connection between light penetration and
photosynthetic activities, these transparency values give an
indication of the extent of the euphotic zones which, during the
period reported averaged about 10 m for Volta, and ranged from
10 - 16 m in Kariba along the long axis of each lake. In lakes
Nasser and Kainji, largely because of extremes of the effects of
the inflowing waters, the range of depth was 1 - 10 m and 1 - 8 m
respectively.

Oxygen from the atmosphere is incorporated into water
largely from the action of winds and waves, and from photosynthesis of aquatic chlorophyllous plants. Water is
deoxygenated through its removal for the respiratory processes
of living organisms, and the break-down of organic matter. In
the three lakes, where much vegetable matter was submerged
during the flooding of the river, the decaying process caused
temporary deoxygenation of the waters as in the Volta. There
was close correlation between dissolved oxygen and temperature
on all four lakes, dissolved oxygen values reaching maximum
towards the end of the circulation period where there was
stratification as on Kariba. Otherwise, as observed in the Volta,
oxygen was generally present throughout the water column.
Values recorded in Lake Nasser in 1970 showed oxygen to be present
right to the bottom waters in the winter largely because of the
mixing of the surface waters into which oxygen had been
incorporated by strong winds.

The chemical factors recorded on all four lakes appeared to
be in line with what is generally found in other African freshwaters. They originate mainly from inflowing waters and only
minimally from rainfall. The concentrations of dissolved salts
held up in the impoundments, also became altered. Phosphates
and nitrates received the most attention in limnological
investigations because of their importance to productivity. The
values and distribution of the two are recorded to have been
influenced by the vegetable matter and dead plankton in the
medium. The vertical distribution of the chemical factors was
affected by the circulation and movement of the water. Not much
else was outstanding except the high production of H_2S at certain

times, which was common to all four lakes.

The waters of lakes Nasser and Kariba, like the Volta, tended
to be alkaline with ranges of pH values 8.67 - 8.74 and 7.5 - 8.5
respectively. The decrease of pH values with depth was similar to
observations on Volta where the presence of large amounts of
decomposing plant and other organic matter in the deeper levels of
the water depressed the pH values. In the surface waters,
photosynthetic activities caused an increase in pH values. With
stratification and deoxygenation of the waters, the pH was also
reduced. Total conductivity recorded for Kainji (40 - 54 μmhos)
was much lower than for Volta (65 - 180 μmho) and for Kariba
(55 - 81 μmho). Nasser, on the other hand, had the high range
140 - 240 μmho.

Certain phytoplankton groups were found to be common to all
four lakes. From pre-impoundment information on the Nile, diatoms,
especially *Melosira* spp., Chlorophytes mainly *Pediastrum* sp.,
and the blue-green *Microcystis, Anabaena* and *Oscillatoria* were
always found in the river. Earlier, Talling (1966) had also
recorded the presence of diatoms *Melosira* spp, and *Nitzschia* spp.
and blue-green *Anabaena* in lakes Victoria and Albert which are in
the Nile system. There were extensive blooms of *Melosira* formed
in Lake Nasser in the early days. *Oscillatoria* and *Microcystis*
were also then recorded and the chlorophytic plankters *Closterium,
Pediastrum* and *Scenedesmus,* the diatoms *Navicula* and *Nitzschia*
were listed. As in the Volta, *Volvox* blooms dominated the algae
population at some times. Similarly, in Kariba, early samples
showed dominant species to include *Microcystis, Melosira, Eudorina*
and *Closterium.*

The Niger river had diatoms and blue-green colonial algae.
In the lake, *Microcystis* and *Anabaena* blooms appeared annually.
The peak phytoplankton samples were 2628 algae per milliliter.

In all four lakes, zooplankton were present with the cladocera
at one time making up 43% of the plankton samples in the southern
end of Lake Nasser. Rotifers were well represented in the
zooplankton in Kainji which, in the period reached peak density
of 4244 organisms per haul. Also interesting at this time, was
the appearance of ostracods and coelenterates (*Hydra* sp. and
Medusae) which had not previously been recorded on the Niger.

Chironomids and other invertebrates appear to have done
reasonably well on all the lakes. A biomass of larvae, pupae and
adults reaching an estimated 100,000 tons were observed in Nasser
in 1970. Mayfly and dragonfly nymphs and corixid bugs were also
present. Bivalves as well as oligochaetes were also recorded
from the bottom samples taken in the northern and central sections.

It is on record that there was an explosive growth of chironomids during the filling of the Kariba and, subsequently, they were the dominant fauna on submerged trees (McLachlan 1970). Molluscs, including *Bulinus* snails, appeared in reasonably quiet littoral zones and were favoured in the four lakes by the presence of abundant plants and vegetation. Kariba alone, had both *Biomphalaria pfeifferi* and *Bulinus globosus*.

The explosion of the water fern *Salvinia auriculata*, eventually covering up to 21.5% of the surface of the Kariba, made history indicating generally the high level of dissolved salts which had entered the forming lake. Kariba had also, *Pistia*, *Potamogeton sweinfarthii*, *Ceratophyllum*, *Najas* sp. and a few other submerged plants. Prior to the formation of the Kainji, 36 genera of plants had been reported in the reservoir area. They included *Pistia*, *Salvinia nymphellula*, *Utricularia* sp. and *Ceratophyllum*. Other plants, *Polygonum senegalense*, *Jussiae repens* and *Utricularia inflexa* which had appeared in the Volta were major weeds.

In addition to the production of electricity, it had been intended from the start that these lakes would support fisheries. Prior to the construction of the Kainji dam, the average composition of the river fishes in the dam area was 148 species per 10,000 specimens. The Citharinidae, Mochokidae, Mormyridae, Characidae and Schielbeidae were the most abundant. The families which became dominant in the early days were the Citharinidae and Characidae. The Mormyridae being mostly bottom feeders, became drastically reduced in population because of the increased water depth. The Centropomidae, on the other hand, appeared and became fairly abundant. Small clupeids such as *Sierrathrissa leonensis* and *Pellonula afzelinsi* became important elements of the fauna, the former being distributed all over the lake. As in the Volta, even early on, the carnivorous *Lates niloticus* seemed likely to be well established. The Lake Nasser fisheries started with the typical lake species. *Tilapia nilotica*, *Lates niloticus*, *Alestes* sp., *Labeo* sp., *Hydrocynus* spp., *Schilbe* spp., *Synodontis* sp., *Mormyrus* sp., and *Clarias* spp., became the most important among the 53 species recorded in the early days. In all four lakes there was a boom in fish production. In Kariba by 1970, the production was 1550 metric tons a year; in Nasser, the catch rose from 746 metric tons in 1966 to 4545 metric tons in 1969.

The boom in production at various levels reflects the reaction of the biota to the change from a lotic to a lentic condition. The shallower the lake, it seems, the greater the overall productivity. In the Volta, fish catch, especially of *Tilapia* spp., was often considerably greater in the shallower waters of the Afram plains. It was in shallow areas also that

aquatic plants did best. The type of substrate and the nutrients
which it held were also contributary factors. Where plants were
cleared and burned before inundation, aquatic weed growth seemed
to have been vigorous. Generally, the enlarged water body attracts
water-loving plants and animals to its shores.

The question of the effect of dams on local climate is
unresolved. It needs more observation. Similarly, it is still
inconclusive whether local rainfall patterns are affected by dams.
Where a dam lies in an arid area with a high rate of evaporation,
conditions are such that it is unlikely that enough moisture can
collect to fall as rain over the area. Lake Nasser has not made
any appreciable impact on the rainfall.

Downstream effects of dam:

The effects of the interference with streams and rivers go
beyond the immediate aquatic environment. Much more attention is
usually given to the upstream impact of the construction of dams
than to changes which occur downstream. This was the case during
the early days of the large dams. However, observations show that
though it may not be so everywhere, there is a noticeable lowering
of the average temperature of the water. Also, the water flow
becomes faster than on the parent river and this causes greater
erosion of the banks. On the Volta, the fast flow favoured the
Simuliidae which continued to occur below the dam. Since much
of the sediment in the impounded water is deposited behind the dam,
downstream nutrients are reduced and this affects the productivity.
On the Volta there was a sharp reduction in both the individual
size and quantities of the crustacean *Atya gabonensis* which was
harvested from the river below the dam. Changes in salinity due
to the reduced flow affect the fauna in the estuaries, and the dam
itself creates a barrier to fish which normally migrate to spawn
in headwaters.

Although the foregoing account has concentrated on the changes
in the aquatic ecosystem, the impact reverberates throughout the
entire basin. In tropical areas, where faster growth causes more
serious imbalances in disturbed media than in temperate zones, the
range of the complex impacts which occur, as indeed observed on
the four lakes, appears to be typical of most artificial dams,
both large and small. The impacts include flooding and land
inundation, changes in water quantity, dislocation of people in
inhabited areas, changes in water quality and biota - especially
increased plankton, fish and weed production, loss of benthic
invertebrates, invasion by other invertebrates and by some disease
vectors. The pressure of the impounded waters also often affects
the hydrology and water table (Table 3).

Table 3. Summary of major changes on dammed rivers and their basins.

Pre-impoundment	Changes during impoundment	Post-impoundment conditions
People	Dislocation/re-location	
Settlements	Submerged/new settlements	
Farms/developed land	Destroyed permanently	
Undeveloped land	Submerged permanently	
Subterranean minerals	Submerged permanently	
Wildlife	Rescued (some perish) rehabilitation	
Water body:		
Depth	increased	affected seasonally by floods
Width	increased	drawn-down changes with rainy seasons
Area	enlarged	fluctuates with rainy seasons
Volume	enlarged	fluctuates with floods
Current and water movement	flow reduced/vertical movement	vertical mixing of water
Temperature	range remains stable	
Colour	turbid	changes with rainy/dry seasons
Suspended matter	increased	sedimentation and siltation

Oxygen	temporarily reduced	generally adequate
Dissolved salts	increased	some fluctuate with seasons
Plankton	increased/blooms	blooms tail off/seasonal
Invertebrates	benthic-submerged littoral-flourish	permanently destroyed plants established
Fish	some destroyed/increased growth boom in catch	boom yield tails off and becomes stabilised
Plants	invasion of aquatic plants	increased growth – littoral, submerged, floating plants
Other aquatic vertebrates	generally disappear	
Climate (local)	not much change	
Water table (basin)	raised – in some places	stabilised
Seismic activity	increased – in some places	less evident
Public health		
Malaria	limited breeding sites	
onchocerciasis	vector breeding sites destroyed	permanently destroyed upstream of dam
schistosomiasis	vector breeding sites created	permanent breeding sites above dam
trypanosomiasis	vector breeding enhanced	

The ecological consequence of dam construction are widespread and complex. In tropical areas they are yet to be fully studied and understood. There is urgent need for such an understanding to contribute to planning and management of the altered ecosystems. The benefits from dams are far too important to development for construction to be stopped because of man's inability to manage the ecological impacts. With sound planning, it is possible, even on the basis of present limited studies to forestall and reduce some of the serious impacts of impounding water.

REFERENCES

Ackermann, W.C., White, G.F. and Worthington, E.B., eds. 1973. Man-made lakes and their environmental effects. American Geophysical Union, Washington.
Balon, E.K. and Coche, A.G. 1974. Lake Kariba, A man-made tropical ecosystem in Central Africa. D.W. Junk Book Publishers, The Hague.
Lowe-McConnell, R.H. ed. 1966. Man-made lakes. Academic Press, London and New York.
Obeng, L.E. ed. 1969. Man-made lakes. Ghana Universities Press, Accra.
Proceedings of the International Conference on Kainji and river basins development in Africa. 1980. Volumes 1 + 2. Kainji Lake Research Institute.
Water Supply and Management. 1980. Vol. 4, Special Edition on the Nile. Pergamon Press, Oxford.

Other References

Brook, A.J. 1954. A systematic account of the phytoplankton of the Blue and White Nile at Khartoum. *Ann. Mag. Nat. Hist.* 12(7): 648-656.
Brook, A.J. and Rzoska, J. 1954. The development of Gebel Aulyia dam on the development of Nile. *J. Anim. Ecol.* 23: 101-114.
Hynes, H.B.N. 1970. The ecology of running waters. Liverpool University Press, Liverpool.
Irvine, F.R. 1949. Fishes and fisheries of the Gold Coast. The Crown Agents, London.
McLachlan, A.J. 1970. Submerged trees as a substrate for benthic fauna in recently created Lake Kariba (Central Africa). *J. App. Ecology* 7: 253-266.
Rzoska, J. ed. 1976. The Nile: biology of an ancient river. D.W. Junk Book Publishers. The Hague.
Talling, J.F. 1957. The longitudinal succession of water characteristics in the White Nile. *Hydrobiologia* 11: 73-89.
Talling, J.F. 1966. The annual cycle of stratification and phytoplankton growth in Lake Victoria (East Africa). *Int. Revue ges. Hydrobiol.* 51: 545-621.

AQUATIC ORGANISMS AND PALAEOECOLOGY: RECENT AND FUTURE TRENDS

Nancy E. Williams

Division of Life Sciences
Scarborough College
University of Toronto
West Hill, Ontario, Canada

INTRODUCTION

The interpretation of past environments has traditionally
been based on the study of certain fossil organisms and parts of
organisms which have been particularly well preserved over time.
Frey (1964, 1969, 1974, 1976) has reviewed the use of various
aquatic plants and animals and the reader should refer to his works
for coverage of this field to 1976. My purpose is not to repeat
his excellent reviews but to discuss the direction of the most
recent palaeoecological work, its contribution to the understanding
of problems in the field and the advantages of studying aquatic
organisms.

For many years, pollen grains were the chief tool employed to
interpret past climatic and edaphic conditions. Pollen's main
component, pollinin, is one of the most inert organic substances
known (Frey, 1976) and so palynology continues to be important to
palaeoecology despite problems and disadvantages associated with
pollen analysis and interpretation (Wright, 1977). In recent
years, however, there has been increasing use of other organisms
whose skeletal materials: silica, chitin, alkaline earth
carbonates and cellulose are also well preserved in sediments.
Coleoptera, Cladocera, Ostracoda, Mollusca, Chironomidae and
Trichoptera are among the groups used most successfully and indeed
it is hardly surprising that four of these six are essentially
wholly aquatic, since most fossil-bearing sediments accumulate as
a result of deposition by water.

SINGLE GROUP STUDIES

A few recent workers have used a single aquatic group to
interpret the past environment of a site. For example, Gasse and
Delibrias (1976) detected three tropical periods within a 100,000
year time period, each with different climatic characteristics,
by analysing the diatoms of a core and geological sections from
an Ethiopian lake basin. Czeczuga and Kossacka (1977) described
post-glacial climate and depth changes in Wigry lake in Poland by
studying indicator species, species diversity and abundance of
Cladocera. Wiederholm and Eriksson (1979) described the eutrophication of a Swedish lake by comparing the chironomid remains
from sediments deposited over a 150 year period with species
found in present-day Swedish lakes.

The potential of a previously little-used aquatic group was
investigated by Williams and Morgan (1977) in a study of fossil
caddis larvae from the Don Formation at Toronto, Ontario. They
identified 22 trichopteran taxa from which they deduced
characteristics of a river and lake and their watershed. The
climatic inferences were in agreement with those made from
studies of other groups and the conclusion was made that immature
Trichoptera would merit further use in palaeoecological studies.
This prediction has been born out as abundant identifiable caddisfly
fragments have been found at many other North American sites
(Williams unpublished data, Williams et al., 1981) and British
workers also have begun to make use of fossil caddisflies. Moseley
(1978) has published a key for the identification of frontoclypeal
sclerites of seven European species in the family Hydropsychidae
while Wilkinson (1981) has described changes in the caddisfly
fauna identified from a late Glacial site in Britain.

It should be useful, then, to include here some drawings and
photographs of caddisfly larval parts in order to publicize the
anatomy of these insects now shown to be both abundant in sediments
and useful in interpretation of past conditions. Figure 1a
illustrates most of the sclerites of the larval head and thorax.
Although sclerites are usually retrieved separately and devoid of
setae, their setal alveoli, muscle scars, shapes and colour
patterns are useful in identification. Figure 2a shows the frontoclypeus of a fossil specimen of the largest caddisfly family, the
Limnephilidae. Members of this group are often the only caddisflies
found in sediments of tundra ponds and other cold climate deposition
sites. Figure 2b shows the frontoclypeus of *Hydropsyche scalaris*
grp. (Hydropsychidae). Members of this species group are usually
inhabitants of deciduous forest streams and rivers. On both the
above specimens, setal alveoli, sensory pits and muscle scars can
be seen. Their use in identification has been described elsewhere
(Williams and Morgan, 1977; Williams et al., 1981). The front
coxa of a phryganeid, with characteristic comb structures and the

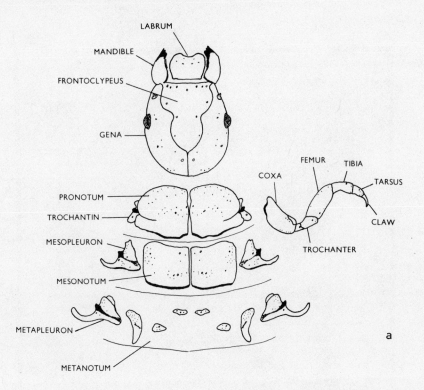

Figure 1. (a) Sclerites of the head and thorax of a limnephilid
caddisfly larva.
(b) Ventral view of the head capsule of a chironomid
larva (Chironomini).

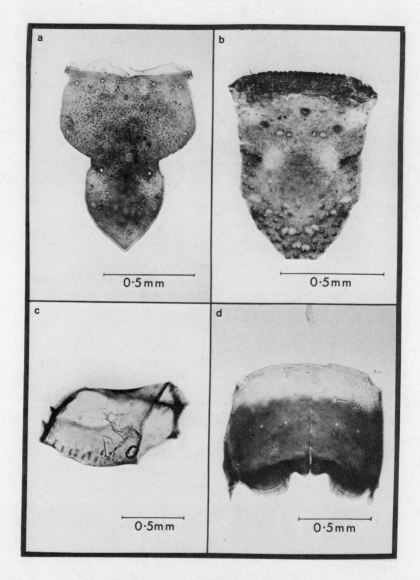

Figure 2.   Fossil caddisfly sclerites (> 50,000 yrs before present).
　　　　(a)   frontoclypeus of ? *Asynarchus* sp. (Limnephilidae).
　　　　(b)   frontoclypeus of *Hydropsyche* nr. *venularis*
　　　　　　　(Hydropsychidae).
　　　　(c)   front coxa of ? *Phryganea* sp. (Phryganeidae).
　　　　(d)   pronotum of *Rhyacophila* ? *melita*
　　　　　　　(Rhyacophilidae).

pronotum of a rhyacophilid (unusual in that right and left halves
are preserved together) are illustrated in Figure 2c and 2d,
respectively.

MULTIPLE GROUP STUDIES

Most palaeoecologists would now agree with Frey (1976) that
"interpretations from a single group of organisms are usually more
suggestive than definitive, and without support from other
evidence lead to open-ended and unsatisfying conclusions".
Indeed the most obvious recent advance in palaeoecology has been
the large number of co-operative studies undertaken.

Advantages of this approach are obvious in studies such as
that of Bell et al. (1972) on mid-Weichselian deposits in
Leicestershire, England. While molluscan and mammalian remains
did not individually yield definitive conclusions about the site,
they did support the climatic inferences indicated by arctic-alpine
plant macrofossils and insects. Similarly Matthews (1974)
maximized information obtained from an Alaskan site by identifying
molluscs, ostracods, insects and plant macrofossils as well as
pollen. Some plant species not indicated by the pollen were
discovered as macrofossils, and indicate that certain boreal or
taiga species may have survived the late Wisconsin in a steppe-
tundra environment, while others became extinct in Alaska at this
time.

The study of Karrow et al. (1975) on Lake Algonquin
sediments in southwestern Ontario shows that one group of organisms
may be extremely useful at one site but poorly represented or
difficult to interpret at another. Molluscs and ostracods were
widespread throughout the sampling sites but pollen and diatoms
had patchy distributions. This meant for example that shallowing
water conditions were indicated at one site by pollen, plant
macrofossils, molluscs and ostracods, at another by pollen, plant
macrofossils and diatoms, and at a third, only by ostracods.

A number of British studies have employed aquatic groups in
combination with other organisms. Shotton et al. (1977) analysed
the molluscs, pollen and beetles of a Flandrian deposit dated at
about 8500-8300 B.P. These three groups provided good evidence of
a marsh, its surrounding woodland and climate. Similarly, the
palaeoecology of interglacial deposits at Sugworth has been
investigated co-operatively, although discussed in three separate
papers. The molluscs (Gilbertson, 1980), ostracods (Robinson,
1980) and beetles (Osborne, 1980) together suggested a warm
temperate climate at the time of deposition, although evidence from
each group considered singly would appear somewhat weak.

Chironomids, beetles and caddisflies from the Pleistocene

Scarborough Formation at Toronto, Ontario were used by Williams
et al. (1981) to detect climatic differences between two sites.
Here again the use of several groups was fortuitous since
identifiable beetle parts were rare at site 1, while chironomids
and caddisflies were rare at site 2. The Chironomidae were the
only group to clearly show amelioration in climate from lowest to
highest of three levels sampled at site 1. Chironomids have been
used extensively to characterize trophic conditions – present and
past (Hofmann 1979, Saether 1971, 1979), but difficulty in
obtaining specific identifications and lack of ecological data
have tended to discourage their use as indicators of climate.
This is unfortunate since the head capsules used for identification
in this diverse family (Figure 1b, 3) are usually very abundant
and well preserved in sediments. As indicated by Oliver (1971,
1979) and Williams et al. (1981), chironomids can be particularly
sensitive indicators of temperature and vegetational change.

QUANTITATIVE STATISTICAL TECHNIQUES

    Aside from multiple group studies, several other trends are
evident in recent palaeoecological work. The use of quantitative
statistical techniques is one of these. In studies of aquatic
groups the work of Delorme et al. (1976) is notable for its
method of calculating palaeoclimatic parameters. A modern collection of molluscs and ostracods from 5,000 sites in western Canada
was used for comparison with fossil assemblages from the lower
Mackenzie River area of northwestern Canada. The limits of
physical, chemical and climatic parameters for each species in
question were recorded and a range calculated for each parameter
on the basis of species present in the fossil assemblage. Brugam
(1980) employed a similar analysis in studying the diatom
stratigraphy of Kirchner Marsh in Minnesota. He compared his
fossil assemblages with surficial assemblages from lakes in
Minnesota and Labrador in a cluster analysis. Although he
recognized the need for an expanded set of surface samples, he was
able to detect eutrophication and changes in lake depth over time.

BIOSTRATIGRAPHIC ZONATION

    Pollen zones have long been used for biostratigraphic correl-
ation but attempts to characterize assemblages of aquatic
organisms for this purpose are just beginning. Shotton et al.
(1977) assigned their mollusc assemblage from Alcester, England
to a zone characterized particularly by the arrival and expansion
of the species *Discus rotundatus*, replacing *D. ruderatus*, as
described by Kerney (1977) for other parts of Britain. In
southwestern Ontario, Miller et al. (1979) found that many of
the molluscs appeared to be stratigraphically limited and
suggested their potential use in zonation of Algonquin, Nipissing

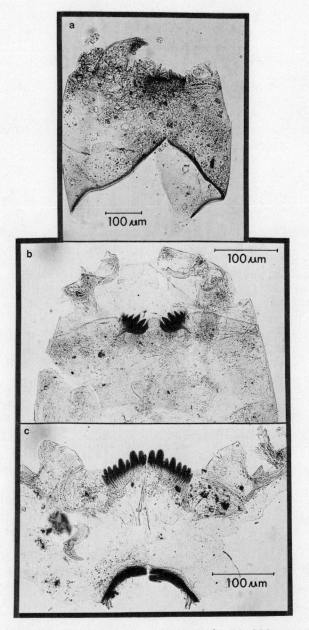

Figure 3. Fossil chironomid sclerites (> 50,000 yrs B.P.)
  (a) ventral view of head capsule (including right
       mandible) of a tanytarsine chironomid.
  (b) head capsule of *Cryptochironomus* sp. showing
       labial plate, both paralabial plates and
       premandibles.
  (c) labial plate and paralabial plates of
       Phaenopsectra sp.

and transitional-age deposits. They proposed using *Pisidium lilljeborgi* and *P. nitidum* cf. *pauperculum* to identify Algonquin stage sediments in the Alliston Embayment area and *Goniobasis liviscens* for Nipissing stage deposits. They stated however that the study of many more sites would be necessary to test the general applicability of molluscs for zoning these sediments.

Osborne (1980) stated that insufficient insect faunas have been described from interglacial times for use in stratigraphic correlation, but he did compare the Sugworth fauna with a few other assemblages and suggested, as did Shotton et al. (1977) that correlation might be possible in future.

SYSTEMATICS

Perhaps the most interesting recent trend has been the use of palaeoecological data to study systematics of the organims themselves. Morgan and Morgan (1980) have pointed out that "A common theme throughout many of the more recent (entomological) papers has been the practice of hypothesizing on the possible origins and shifts of species based entirely on the patterns of modern distribution. However, with the accumulation of fossil insect records it is now becoming possible to raise above the realm of speculation our knowledge of faunal histories and insect speciation." This applies equally well to many non-insect groups.

Hofmann (1977) studied the evolution of *Bosmina* (Chydoridae) populations found in late glacial and postglacial sediments of the Grosser Segeberger See in Schleswig-Holstein. His data on morphological transition and co-occurrence suggested that the two forms *kessleri* and *coregoni* are two morphs of one species rather than separate species or subspecies as suggested by other taxonomists.

Several papers have contributed to the knowledge of mollusc distributions and migrations. Cvancara (1976) discussed the distribution of aquatic molluscs in North Dakota during the last 12,000 years and suggested causes for presumed regional extinctions of species, and other species changes. Mollusc fossils from Pleistocene deposits in the Old Crow Basin, Yukon Territory enabled Clarke and Harington (1978) to detect the past presence of two Asian species not previously recorded in North America. They speculated on reasons why some Asian migrants have become widespread in North America while others such as these two, failed to become established. They suggested that among dioecious molluscs, only the smaller species appeared able to migrate easily between adjacent, non-confluent river systems and that the severe arctic climate of Beringia during the Wisconsin glaciation meant that only tundra-adapted species could survive in this refugium.

The long-range objective of the study of Miller et al. (1979)
was to document the presence of various molluscan species in lakes
of various ages and deduce therefrom the timing and patterns of
migration of molluscs as climate and palaeogeography changed.
The distributions of the aquatic species from the 17 sites within
the study area were compared with dated records outside the area
to detect three different migration patterns.

Returning to insects, it should be noted that organized study
of fossil insect sites began only during the fifties in Britain
and even later in North America. Morgan and Morgan (1980) listed
362 fossil beetles from 57 sites in Canada and the neighbouring
United States and discussed distributional shifts through time
for selected species. Matthews (1976, 1977) has discovered
Tertiary Coleoptera fossils from northern North America many of
which are apparently extinct but closely related to variously
distributed modern species. They show great promise for evolutionary
studies when more is known about the present northern fauna. That
this type of data is already becoming available for the Coleoptera
surely bodes well for other groups whose study is just beginning.
As Matthews (1979) predicted, "it may be feasible within the next
two decades to discuss the development of the Canadian insect fauna
entirely in terms of insects rather than by analogy with
vegetation history."

AQUATIC VERSUS TERRESTRIAL ORGANISMS

As previously mentioned the most obvious reason for using
aquatic organisms in the study of past environments is that most
fossil-bearing sediments are waterlaid. This means that aquatic
plant and animal remains are usually very abundant and in relatively
good condition. One can assume, with reasonable confidence, that
the majority have not been transported from great distances to the
deposition site although some specimens may come from upstream
tributaries or another part of a lake. It can also be assumed that
organisms from the time of deposition will predominate over reworked
material which would in any case tend to be broken down by the
reworking process. On the other hand terrestrial organisms in
waterlaid sediments enter the water by a variety of methods and
comprise allochthonous assemblages derived from a wide variety of
habitats and perhaps times. Aquatic organisms have the additional
advantage that they reflect not only the limitations imposed upon
them directly by their aquatic environment but also the indirect
effects of the nature of the surrounding watershed in that the
soil and vegetation affect water chemistry and provide food for
aquatic organisms. For example, Ross (1963) has indicated that
many caddisfly species, subgenera and genera are correlated with
terrestrial biomes e.g. eastern temperate deciduous forest, montane
coniferous forest or northern boreal coniferous forest.

Table 1.   Immigration rates of various organisms in response to climate change.

Group	Immigration Rate	Relevant Factors	References
Ostracoda	immediate	rapidly produced, easily dispersed resistant eggs	Delorme, 1969; Delorme et al., 1977
Cladocera	immediate	rapidly produced, easily dispersed resistant eggs	Czeczuga & Kossacka, 1977; Decosta, 1964; Frey, 1976; Whiteside, 1969
Terrestrial insects	rapid	rapid reproduction, high mobility-wings, can take advantage of microclimate	Coope, 1969, 1970, 1975; Morgan & Morgan, 1980; Ross, 1965
Aquatic plants	rapid	rapidly produced and dispersed seeds, no maturing of soil needed	Iversen, 1960, 1964
Aquatic insects	fairly rapid	rapid reproduction, high mobility	Ross, 1963, 1965, 1967; Williams & Morgan, 1977
Small terrestrial plants	slow but variable	can depend on microclimate	Iversen, 1964
Mollusca	same as trees, or faster	many depend on plants for food	Clarke & Harington, 1978; Frey, 1964; Miller et al., 1979; Newman, 1975; Taylor, 1960, 1965
Trees	slow lags of 1000 -2000 yrs after suitable climate; migration rates - 100-500 m/yr	slow to mature and reproduce, dependent on soil conditions	Davis, 1976; Iversen, 1958, 1960, 1964; Newman, 1975
Mammals	same as or following terrestrial plants	must follow food organisms	Hibbard et al., 1965; Semken, 1975
Diatoms	usually slower than terrestrial plants but variable	need silica leached from soil	Frey, 1974; Pennington et al., 1972; Round, 1964

In interpreting data acquired from different plant and animal groups one must consider the different ways of responding to climate change. These are:
1. avoidance by making use of microclimate
2. change in rates of growth and reproduction
3. migration to a more suitable area
4. evolution
5. lack of response.

One should consider that the same organism may have different means of responding to improving and deteriorating climate. Small terrestrial organisms are most able to make use of microclimate, while organisms in very large water bodies have a more uniform environment buffered from changes in precipitation, wind and temperature. River assemblages are perhaps most likely to reflect macroclimate since the river responds to macroclimatic changes but offers little variation in microclimate to its fauna.

Since responses are governed by a complex set of interacting variables it is perhaps not surprising that there are very little comparative data on another important consideration, the response rates of organisms used to make climatic deductions. However, recent, studies suggest some relative immigration rates (Table 1). It should be remembered that differences within groups are also important.

It should be clear from the above discussion that the greatest information will be acquired from studies which consider physical barriers, available refuges and other non-climatic factors which may affect distributions, morphology and numbers of organisms as well as species composition of assemblages.

THE FUTURE

The future will certainly see more palaeoecological studies involving multiple groups of organisms and more contributions to systematics by these studies. One should not overlook the contribution to modern taxonomy which can be made by palaeo-ecologists struggling to find means of identifying fossil fragments. My own work has certainly made me aware that diagnostic characters can sometimes be noticed more readily on clean fossil skeletons than on modern specimens (Williams et al., 1981).

There is also room for the introduction of new aquatic groups to this field. The requirements include a diversity of specialized taxa, vulnerability to environmental impact, suitable taxonomic methods and abundance in fossil assemblages. One possibility as suggested by Williams (1980) might be the Ephemeroptera. Are there any takers?

REFERENCES

Bell, F.G., Coope, G.R., Rice, R.J. and Riley, T.H. 1972.
    Mid-Weichselian fossil-bearing deposits at Syston,
    Leicestershire. *Proc. Geol. Ass.* 83: 197-211.
Brugam, R.B. 1980. Post-glacial diatom stratigraphy of Kirchner
    Marsh, Minnesota. *Quat. Res.* 13: 133-146.
Clarke, A.H. and Harington, C.R. 1978. Asian freshwater
    molluscs from Pleistocene deposits in the Old Crow Basin,
    Yukon Territory. *Can. J. Earth Sci.* 15: 45-51.
Coope, G.R. 1969. The response of Coleoptera to gross thermal
    changes during the Mid-Weichselian interstadial. *Mitt. Int.
    Verein. Limnol.* 17: 173-183.
Coope, G.R. 1970. Interpretations of Quaternary insect fossils.
    Ann. Rev. Ent. 15: 97-120.
Coope, G.R. 1975. Climatic fluctuations in northwest Europe
    since the last interglacial, indicated by fossil assemblages
    of Coleoptera. *In*: Ice Ages: Ancient and Modern. Wright,
    A.E. and Mosely, F. eds. Liverpool University Press.
Cvancara, A.M. 1976. Aquatic molluscs in North Dakota during
    the last 12,000 years. *Can. J. Zool.* 54: 1688-1693.
Czeczuga, B. and Kossacka, W. 1977. Ecological changes in
    Wigry Lake in the post-glacial period. Part II.
    Investigations of the cladoceran stratigraphy. *Pol. Arch.
    Hydrobiol.* 24: 259-277.
Davis, M.B. 1976. Pleistocene biogeography of temperate
    deciduous forests. *Geoscience and Man* 13: 13-26.
DeCosta, J. 1964. Latitudinal distribution of chydorid
    Cladocera in the Mississippi Valley, based on their remains
    in surficial lake sediments. *Invest. Indiana Lakes,
    Streams* 6: 65-101.
Delorme, L.D. 1969. Ostracods as Quaternary paleoecological
    indicators. *Can. J. Earth Sci.* 6: 1471-1476.
Delorme, L.D., Zoltai, S.C. and Kalas, L.L. 1976. Freshwater
    shelled invertebrate indicators of paleoclimate in north-
    western Canada during the late glacial. *Paleolimnology of
    Lake Biwa and the Japanese Pleistocene* 4: 605-657.
Delorme, L.D., Zoltai, S.C. and Kalas, L.L. 1977. Freshwater
    shelled invertebrate indicators of paleoclimate in north-
    western Canada during late glacial times. *Can. J. Earth
    Sci.* 14: 2029-2046.
Frey, D.G. 1964. Remains of animals in Quaternary lake and bog
    sediments and their interpretation. *Arch. Hydrobiol. Beih.
    Ergebn. Limnol.* 2: 1-114.
Frey, D.G. 1969. The rationale of paleolimnology. *Mitt. Int.
    Verein. Limnol.* 17: 7-18.
Frey, D.G. 1974. Paleolimnology. *Mitt. Int. Verein. Limnol.*
    20: 95-123.

Frey, D.G. 1976. Interpretation of Quaternary paleoecology from
    Cladocera and midges, and prognosis regarding usability of
    other organisms. *Can. J. Zool.* 54: 2208-2226.
Gasse, F. and Delibrias, G. 1976. Les lacs de l'Afar Central
    (Ethiopie et I.F.A.I.) au Pleistocene superieur.
    Paleolimnology of Lake Biwa and the Japanese Pleistocene
    4: 529-575.
Gilbertson, D.D. 1980. The palaeoecology of middle Pleistocene
    Mollusca from Sugworth, Oxfordshire. *Phil. Trans. R. Soc.
    London Series B.* 289: 107-118.
Hibbard, C.W., Ray, D.E., Savage, D.E., Taylor, D.W. and Guilday,
    J.E. 1965. Quaternary mammals of North America. *In*:
    The Quaternary of the United States, Wright, H.E. and Frey,
    D.G. eds. Princeton Univ. Press.
Hofmann, W. 1977. *Bosmina (Eudosmina)* populations of the Grosser
    Segeberger See during late glacial and postglacial times.
    Arch. Hydrobiol. 80: 349-359.
Hofmann, W. 1979. Chironomid analysis in Palaeohydrological
    changes in the temperate zone in the last 15,000 years.
    Subproject B. Lake and mire environments. Project Guide
    Vol. 2: specific methods. Berglund, B.E. ed. International
    Geological Correlation Programme Project 158.
Iversen, J. 1958. The bearing of glacial and interglacial
    epochs on the formation and extinction of plant taxa.
    Uppsala Univ. Arsskr. 6: 210-215.
Iversen, J. 1960. Problems of the early post-glacial forest
    development in Denmark. *Danmarks Geol. Undersøgelse IV*:
    4:3:32 pp.
Iversen, J. 1964. Plant indicators of climate, soil, and other
    factors during the Quaternary. *Rept. 6th Int. Congr. Quat.
    Warsaw 1961. II Palaeobotanical Section*: 421-428.
Karrow, P.F., Anderson, T.W., Clarke, A.H., Delorme, L.D. and
    Sreenivasa, M.R. 1975. Stratigraphy, paleontology, and
    age of Lake Algonquin sediments in Southwestern Ontario,
    Canada. *Quat. Res.* 5: 49-87.
Kerney, M.P. 1977. British Quaternary non-marine Mollusca:
    A brief review. pp. 31-42 *In*: British Quaternary Studies:
    Recent Advances, Shotton, F.W. ed. Clarendon Press, Oxford.
Matthews, J.V. 1974. Wisconsin environment of interior Alaska:
    pollen and macrofossil analysis of a 27 meter core from the
    Isabella Basin (Fairbanks, Alaska). *Can. J. Earth Sci.*
    11: 828-841.
Matthews, J.V. 1976. Insect fossils from the Beaufort Formation:
    geological and biological significance. *Geol. Surv. Can.
    Paper* 76-1B: 217-227.
Matthews, J.V. 1977. Tertiary Coleoptera fossils from the North
    American Arctic. *Coleopterist's Bull.* 31: 297-308.
Matthews, J.V. 1979. Tertiary and Quaternary environments:
    historical background for an analysis of the Canadian insect
    fauna. pp. 31-86. *In*: Canada and its Insect Fauna. Danks,
    H.V. ed. *Mem. Ent. Soc. Can.* 108.

Miller, B.B., Karrow, P.F. and Kalas, L.L. 1979. Late Quaternary mollusks from glacial Lake Algonquin, Nipissing and transitional sediments from southwestern Ontario, Canada. *Quat. Res.* 11: 93-112.

Morgan, A.V. and Morgan, A. 1980. Faunal assemblages and distributional shifts of Coleoptera during the late Pleistocene in Canada and the northern United States. *Can. Ent.* 112: 1105-1128.

Moseley, K.A. 1978. A preliminary report on Quaternary fossil caddis larvae (Trichoptera). *Quat. Newsletter* 26: 2-12.

Newman, W.S. 1975. Late Quaternary palynology- delay en route? *Quat. Non-Marine Pal. Conf., May 12-13, 1975, Program and Abstracts. Univ. of Waterloo, Ontario.*

Oliver, D.R. 1971. Life history of the Chironomidae. *Ann. Rev. Ent.* 16: 211-230.

Oliver, D.R. 1979. Contribution of life history information to taxonomy of aquatic insects. *J. Fish. Res. Bd. Can.* 36: 318-321.

Osborne, P.J. 1980. The insect fauna of the organic deposits at Sugworth and its environmental and stratigraphic implications. *Phil. Trans. R. Soc. London Series B.* 289: 119-133.

Pennington, W., Haworth, E.Y., Bonny, A.P. and Lishman, J.P. 1972. Lake sediments in Northern Scotland. *Phil. Trans. R. Soc. London, Series B.* 264: 191-294.

Robinson, J.E. 1980. The ostracod fauna of the interglacial deposits at Sugworth, Oxfordshire. *Phil. Trans. R. Soc. London, Series B.* 289: 99-106.

Ross, H.H. 1963. Stream communities and terrestrial biomes. *Arch. Hydrobiol.* 59: 235-242.

Ross, H.H. 1965. Pleistocene events and insects. *In*: The Quaternary of the United States. Wright H.E. and Frey, D.G. eds. Princeton Univ. Press.

Ross, H.H. 1967. The evolution and past dispersal of the Trichoptera. *A. Rev. Ent.* 12: 169-206.

Round, F. 1964. The diatom sequence in lake deposits: some problems of interpretation. *Ver. Int. Verein. Limnol.* 15: 1012-1020.

Saether, O.A. 1975. Nearctic chironomids as indicators of lake typology. *Verh. Int. Verein Limnol.* 19: 3127-3133.

Saether, O.A. 1979. Chironomid communities as water quality indicators. *Holarctic Ecol.* 2: 65-74.

Semken, H.A. 1975. Vertebrates as paleoecological indicators of Pleistocene environments. *Quat. Non-marine Pal. Conf., May 12-13, 1975. Program and Abstracts. Univ. of Waterloo, Ontario.*

Shotton, F.W., Osborne, P.J. and Greig, J.R.A. 1977. The fossil content of a Flandrian deposit at Alcester, Warwickshire. *Coventry Nat. Hist. Sci. Soc.* 5: 19-32.

Taylor, D.W. 1960. Late Cenozoic molluscan faunas from the High Plains. *U.S. geol. Surv. Prof. Pap.* 337: 1-94.

Taylor, D.W. 1965. The study of Pleistocene nonmarine mollusks
    in North America. *In*: The Quaternary of the United States.
    Wright, H.E. and Frey, D.G. eds. Princeton Univ. Press.
Whiteside, M.C. 1969. Chydorid (Cladocera) remains in
    surficial sediments of Danish lakes and their significance
    to paleolimnological interpretations. *Mitt. Int. Verein.
    Limnol.* 17: 193-201.
Wiederholm, T. and Eriksson, L. 1979. Subfossil chironomids as
    evidence of eutrophication in Ekoln Bay, central Sweden.
    Hydrobiologia 62: 195-208.
Wilkinson, B. 1981. Quaternary sub-fossil Trichoptera larvae
    from a site in the English Lake District. *Proc. 3rd Int.
    Symp. Trichoptera.* July 28-Aug. 2, 1980. *Series Entomologica,*
    20. Moretti, G.P. ed. Junk, The Hague.
Williams, D.D. 1980. Applied aspects of mayfly biology. *In*:
    Advances in Ephemeroptera Biology, Flannagan, J.F. and
    Marshall, K.E. eds. Plenum, N.Y.
Williams, N.E. and Morgan, A.V. 1977. Fossil caddisflies
    (Insecta: Trichoptera) from the Don Formation, Toronto,
    Ontario, and their use in palaeoecology. *Can. J. Zool.*
    55: 519-527.
Williams, N.E., Westgate, J.A., Williams, D.D., Morgan, A. and
    Morgan, A.V. 1981. Invertebrate fossils (Insecta:
    Trichoptera, Diptera, Coleoptera) from the Pleistocene
    Scarborough Formation at Toronto, Ontario, and their
    palaeoenvironmental significance. *Quat. Res.* (in press).
Wright, H.E. 1977. Quaternary vegetation history - some
    comparisons between Europe and America. *A. Rev. Earth
    Planet. Sci.* 5: 123-258.

ORGANIZATIONAL IMPEDIMENTS TO EFFECTIVE RESEARCH ON RUNNING WATERS

Ron R. Wallace

Canstar Oil Sands Limited
Calgary, Alberta, Canada

> From the beginning we have called attention to the
> kinds of information a co-operative study might
> produce, but always we have encouraged individuality
> in the design and execution of research. We deem
> this individual research freedom one of the greatest
> assets of the Hubbard Brook Study.
>
> Likens et al. 1977

During the past decade, environmental impact research, of
which studies in running waters form an important part, has
received increasing public and scientific attention. Indeed,
there has been a phenomenal growth in the amount of funding
which has been directed toward environmental research in existing
Canadian institutions, or toward the creation of new government-
based institutes for such research. The time may be opportune to
question whether or not the increased amount of public attention,
and the resultant changes in patterns of funding for research,
have made limnological research in Canada more productive.*

There have been a number of reviews of this topic. In 1974,
Hamilton wrote a general overview of the problems facing
scientists employed by the Federal government who were attempting
to carry out environmental impact research. Vallentyne (1978)
provided a much more detailed view of events and decisions that
confirmed many of the trends in the rise of impediments to

---

*Although most of the following discussion is in terms of
Canadian research trends, similar problems are being encountered
by many of the industrialized nations.

effective research noted in the earlier review. Indeed, the overall
concensus was one of a perceived erosion of the climate for
Canadian environmental research, a theme which has been echoed
elsewhere (Wallace, 1981).

During the 1970's, interest in what the general public and
media have come to call "the environment" led to increasing
pressures on government agencies to bring about more effective
pollution monitoring and control policies. Government subsequently
greatly enlarged its legislative base which, in turn, gave rise to
an unprecedented increase in government regulatory agencies needed
to enforce, monitor and advise upon the new environmental
legislation. Unfortunately, the new organizational structures
meant a concurrent increase in the mechanisms for budget allocation
and control for many established centers of research. It was the
sudden increase in management of science by individuals or agencies,
which were not accountable for the quality or quantity of
scientific output, which was viewed by some as a counterproductive
trend.

The question remains: Has the development of highly visible
Provincial and Federal departments of the Environment enhanced our
scientific understanding about environmental impact assessment?
Vallentyne's (1978) study documents the widespread decline of
environmental (chiefly limnological) research in Canada. A more
general study of the Federal Public Service by Laframboise (1977)
singled out a prime underlying problem as the growing class of
officials which make and impose vital managerial decisions on
operational units. The implication is that when decisions are
made by those who are not responsible for the performance of
research laboratories there may be profound and lasting
consequences for those units. As the influence of distant
managers increases, and as the discretionary powers of those
responsible for scientific output decreases, the quality and
quantity of scientific output must, inevitably, decrease.
Historically, scientific research and development has been a type
of "free enterprise" system, in that many small laboratories
throughout the world have followed their pursuits of new, or
promising, lines of enquiry. The researchers determined, in
collaboration with their peers, directions for research. The
published scientific literature served as the intellectual
"currency" between laboratories. Funding was commonly a problem
but, once obtained, researchers were free to make the decisions
controlling lines of enquiry. By contrast, authors such as Hayes
(1973) provide details about the consequences of imposing national
controls in a system of scientific or technical interchange.

The trend toward larger, more centralized centers for
government research is a reflection of an overall increase in the
national, institutional controls over research science. While

cost-effectiveness may be served by the creation of such institutions, the consequences for productivity of research may be profound. Policy-makers often contend that large research groups are a good idea. This has led to the concentration of scientific resources into fewer but larger groups, especially in government.

Blume (1980) dealt with this situation in his review of Stankiewicz (1979). The latter found that there was an increase in effectiveness in groups of up to 3 to 5 or 5 to 7 scientists, but thereafter performance declined. Interestingly, the study further noted that highly experienced scientists profit more from large research groups than do the less experienced.

Although such studies have exceedingly important ramifications for the management of research science, they are all too rare. This is the more surprising in light of the relatively greater amounts of time and money which have been devoted to studying organizational theory and development, particularly in the larger, government institutions. Unfortunately, it is in the large, centralized laboratories that the vital concept of a community of science has been sacrificed to the idea of uniform management theories. As Blume (1980) noted, the few detailed studies of scientific productivity suggest that "policies and managerial practises to be adopted in furthering research in one specialty will not necessarily be the same as those appropriate to another ..." These important distinctions are often not visible to distant granting agencies, especially when very large disbursements for funds are necessary. Such problems, unfortunately, are often the hallmarks of Canadian environmental impact research programs. As Hamilton (1974) discussed, there are four characteristics of those major programs:

1. The overall program planning and financing is done outside the established research agencies.

2. The program of research is short-term and crisis-oriented with a research component designed to answer management-derived questions that are applicable to the specific area and problem.

3. The impacts are often awesome in scope. Co-ordination problems associated with large multi-agency, multi-disciplinary studies are immense, as are the problems of operating research crews in remote and very large geographic areas.

4. The external resources allotted are substantial as compared with the internal support provided to non-impact programs; however, limitations in manpower, time and funds are still very restrictive.

As examples, these essential characteristics apply to the 1969
Canada-British Columbia Agreement for the Okanagan Basin Study,
the 1971 Canada-Manitoba Agreement on the Lake Winnipeg, Churchill
and Nelson Rivers Study, the 1972 Northern (Mackenzie River)
Pipeline Study, and the 1973 Alberta Oil Sands Environmental
Research Program (AOSERP).

The four criteria cause fundamental problems for research
scientists concerned with the fulfillment of basic research goals.
Major environmental impact studies often lack guidelines which
are useful to research scientists. Often, the agreements which
set out the program objectives are written as legal documents
which provide scant guidance in the design of experimental or
survey research. Further, conflicts between program managers and
scientists invariably develop as commitments for funding change
(often times during the early life of the program of research)
and because of the administrative uncertainty which often
surrounds large research programs. In short, the larger the
research program the more vulnerable it is to financial cutbacks
and distant management controls.

There may be other consequences for the practising scientist
who ignores the administrative conflicts which surround the
direction and support of research programs. Scientists and their
work will inevitably be influenced by such conflicts. Yet if they
are able to retain credibility as scientists they are tied by
professional peer-reviewers and personal motivations to
producing publishable, scientific research. Typically, this
objective is further hampered because study agreements may
explicitly restrict the freedom to publish scientific findings. In
most cases this feature of environmental impact programs has been
more of an annoyance than a hardship; however, the implications are
serious.

A related concern voiced by Schindler (1976) is that peer
review, a vital mechanism in upgrading the quality of any research,
is often missed in environmental studies because the results are
too infrequently published in reputable scientific journals.
Many, therefore, view much of the current environmental impact
research with disdain.

In short, it could be said that good administrative support
for field and laboratory research is essential but not sufficient
to guarantee the realization of productive science. It could also
be argued that too much attention has been devoted to the concerns
of organization and management instead of to the facilitation of
research. Clearly, the essential climate of research will depend
upon the scientific maturity and sensitivity of senior management.
If those managers are, in turn, not held responsible for the
quality of research produced by their units, efficiency and quality
will decline.

A good scientist usually is far better qualified to recognize an existing area for potential advances than is an administrator far removed from the laboratory and scientific literature. Yet, as environmental research programs have become larger and orientated to "crash studies", the gaps between the scientists and the policy makers have also become wider. This phenomenon is well documented by Vallentyne (1978) but it is not limited only to government laboratories. Laframboise (1977) stated that "The distance between Ministers of the Crown and the vast units of administrative machinery for which they are individually or collectively responsible appears to be growing. Powers external to the line of command from a Minister through his deputy and his other senior officials effectively temper or frustrate clear-cut initiatives and flexible decision-making .... The principle that ..... a manager commands and is responsible for his resource is going by the board." This type of administrative alienation also promotes the problems noted by Schindler (1976) and Vallentyne (1978).

The conflicts over the control and administration of environmental impact research have created a great deal of frustration and disillusion for administrators and scientists alike. In some cases these conflicts are kept beneath the surface. In others, like the Alberta Oil Sands Environmental Research Program (AOSERP), deep conflicts between program administrators and scientists were frequently voiced in the media.

Perhaps the crucial question is this "is environmental impact research too important and too sensitive to allow research scientists a major input into the design and management of an environmental impact program?" All too often research design and, to an even greater extent, program management have become the battleground for bureaucrats and politicians. When various levels of government, line agencies, individual scientists and the program administrators are all competing over resources and jurisdiction, the "rule of reason" seems irrelevant. If a climate of competition develops, effective inter-agency, interdisciplinary environmental research is simply impossible. The "top-down" control of science and scientists that emerges typically leads to the assignment of work activities in accordance with jurisdictional mandates and immediate political realities - not on the basis of either the problem or scientific competence. At best the end result is a patchwork that treats the pieces and not the inter-relationships.

It is clear that environmental issues extend across scientific disciplines. If these issues are to be addressed in a manner consistent with that which exists in nature the research must be designed appropriately. Is it possible to establish a co-operative climate for co-ordinating interdisciplinary research on environmental problems? I think that it is, but first we have

to acknowledge the importance of open scientific debate and peer
review. As an example, social research studies almost invariably
show that organizations that allow for effective, meaningful worker
input into policy and administrative decisions are much more
productive than those that follow more autocratic approaches
(Likert and Likert, 1976; Bennis, 1969). This should not be
misinterpreted as a call for unlimited funding for research
scientists. Quite the contrary. Science, including environmental
research, cannot and should not be exempted from sensible fiscal
controls or from public spending reductions in hard times.
However the unfortunate aspect of "top-down" fiscal control is
that when cuts in spending are made they are often made crudely.
Hence, the larger and more concentrated the research institution,
the more vulnerable it may be to such crude fiscal management.
A second characteristic of large institutes is that they tend to
be less flexible in their ability to respond to changing times.
Therefore, the concept of a stable centre of research resulting
from a large institute is possibly quite invalid.

     University laboratories are prone to their own, distinct
problems as well. There, older, tenured scientists often enjoy
job security while the younger, often more productive,
researchers are allowed to seek other possibilities for
employment. The universities, however, face the inexorable
realities of demography over the next several decades because of
the declining birth rate. Universities expanded rapidly during
the 1960's and many scientists who were hired at that time won
academic posts which led to tenured positions. It has been
suggested (Anon, 1981) that this trend presently threatens
science because a disproportionate number of the most important
discoveries have, historically, been made by younger scientists.
For the next several decades there may not be room in the existing
university community for the brillant, young scientists who shall
come to market. Furthermore, in a climate of fiscal restraint
the ability of scientists to pursue unexpected insights and
opportunities in their research is diminished. Inevitably, major
advances in science occur when previously unsolvable problems
are approached by a new, innovative technique or are made
susceptible to new discoveries.

     Ironically, it will fall more and more to the research
scientists themselves to agree to sensible, co-ordinated
economies of scale for their research. Without question, should
scientists fail to adopt sensible economies in their work,
decisions as to funding will, as in the past, be made for them
by distant administrators.

     In general, the Canadian research infrastructure that once
provided a reasonably stable, secure base for research has now
been replaced with a bureaucratic structure which has more political,

immediate objectives: it is often more concerned with the form of
the process than with the substance of research. Centralized
organizations for research funding exert a profound effect not
only on government laboratories but on universities as well. It
is argued that highly centralized control of environmental
research may be an extreme and highly counterproductive management
solution to the control of costs. It is further argued that it is
highly unlikely that new discoveries in aquatic research can be
facilitated by a strict "management of objectives" approach.
Indeed, if it were possible to foresee a discovery clearly
enough to formulate a precise management plan, the "discovery"
would, by definition, not constitute new science.

In order to reverse the decline of substantive research it is
recommended that a detailed review be carried out of existing
institutional arrangements for aquatic research in Canada, with
a view to evaluating the relative successes of past environmental
impact research programs and existing laboratories. Strong
consideration should be given to the possibility of establishing
more, smaller, stable research laboratories rather than to
concentrating funds and attention on fewer, larger institutional
facilites. In this vein, it is clear that there are far too few
financially-independent research foundations in Canada. Such
independent bodies could provide a much needed alternative avenue
for the furtherance of Canadian research.

It could be suggested that should administrative problems and
allocative conflicts for research science not be resolved, and
soon, the public purse which endows the greatest majority of
research science in Canada may become ever more inaccessible to
scientists. Sir Adam Patrick Herbert put it this way:

> "Fancy giving money to the Government.
> Nobody will see the stuff again.
> Well, they've no idea what money's for −
> Ten to one they'll start another war.
> I've heard a lot of silly things, but, Lor'!
> Fancy giving money to the Government!"

REFERENCES

Anon, 1981. Scientist, heal thyself. *The Economist*   Mar. 7,
      1981. p. 18.
Bennis, W.G. 1969. New patterns of leadership for adaptive
      organizations. *In*: The Temporary Society, Bennis, W.G. and
      Slater, P.E. eds. Harper and Row, N.Y.
Blume, S.S. 1980. A managerial view of research. *Science*
      207: 48-49.

Hamilton, A.L.   1974.   Organizational impediments to effective
     environmental impact research. *In*: The allocative conflicts
     in water resource management. Agassiz Centre for Water
     Studies, Winnipeg, Manitoba.

Hayes, F.R.   1973.   The Chaining of Prometheus: Evolution of a
     Power Structure for Canadian Science. University of
     Toronto Press, Toronto, Ontario.

Laframboise, H.L.   1977.   Counter-Managers: The abandonment of
     Unity of Command. *Optimum*   8 (4): 18-28.

Likens, G.E., Bormann, F.H., Pierce, R.S., Eaton, J.S. and
     Johnson, N.M.   1977.   Bio-geo-chemistry of a forested
     ecosystem. Springer-Verlag, N.Y.

Likert, R. and Likert, J.G.   1976.   New ways of managing conflict.
     McGraw-Hill, N.Y.

Schindler, D.W.   1976.   The impact statement boondoggle.
     Science   192: 4239; 509.

Stankiewicz, S.   1979.   *In*: The effectiveness of research groups
     in six countries. Andrews, F.M. ed. Cambridge University
     Press, New York, and UNESCO Paris.

Vallentyne, J.R.   1978.   Facing the long term: an enquiry into
     opportunities to improve the climate for research with
     reference to limnology in Canada. *J. Fish. Res. Bd. Canada*
     35: 350-369.

Wallace, R.R.   1981.   Environmental Impact Research: A Time for
     Choices. *Alternatives*   9 (4): 42-48.

SECTION IV

REGIONAL RUNNING WATER ECOLOGY

ECOLOGY OF STREAMS AT HIGH LATITUDES

P.P. Harper

Département de Sciences Biologiques
Université de Montréal
Montréal, Québec, Canada

INTRODUCTION

Much of our current knowledge of the structure and the function of lotic ecosystems has arisen from studies conducted in the north of Europe and in temperate North America; this has given rise to a conception of what must be a typical running water habitat, its biota and their interrelationships. Investigations at more polar or equatorial latitudes have been much fewer, and by necessity, have often been comparisons with temperate systems, rather than studies in their own right. The situation is usually due to logistic problems which plague such endeavours. Not too surprisingly, limnological history is repeating itself at these latitudes and a much greater interest is being shown for the study of standing waters, lakes, ponds, and pools, than for running waters.

The purpose of this paper is to attempt to review some of the current knowledge of running water ecosystems at high latitudes; little is said of such environments in current textbooks (Hynes, 1970; Whitton, 1975) and even specialized reviews of arctic or antarctic limnology scarcely mention streams (Rawson, 1953; Livingstone, 1963; Goldman, 1970; Hobbie, 1973). The review will have a strong biological bent, but will exclude much of the fisheries and fisheries management literature; the author's linguistic limitations will as well restrict the coverage of soviet work and give the paper a nearctic bias. High latitudes will be taken in a rather large sense and will cover the ill-defined categories of arctic, subarctic, north boreal and their southern hemisphere equivalents (Ives and Barry, 1974). Little will be said of antarctic systems, not only for lack of data, but because few streams exist in the area, either through inadequate conditions or the absence of land masses

313

at suitable latitudes, except for scattered subantarctic islands,
parts of the Antarctic coast and some alpine situations in southern
Patagonia.

In a way, streams at high latitudes are but special cases of
the classical running water ecosystem, but set in particularly
extreme conditions, nonetheless there occurs a variety of running
water habitats colonized by a somewhat specialized biota.

THE PARTICULAR CONDITIONS

The living conditions at high latitudes are marked both by
extreme photoperiods and by a variety of temperature conditions
(Bownes, 1964). There thus exists an array of combinations of light
and temperature regimes, ranging from the relatively mild conditions
of parts of Fennoscandia and Alaska, to the extremely harsh
situation of eastern Siberia and the persistant cold of the Canadian
Arctic Archipelago.

Extreme photoperiods: photoperiods change regularly with latitude;
strong seasonality exists already at mid-latitudes, but it becomes
extreme polar-wise especially beyond the polar-circle (66°30'N) where
it gives rise to continuous daylight during the polar summer.

Low rates of inflow of radiant energy: because temperatures do not
follow latitude as closely as photoperiod, many latitudes exhibit a
variety of climatic conditions, which range from boreal to arctic and
can be best recognized by the plant cover of the terrestrial ecosystem
(coniferous forest to polar desert).

Though extremely low temperatures are recorded for short periods
in the continental temperature zone, streams are often well protected
by a heavy blanket of snow. In higher latitudes not only is the cold
more persistant, but the snow-cover is generally surprisingly thin,
and some areas are kept virtually snow-free by the wind even in winter
Streams are thus directly exposed to prolonged cold spells, and lose
much of their accumulated heat; the presence of an underlying permafrost layer causes heat loss at the stream bottom as well.

In summer, much of the absorbed radiation is used for melting
the snow and ice and the thawing of an active layer in the soil.
Photoperiods and incoming radiation are already on the decline when
water bodies start to warm up. This results in a very short summer,
which is in part compensated by the still extended periods of daylight.

Other features: in other respects, streams at high latitudes do not
differ markedly from others around the world. Discharge is related to
climate, soil and vegetation and will be dealt with below. The
chemical composition of the water reflects the lithology of the
catchment area, the terrestrial biogenic processes, and the
precipitation patterns; the resulting composition is not particularly

different from that at temperate latitudes (Brown et al., 1962; Livingstone, 1963; Kalff, 1968; Parker and Simmons, 1978), though low nutrient levels are often reported, especially in smaller watersheds (Brunskill et al., 1975; Grobbelaar, 1968); a large percentage of the nitrogen and the phosphorus may be in particulate form and not readily available to the plants (Brunskill et al., 1975).

A VARIETY OF HABITATS

Running water environments vary considerably and though a definite classification is still premature, the following breakdown appears useful for our discussion. Most stream systems are composites of two or more of the categories; the distinctions are thus somewhat artificial, particularly in regard to the "river continuum concept" (Vannote et al., 1980).

Large rivers: these are characteristic features of the north slope of the North American and Asian continents and indeed some of the largest rivers of the world, Yenisei, Lena, Ob-Irtysh, and Mackenzie, flow into the Arctic Ocean. As they originate in temperate forests and steppes, their drainages are not characteristically arctic and they are often seen as extensions of the boreal forest biome towards the north (Mackay and Løken, 1974; Snow and Chang, 1975).

The heat content of the river thaws the underlying permafrost to a considerable depth and for some distance laterally; up to some 40 m under the River Shaviovik of Alaska (Brewer, 1958) and to some 100 m of the shore of the Siberian River Yana (Grigor'ev, 1959). The river flows over an unfrozen zone which has the shape of an extended trough, the width of which is proportional to that of the river (Johnston and Brown, 1964).

Typically there is a period of high flow in the early summer which accounts for 14-46% of the annual discharge and is 4-5 times the minimum discharge (Bird, 1967); the lower values are found in rivers, such as the Mackenzie, in which large lakes, in this instance Great Slave Lake, stabilize the flow; the higher values occur in rivers which have a more regular drainage such as the Back River of Canada (Mackay and Løken, 1974). Rivers which arise in high mountain ranges, such as the Ob which drains the Altai mountains of Central Asia, have a second peak of high discharge in mid-summer due to ice-melt in the mountain glaciers (Zhadin and Gerd, 1961).

In autumn, ice forms on the river following the usual processes, clear ice on the slow regions and an accumulation of frazil ice in the turbulent reaches; maximum thickness of 1-2 m is reached in late winter (Mackay and Løken, 1974), and some supercooling of the water may occur under the ice (Brunskill et al., 1973). Even in very high latitudes, some sections remain open throughout the winter, usually in fast reaches, at the outlet of large lakes or near groundwater infiltrations (Oliver and Corbet, 1966; Power, 1969); during cold

spells, large banks of frazil and anchor ice undoubtedly develop,
though this is not well documented.

In most instances flow is maintained throughout the winter,
though at a much reduced rate. Smaller rivers have no discernable
flow in late winter (Arnborg et al., 1966) and Power (1969) indicates
that "even quite large rivers (in the Ungava peninsula) are reduced
to a mere trickle or a series of frozen pools with the intervening
riffles quite dry".

The spring thaw generally starts in the south in the upper
basin and gradually proceeds downstream. The greatly increased
water level in the lower reaches detaches the ice cover from the
sides of the channel, breaks it into large slabs and produces ice
jams which further increase the level; large plates of ice are
carried downstream and many are left stranded on the shores when
the water recedes (Mackay and Løken, 1974). The lower course of
the river leaves its bed and forms a large number of delta lakes,
side channels, overflows, and back eddies; these are of considerable
importance in the ecology of the river and they come into contact
with the main channel only in periods of high water.

The high discharges of spring and summer together with the
eroding effect of the spring ice run and the "thermal erosion"
consecutive to the thawing of the stream bank (Bird, 1974) render
the waters turbid. Wynne-Edwards (1952) qualifies the rivers as
extremely turbid, an exception being rivers in which a large lake
acts as a sediment trap (e.g. Great Slave Lake on the Mackenzie and
Lake Baikal on the Angara); on the other hand, Zhadin and Gerd
(1961) consider the Asian rivers as only moderately turbid. In
fact turbidity varies seasonally and the turbid Liard River in the
Mackenzie drainage carries a sediment load of up to 2000 g/m^3 in
the early summer, but becomes quite clear (1-10 g/m^3) in the winter
(Brunskill et al., 1973).

Summer temperatures are moderate (14-16°C), but sometimes higher
than those of lakes at the same latitude; this is the influence of
the warmer waters flowing from the south (Wynne-Edwards, 1952).

Oxygen levels are generally high and remain so most of the year;
somewhat reduced concentrations may occur under the ice (Brunskill
et al., 1973; Moore, 1978a), but rarely to the point of deoxygenation
commonly observed in the Ob which drains extensive peatlands
(Yukhneva, 1971).

Nutrient transport systems have seldom been studied, but appear
similar to those in the temperate zone (Brunskill et al., 1975).

Mountain streams: the boreal mountain streams of Scandinavia, such
as the Kaltisjokk (Müller et al., 1970) the Vindelälven (Ulfstrand,
1968b) and the Ricklean, (Göthberg and Karlström, 1975; Karlström
and Backlund, 1977), are among the world's best known stream ecosystems
and behave in a "classical" manner' indeed much of our understanding
of streams stems from these studies. Many however dry up in summer
(Brinck and Wingstrand, 1949) or freeze to the bottom in the winter
(Mendl and Müller, 1978).

Of special interest are the watercourses found in lake areas;
the presence of the lakes maintains higher stream temperatures even
in the winter (Melin, 1947; Müller, 1954b) and provides a distinctive
habitat for the so-called "lake-outlet biocoenosis" (Müller, 1955;
Illies, 1956; Ulfstrand, 1968a).

Mountain streams beyond the tree-line are markedly different
(Stocker and Hynes, 1976; Slack et al., 1977, 1979). They flow only
during the summer months, often even less (Johansen, 1911). The
spring thaw causes extreme flooding for a fortnight or so during
which flows some 80% of the annual discharge; the flow then decreases in a roughly asymptotic fashion during the summer, except for
an occasional freshet following precipitation (McCann and Cogley,
1972); when the surrounding terrestrial vegetation is sparse, peak
flows are short and high.

By late summer, the stream freezes in its bed. Early thawing
may cause some of the meltwater to flow under the snow over the still
frozen streambed, thus preserving it from erosion (Pissart, 1967).
The bed will eventually thaw and considerable erosion may occur, up
to 90% of the yearly total during the spring flood (Cogley, 1971;
Stocker and Hynes, 1976). The frost shattering and the weathering
of the stream bed in the winter combined with the removal of the
weathered material in the summer often produces a deeply entrenched
channel, particularly in the middle course of streams in sedimentary
areas; the upper valley often remains open and shallow as it
receives little water after snow-melt and usually dries up (Bird,
1967). Hynes et al. (1974) believe that in some streams the top 5 cm
of the substrate may be lost during the spring flood.

Some streams retain flow in the winter and even ice-free areas
in less extreme climates. A cold spell can bring about the formation
of sufficiently large amounts of frazil and anchor ice in the highland
sections to reduce the flow in lowland sections; in such an instance,
an icelandic river has had its discharge decrease from 340 m^3 to
20 m^3 (Eythorsson and Sigtryggsson, 1971).

In the high polar desert, streams much resemble the desert
streams of middle latitudes to the extent of forming the characteristic alluvial fans (Bird, 1967).

Temperatures remain low throughout the summer, they reach a maximum of 5-15°C in late July and decrease to freezing levels within 4-6 weeks (Oliver and Corbet, 1966; Stocker and Hynes, 1976).

Streams which retain flow have low concentrations of oxygen in winter and deoxygenation may occur; concentrations are high in summer, except where the flow is sluggish and high levels of POM are present (Frey et al., 1970; Brunskill et al., 1973).

Water chemistry varies seasonally. The early flow is essentially meltwater and is very poor in minerals; conductivity increases steadil during the summer as the flow is reduced and the water drains from a deeper active layer in the soil (Oliver and Corbet, 1966; Schindler et al., 1974); there in fact appears to be an inverse correlation between TDN, TDP, and DOC and the flow (De March, 1975). In areas of sparse plant cover, there is little retention of the water in the soil and little contribution of the soil to the water quality (De March, 1975); inversely, a more developed terrestrial vegetation has a stabilizing effect on the flow of water (Bird, 1967) and affects the water quality (Woo and Marsh, 1977). The role of these streams in the nutrient cycles of the lakes is starting to be assessed both in the Arctic (Bliss et al., 1973; Schindler et al., 1974; De March, 1975) and the Antarctic (Samsel and Parker, 1972). In areas of volcanic activity, volcanic ash can provide an important input of nutrients (Eicher and Rounsefell, 1957; Dugdale and Dugdale, 1961).

Glacier streams: Steffan (1972) considers glacier-fed streams to be sufficiently different from other running-water habitats to be set in a particular category he calls the kryon. These brooks are charact- erized by both low temperatures and low nutrient concentrations. The suspended load of sediment is variable, but is highest in summer (Church, 1974); it can contribute to the ionic content of the water (Slatt, 1970).

Flow is much reduced in winter and the stream freezes over except at the outlet of the glacier where considerable anchor ice may develop. There is marked daily fluctuation in flow, with a maximum discharge in late afternoon and a minimum in early morning; the development of a lagoon at the foot of the glacier sometimes stabilizes the flow to some extent (Eythorsson and Sigtryggsson, 1971; McCann and Cogley, 1977). There is also a daily fluctuation in stream temperatures, the amplitude of which increases downstream (Oliver and Corbet, 1966). Much of the literature on alpine glacial streams (Steinböck, 1934; Saether, 1968; Bretschko, 1969) is doubt- less applicable to polar glacial streams.

A peculiar situation is that of lakes dammed by glaciers; the erosion of the dam during periods of warm weather or high precipita- tion may cause a rapid flushing of the lake into the downstream river (= Jökulhlaup) (Alseth, 1952; Post and Mayo, 1971; Cogley and

McCann, 1976; Young, 1977. For example, the rupture of the ice dam
(Knik Glacier) restraining the Alaskan Lake George can cause a drop
of some 35 m in the water level of the lake and raise the Knik River
from 2 to 6 m and widen it from 50 to 1300 m (Stone, 1963).

Spring-fed streams: springs can occur even in the permafrost area.
Water may come from a suprapermafrost source in the active layer of
the soil and give rise to seasonal seeps and springs which freeze
over with the permafrost in the fall. Subpermafrost water often
exists in aquifers particularly in sedimentary terrains and surfaces
as permanent springs with rather constant flow and temperature; in
the subarctic where discharges are greater, extensive karst can
occur, particularly in Alaska, the Lena basin of Siberia (Bird, 1967),
and the "moberg" region of Iceland (Eythorsson and Stigtryggsson,
1971). Water also exists within the permafrost as thaw pockets
(= taliks) and pipes (Mackay and Løken, 1974).

The permanent springs may surface individually or arise in
rivers and lakes; they help maintain flow and open water in many
streams during the winter. Downstream from the infiltration of the
spring, the water eventually freezes and there is a buildup of large
quantities of ice (= icings, Aufeis, wadeli) which do not thaw
completely until late into the summer; the presence of such icings
has been used to locate springs (Lewis, 1962; Bird, 1967; Post and
Mayo, 1971; Craig and McCart, 1974). The spring will maintain a
relatively milder microclimate in its immediate vicinity and often
will allow the establishment of a more temperate vegetation; such
areas have been likened to cases in a polar environment (Craig and
McCart, 1975).

In regions of volcanic activity such as Alaska and Iceland,
there are numerous hot springs with temperatures ranging up to $54^{\circ}C$
(Nava and Morrison, 1974). These highly specialized habitats occur
at all latitudes and comprehensive reports on them have been given
by Tuxen (1944) and Castenholz and Wickstrom (1975).

Lowland streams: these are undoubtedly the most characteristic of
high latitude lotic habitats, but they are also among the least
studied; data are available only on the Alaskan Ogotoruk Creek
(Watson et al., 1966) and a few Siberian streams (Popova, 1957;
Levanidov, 1977). This category groups foothill and floodplain
streams with reduced flow and low slope.

In coastal plains where there is considerable ground ice, the
stream may thaw some of this ice and form a series of pools along
its course, thus becoming a so-called "beaded stream" (Bird, 1967,
1974). Streams often have a number of parallel and intertwining
channels and are called "braided streams" (Craig and McCart, 1975).

Flow is restricted to the summer months, and flooding is less severe than in mountain streams particularly if the surrounding tundra vegetation is dense (Dingman, 1966). Summer temperatures tend on the whole to be higher than in mountain streams, especially in beaded streams (Levanidov, 1977), and turbidity is distinctly lower as the tundra acts as a filter and the ponds as sediment traps. The extensive flatlands give the water its distinctive dark colour, low pH, low calcium and nutrient concentrations, low conductivity, and high oxydability (Popova, 1957; Church, 1974; Craig and McCart, 1975). The streams appear to play an important role in the nutrient cycling of surrounding terrestrial ecosystems (Bliss et al., 1973).

In winter the streams freeze completely to the bottom (Watson et al., 1966) as there is little ground water resurgence in the coastal plains, except sometimes near the foothills, where flowing water may occur for a short distance before icings eventually build up (Lewis, 1962).

A lowland stream of similar character has been intensively studied in a boreal lowland of Northern Alberta (Clifford, 1969, 1978). This differs in many respects from more polar streams by its permanent flow, milder climatic setting as well as by a high pH (from a sedge and reed peatland).

A SPECIALIZED BIOTA

Polar biotas usually comprise the most robust elements of the temperate flora and fauna together with a distinctive element; the resulting biogeographical implications have been discussed else-where, though much work remains to be done in the case of the stream biota which is still not well known taxonomically. The following review will restrict itself to ecological topics relating to the major plant and animal groups and to stream productivity.

Plants and algae: mosses and herbaceous plants are very abundant in tundra ponds and lakes, but little is reported of these in running waters; more often than not, only taxonomic details are given and species are listed only as having been collected from a stream (e.g. Polunin, 1948; Holmen and Scotter, 1967). Mosses are commonly found in mountain streams whereas aquatic macrophytes dominate in the lower sections of large rivers (Zhadin and Gerd, 1961) particularly in the flood-plain lakes and overflows (Snow and Chang, 1975); ice scouring in the spring often limits plant growth in the main channel. Johnson (1978) has estimated the contribution of aquatic mosses to stream metabolism in the boreal river Ricklean, as part of a general investigation (Kalström, 1978).

Reports on algae are patchy and, as can be expected, range from indications of lush growths to virtual absence (Watson et al., 1966; Stocker and Hynes, 1976). Phytoplankton develops only in the slower sections of large rivers. It is poorly represented in turbid waters, as were those of the Ob before the construction of hydroelectric dams; it often increases in biomass and diversity downstream, as in the Yenisei, to reach a maximum in the delta. In other rivers, such as in the Lena, it is poorly developed (Zhadin and Gerd, 1961). Diatoms often dominate (Hooper, 1947). The floodplain standing waters do support highly productive phytoplankton communities and their contribution to the main river in periods of high water are probably substantial (Zvereva, 1957; Snow and Chang, 1975). Moore (1978b) in a comparison of phytoplankton of ponds, lakes and rivers of the Canadian Northwest Territories found important populations only in the large wide rivers; their composition was similar to that of the lakes of the area. In the Subarctic Yellowknife River, the phytoplankton is marked by a heavy bloom in April of *Chlamydomonas lapponica* with *Dinobryon* spp. developing during the summer, but at a lower population level. Diatoms are rare in the summer due to low SiO_2 concentrations and there is no autumn bloom, due doubtless to the low temperatures and the lack of nutrients (Moore, 1977a, 1979a).

Attached algae are more widely distributed. Glacier brooks, however, seem to lack large populations, because of low temperatures, high turbidity and low nutrient concentrations (Steffan, 1972). Other rivers possess variable growths according to the transparency of the water, the scouring effect of the flow, the presence of suitable substrates and the availability of nutrients. Moore (1979b) has compared epilithic, epipelic, epiphytic and epipsammic algae in three streams (and 18 lakes) of the Canadian Northwest Territories. The species composition of the algal communities growing on stones, mud and plants were very similar and contained a distinctive northern element. The epipsammic algae were more cosmopolitan and resembled similar associations in warmer latitudes, which probably reflects a high degree of specialization. Populations developed during the summer and reached a maximum in late July and then decreased till freezing. Epipelic communities in Baffin Island streams (Moore, 1974a) were highly developed showing a notable species richness (240 taxa, 200 of which were diatoms) and a density growth rate comparable to those of temperate communities. Temperature and light were seen as the major limiting factors, whereas grazing, nutrients and flooding seemed to have little effect. In epilithic and epiphytic communities (Moore, 1974b), densities were much lower, and nutrients, particularly NO_3-N, were limiting. There was a definite seasonal succession of diatoms (in June) and Chlorophyta and Chrysophyta (in August); Cyanophyta tended to dominate throughout by numbers and volume. There was some indication that continuous light in mid-summer may have caused some inhibition as growth rates were superior in late summer. A sharp drop in densities in late August was attributed to a change in species composition from planktonic algae to attached species.

Müller-Haeckel (1967, 1970a, 1970b, 1973) has extensively studied the drifting of algae in streams of northern Sweden. She has correlated the intensity of the drift with periods of high metabolic activity; most species reach maximum drift in mid summer and at mid-day, but night drifting species are also known. Diatom drift and colonization rates seem to indicate that most species have an endogenous potential for two periods of blooming each year, but that local conditions may often be limiting (Müller-Haeckel and Håkansson, 1978).

Zooplankton: this community develops mainly in lakes, but slow reaches in large rivers often harbour important populations (Bardach, 1954). The community is composed mainly of rotifers, cladocerans and some protozoans (Zvereva, 1957; Zhadin and Gerd, 1961). Moore (1977b, 1978a) has found that temperature and seston are the limiting factors for herbivorous plankters, and prey availability for the predaceous species.

Zoobenthos: benthic invertebrates are widely distributed along river courses and have been relatively better studied than other parts of the ecosystem despite inherent taxonomic problems.

Glacier streams support little benthos: typically only detritivorous species occur and Diamesinae dominate near the glacier while *Prosimulium* and Orthocladiinae appear downstream (Steffan 1971, 1972).

Springs on the other hand often support a very dense and diverse fauna, compared to other streams in the same area, especially at high latitudes, the difference often being of one order of magnitude (Craig and McCart, 1975); this is related to the stability of the flow and of the climatic conditions as well as the maintenance of flowing water in the winter. Hot springs harbour the usual highly specialized fauna, such as Ephydridae (Tuxen, 1944), though Peterson (1977) found larvae of *Simulium vittatum* living in warm sulphurous water originating from the condensation of volcanic steam.

Mountain streams below the treeline contain a very diverse and abundant benthos, as the classical studies of Thienemann (1941) and Ulfstrand (1967, 1968b) indicate. At higher latitudes, the fauna becomes rapidly poorer in both species number and population densities. Slack et al. (1977, 1979) have described the zonation pattern of the fauna in northern Alaskan streams: Diamesinae, Orthocladiinae and *Prosimulium* dominate the headwaters (cf. glacier brooks); Plecoptera appear in streams of order 2-3 and represent a significant portion of the community; Ephemeroptera live only in the large sections of order 5 or more. Trichoptera are rare or absent. Numerically the Orthocladiinae dominate throughout most of the course.

The overall densities are often very low (Craig and McCart, 1975) and
reflect the high instability of the substrate and the flow. Many
groups of insects are absent at higher latitudes, such as Odonata,
Plecoptera, Ephemeroptera, and Simuliidae (Oliver et al., 1964) and
often the only insects in streams are Chironomidae, particularly
Orthocladiinae (McAlpine, 1965). Stocker and Hynes (1976) describe
the benthos of an arctic stream as comprising only Chironomidae
(55-95%), Enchytraeidae (1.5-40%), Ostracoda (0-16%) and Copepoda
(0-15%); all the species are detritivores, except the copepod
Cyclops magnus. Antarctic streams have even more reduced faunas, no
doubt for zoogeographical reasons: only Chironomidae, Acari and a
dytiscid beetle have been reported (Gressitt and Leech, 1961; Brundin,
1970; Gressitt, 1970).

Streams flowing from lakes tend to have extremely high populations of filter feeders, principally Trichoptera, Simuliidae and
Chironomidae (Müller, 1955; Illies, 1956; Ulfstrand, 1968a); many of
these groups are however absent at higher latitudes.

Lowland streams have benthos densities which are intermediate
between those of springs and those of mountain streams (Craig and
McCart, 1975). Levanidov (1977) in Eastern Siberia and Watson et al.,
(1966) in Alaska describe much similar faunas in the streams they
investigated; chironomids dominate in numbers, followed by Plecoptera,
Ephemeroptera and Oligocheta; water mites, planarians, Trichoptera,
and Tipulidae also occur. Clifford (1972b) has shown that drift from
the surrounding marshes may bring a significant input of invertebrates, mainly microcrustaceans, into the main stream.

Large rivers have seemingly few invertebrates and the benthos
is dominated by Chironomidae, Ceratopogonidae, Oligocheta;
Ephemeroptera, Trichoptera, Nematoda, Crustacea, and Mollusca are
sometimes mentioned (Popova, 1957; Zvereva, 1957; Yukhneva, 1971;
Brunskill et al., 1973; Snow and Chang, 1975). Higher populations
occur in the slow lateral reaches, particularly in the standing
waters of the floodplain; the central channel may only contain small
psammophilic species of Chironomidae, Oligocheta and Nematoda
(Zvereva, 1957).

Much information is available on the life histories and
adaptations of the benthic organisms. Drift patterns have been
extensively studied, and animals classified as day-active or night-
active (Müller, 1954a, 1970; Sodergren, 1963; Schmidt and Müller,
1967; Clifford, 1972a; Steine, 1972; Gallup and Clements, 1975).
Some invertebrates show little periodicity in their drifting patterns
(Chironomidae, Simuliidae) while others are highly synchronized
except for a period of asynchronization during the polar summer
(Müller and Benedetto, 1970; Müller and Thomas, 1972; Müller, 1973b).
Müller (op. cit.) indicated that this asynchronization is a response

to high light intensities throughout the day in summer, but
Benedetto (1976) showed that it is also related to changes in daily
patterns of activity with the size of the insects. As well, rhythms
of the activity of the adults have been investigated; generally,
clear periodicities of emergence flight and mating are noted (Kureck,
1966, 1969; Tobias and Thomas, 1967; Thomas, 1970; Kaiser and Müller,
1971; Göthberg, 1973). In the high arctic, however, insects are
more opportunist and fly and mate when conditions are adequate
(Downes, 1965; Syrjamaki, 1969) or, in some highly specialized cases,
do away with many of the activities of adult life by various
adaptations such as parthenogenesis, maturation of the eggs during
the immature stages or mating on the ground (Davies, 1954; Downes,
1962, 1964, 1965). Müller (1954a) has proposed the existence of a
colonization cycle by which upstream flight by adult insects
compensates the downstream drift of the aquatic immatures; interesting
modifications of the cycle are noted by Müller et al. (1976) and
Mendl and Müller (1978) in which Plecoptera emerging from a stream
drop their eggs in an upstream lake where these will hatch; the
larvae spend the winter in the lake and drift back into the stream
in spring, thus escaping the river ice.

Life cycles of some insects have been described and little
growth seems to occur during the winter (Svensson, 1966; Ulfstrand,
1968a; Clifford, 1969, 1976; Benedetto, 1973; Rosenberg et al.,
1977b) although Plecoptera have been observed to continue feeding
(Moore, 1977c); life cycles are sometimes accelerated in the warmer
waters at lake outlets (Ulfstrand, 1968a). In the Chironomidae,
cycles can extend for more than a year in high latitudes (Oliver,
1968), but this has not been reported in other groups, except in
some alpine Trichoptera (Decamps, 1967). Nothing is known of the
overwintering benthos in streams which freeze to the bottom; as
Brinck and Wingstrand (1949) observed little difference between the
fauna of Lapland streams which maintain flow and those which do not,
a fairly efficient mechanism must exist. Some observations made on
lentic Chironomidae which survive freezing probably apply as well to
lotic species (Scholander et al., 1953; Salt, 1961; Danks, 1971a,
b, c; Grimas and Wiederholm, 1979); some species also build cocoons
in which to spend the winter (Madder et al., 1977). In rivers which
lose much oxygen in winter, the benthos is composed mainly of resistant forms (*Procladius, Chironomus, Cryptochironomus*) and others which
congregate in the more favourable conditions near the mouths of
tributaries (Yukhneva, 1971).

Fish: an enormous literature has accumulated on the fisheries
management of northern waters, but no attempt will be made to overview it, and only a few points will be stressed here.

Arctic rivers harbour considerable populations of fish, though
exact censuses hardly exist. Commercial fisheries exist for native

and local populations, but yields are generally low (Regier, 1976).

The main river channels contain little food for fish and Zvereva (1957) has commented on conditions being "apparently adequate in river beds seemingly without food". In fact, overflows of the main river are much more productive and provide much of the energy input (Zhadin and Gerd, 1961; Snow and Chang, 1975). The large river channels also provide migration routes and refuges from the ice in the winter (Popova, 1957). The tundra and mountain streams are used in the summer for food (Müller, 1962). Some species, such as the grayling, reproduce earlier in the season in tundra streams and are able to move back to the main channel before the fall freeze up, others, such as arctic char, breed in the fall, and as their eggs cannot withstand freezing, they must spawn in or near springs which maintain open water (Kogl, 1965; McCart and Craig, 1973; Craig and McCart, 1974; Ridder and Peckam, 1978). Springs thus become an important resource in the biology of the fish, particularly since they also harbour large populations of invertebrates, which constitute the main diet of most fish (Loftus and Lenon, 1977; McKinnon et al., 1978).

Another aspect of fish biology which has been much investigated in high latitude rivers is that of activity patterns. Particularly noteworthy are the inversion of diel rythms in some species (from day-active in winter to night-active in summer), and the desynchronization during polar summer (Müller 1968, 1973a, 1976; Andreasson and Müller, 1969; Andreasson, 1972; Figala and Müller, 1972; Eriksson, 1973).

Productivity: although many high latitude stream systems seem to contain populations of algae and invertebrates in densities comparable to those of more temperate zones (Moore, 1974a; Craig and McCart, 1975; Levanidov, 1977), it should be remembered that time scales are much longer (Brown, 1975). Life cycles of both fish and their prey are undoubtedly much extended so that low production rates are to be expected. The long photoperiods in the summer which should compensate the short ice-free period may also have an inhibitory effect (Moore, 1974b).

A global study of the productivity processes in a high latitude lotic ecosystem is still to be done.

CONCLUDING REMARKS

From the preceeding discussion, it is evident that much information is still needed before we get a clear picture of these running water ecosystems.

Present studies: much of the early information on northern systems
in North America was gained through expeditions such as the Harriman
Alaska Expedition of 1899 and the Canadian Arctic Expedition of
1913-1918. These were followed after the Second World War by the
Canadian Northern Insect Survey of 1947-1957 (Freeman, 1959), the
Northern Biting Fly Program of the Canada Defence Research Board
(Twinn, 1952) and, more recently, by the Lake Hazen Project of
1961-1963 (Oliver et al., 1964) and the Char Lake Project during
the International Biological Program (Schindler et al., 1974). In
recent years, studies in the north have taken on a more practical
aspect, and are mainly impact studies related to northern
development projects, to highway construction (Rosenberg and Snow,
1975b; Rosenberg and Wiens, 1975, 1978), forest management
(Sheridan and McNeil, 1968), oil sands exploitation (Barton and
Wallace, 1979a, b) oil and gas exploration and transportation
(Brown, 1975; Didiuk and Wright, 1975; Roeder et al., 1975;
Rosenberg and Snow, 1975a; Snow and Rosenberg, 1975; Snow et al.,
1975; Rosenberg and Wiens, 1976; Rosenberg et al., 1977a),
hydroelectric development (Magnin, 1978), and flood and pollution
control (Frey et al., 1970).

Undoubtedly the same course of events occurs in other countries.
This current trend has allowed considerable research to be conducted at not easily accessible latitudes and has increased our
knowledge of these extreme ecosystems. Much of the information
however is stored in internal reports of the sponsoring agencies,
and thus not readily available to the scientific community at large.
It would be desirable that scientists involved in these endeavours
take the time to condense their findings and report them in
international journals; indeed many interesting papers have arisen
as direct outputs or byproducts of these impact studies.

The need for comprehensive studies: as always happens in limnology,
rivers and streams have received much less attention than lakes at
these latitudes; we still lack a comprehensive investigation of a
running-water ecosystem of the magnitude and value of the Char Lake
Project. Obviously rivers are not as easy to study and do not form
the compact microcosm that a lake does. Further, rivers are perhaps
not as vulnerable as other polar habitats to man-made environmental
changes and have a higher power of recuperation. Nonetheless, many
of the impact studies are centered on river systems which remain
the least known of high latitude systems.

The lack of basic information: despite the growing interest for
northern ecosystems, considerable gaps in our basic knowledge still
exist. There is no comprehensive review of stream faunas and floras;
older syntheses such as Fauna Arctica are of some value, but
considerably out of date. Life history information is lacking for
most plants and animals of the system and little is known of what

happens to them in the winter. Similarly, we do not know what
factors limit the polar-wise extension of many groups or the
adaptations that allow other groups to survive and even to flourish.

   The worldwide economic stagnation and the current aversion for
"pure" research will probably not allow the establishment of large
scale programmes for the study of these stream systems; it is hoped
that scientists will use the opportunities offered by applied
research linked to northern development to gather and make available
the information we need for a fuller comprehension of these extreme
but important running-water ecosystems.

REFERENCES

Alseth, I.B. 1952. Self-emptying lake. *Nat. Hist.* 61: 8-13.

Andreasson, S. 1972. Seasonal changes in diel activity of *Cottus
    poecilopus* and *C. gobio* (Pisces) at the arctic circle. *Oikos*
    24: 16-23.

Andreasson, S. and Müller, K. 1969. Persistence of the diel
    activity phase shift in *Cottus poecilopus* Heckel (Pisces)
    after transfer from 66° to 55° northern latitude. *Oikos*
    20: 171-174.

Arnborg, L., Walker, H.J., and Peippo, J. 1966. Water discharge
    in the Colville River, Alaska, 1962. *Geografiska Annal.*
    48A: 195-210.

Bardach, J.E. 1954. Plankton crustacea from the Thelon watershed,
    NWT. *Can. Field-Nat.* 68: 47-52.

Barton, D.R. and Wallace, R.R. 1979a. Effects of eroding oil sand
    and periodic flooding on benthic macroinvertebrate communities
    in a brown-water stream in northeastern Alberta, Canada.
    Can. J. Zool. 57: 533-541.

Barton, D.R. and Wallace, R.R. 1979b. The effects of an
    experimental spillage of oil sands tailings sludge on
    benthic invertebrates. *Environ. Pollut.* 18: 305-312.

Benedetto, L.A. 1973. Growth of stoneflies nymphs in Swedish
    Lapland. *Ent. Tidskr.* 94: 15-19.

Benedetto, L.A. 1976. Annual and diurnal periodicity of stonefly-
    drift at the arctic circle. *Limnologica* 11: 47-55.

Bird, J.B. 1967. "The physiography of arctic Canada, with special
    reference to the area south of Parry Channel." John Hopkins
    Press, Baltimore.

Bird, J.B. 1974. Geomorphic processes in the arctic. *In*
    "Arctic and alpine environments," Ives, J.D. and Barry, R.G.
    eds, Methuen, London.

Bliss, L.C., Courtin, G.M., Pattie, D.L., Riewe, R.R., Whitfield,
    D.W. and Widden, P. 1973. Arctic tundra ecosystems.
    Ann. Rev. Ecol. Syst. 4: 359-399.

Bretschko, G. 1969. Zur Hydrobiologie zentralalpiner
    Gletscherabflusse. *Verh. dt. Zool. Ges. (Deut).* 1968: 741-
    750.

Brewer, M.C.   1958.   Some results of geothermal investigations of permafrost in northern Alaska.   *Trans. Amer. Geophys. Union.* 39:   19-26.

Brinck, P. and Wingstrand, K.G.   1949.   The Mountain fauna of the Virihaure area in Swedish Lapland.   I.   General account. *Lund Universitets Årsskrift N.F.*   2 (2):   1-69.

Brown, J., Grant, C.L., Ugolini, F.C. and Tedrow, J.C.F.   1962. Mineral composition of some drainage waters from Arctic Alaska.   *J. Geophys. Res.*   67:   2447-2453.

Brown, S.   1975.   Environmental impacts of arctic oil and gas development.   Inuit Tapirisat of Canada Renewable Resources Project, 6.

Brundin, L.   1970.   Diptera: Chironomidae of South Georgia. *Pacific Insect Monog.*   25:   276.

Brunskill, G.J., Rosenberg, D.M., Snow, N.B. and Wageman, R.   1973. Ecological studies of aquatic systems in the Mackenzie-Porcupine drainages in relation to proposed pipeline and highway development.   I and II.   Canada Department of the Environment, Fisheries and Marine Service, Report of the Task Force on Northern Oil Development.   73-40 and 73-41.

Brunskill, G.J., Campbell, P., Elliott, S., Graham, B.W. and Morden, G.W.   1975.   Rates of transport of total phosphorus and total nitrogen in Mackenzie and Yukon River watersheds, N.W.T. and Y.T., Canada.   *Verh. int. Verein. theor. angew. Limnol.*   19:   3199-3203.

Castenholz, R.W. and Wickstrom, C.E.   1975.   Thermal streams. *In* River Ecology.   Whitton, B.A. ed., University of California Press.

Church, M.   1974.   On the quality of some waters on Baffin Island, N.W.T.   *Can. J. Earth Sci.*   11:   1676-1688.

Clifford, H.F.   1969.   Limnological features of a northern brown-water stream, with special reference to the life histories of the aquatic insects.   *Amer. Midl. Nat.*   82:   578-597.

Clifford, H.F.   1972a.   A year's study of drifting organisms in a brown-water stream of Alberta, Canada.   *Can. J. Zool.* 50:   975-983.

Clifford, H.F.   1972b.   Drift of invertebrates in an intermittent stream draining marshy terrain of west-central Alberta. *Can. J. Zool.*   50:   985-991.

Clifford, H.F.   1976.   Observations on the life cycle of *Siphloplecton basale* (Walker) (Ephemeroptera: Metretopodidae). *Pan-Pacific Entomol.*   52:   265-271.

Clifford, H.F.   1978.   Descriptive phenology and seasonality of a Canadian brown-water stream.   *Hydrobiologia*   58:   213-231.

Cogley, J.G.   1971.   Hydrological and geomorphological investigations on a high latitude drainage basin: "Jason's Creek" Southwest Devon Island, N.W.T.   M.Sc. Thesis, McMaster University, Hamilton, Canada.

Cogley, J.C. and McCann, J.B. 1976. An exceptional storm and its
     effects in the Canadian High Arctic. *Arctic and Alpine Res.*
     8: 105-115.
Craig, P. and McCart, P.J. 1974. Fall spawning and overwintering
     areas of fish populations along routes of proposed pipeline
     between Prudhoe Bay and the Mackenzie Delta 1972-1973.
     Arctic Gas Biol. Report Ser. 15 (3): 1-34.
Craig, P.C. and McCart, P.J. 1975. Classification of stream
     types in Beaufort Sea drainages between Prudhoe Bay, Alaska,
     and the Mackenzie Delta, N.W.T., Canada. *Arctic and Alpine
     Res.* 7: 183-198.
Danks, H.V. 1971a. Overwintering of some north temperate and
     arctic Chironomidae. I. The winter environment.
     Can. Ent. 103: 589-604.
Danks, H.V. 1971b. Overwintering of some north temperate and
     arctic Chironomidae. II. Chironomid biology. *Can. Ent.*
     103: 1875-1912.
Danks, H.V. 1971c. Spring and early summer temperatures in a
     shallow arctic pond. *Arctic*   24: 113-123.
Davies, L. 1954. Observations on *Prosimulium ursinum* Edw. at
     Holandsfjord, Norway. *Oikos* 5: 94-98.
Decamps, H. 1967. Ecologie des Trichoptères de la Vallée d'Aure
     (Hautes-Pyrénées). *Ann. Limnologie* 3: 399-577.
deMarch, L. 1975. Nutrient budgets for a high arctic lake.
     (Char Lake, N.W.T.). *Verh. int. Verein. theor. angew.
     Limnol.* 19: 496-503.
Didiuk, A. and Wright, D.G. 1975. The effect of a drilling waste
     on the survival and emergence of the chironomid *Chironomus
     tentans* (Fabricius). *Canada Department of Environment,
     Fisheries and Marine Service, Technical Report*   586: 1-18.
Dingman, S.L. 1966. Characteristics of summer runoff from a small
     watershed in central Alaska. *Wat. Resources Res.* 2: 751-754.
Downes, J.A. 1962. What is an arctic insect? *Can. Ent.* 94:
     143-162.
Downes, J.A. 1964. Arctic insects and their environment.
     Can. Ent. 96: 279-307.
Downes, J.A. 1965. Adaptations of insects in the arctic.
     Ann. Rev. Ent. 10: 257-274.
Dugdale, R.C. and Dugdale, V.A. 1961. Sources of nitrogen and
     phosphorus for lakes on Afognak Island. *Limnol. Oceanogr.*
     6: 13-23.
Eicher, G.J. and Rounsefell, G.A. 1957. Effects of lake
     fertilization by volcanic activity on abundance of salmon.
     Limnol. Oceanogr. 2: 70-76.
Eriksson, L.O. 1973. Spring inversion of diel rhythm of the
     locomotor activity in young sea trout *(Salmo trutta trutta L.)*
     and Atlantic salmon *(Salmo salar L.) Aquilo, Ser. Zoologica*
     14: 68.

Eythorsson, J. and Sigtryggsson, H. 1971. The climate and weather of Iceland. *Zool. Iceland* 1: 1-62.

Figala, J. and Müller, K. 1972. Jahresperiodik and Phasenlage eines tropischen Fisches *(Barbus partipentazona)* am Polarkreis. *Aquilo, Ser. Zoologica* 13: 13-20.

Freeman, T.N. 1959. The Canadian Northern Insect Survey, 1947-1957. *Polar Record.* 9: 299-307.

Frey, P.J., Mueller, E.W. and Berry, E.C. 1970. The Chena River. The study of a subarctic river. United States Environmental Protection Agency, Federal Water Quality Administration, Project 1610-10/70, 96 pp.

Gallup, D.H. and Clements, A. 1975. Base-line studies on the invertebrate drift populations of a section of the Aishinik River. *Canada Department of the Environment, Fisheries and Marine Service. Tech. Rep.* PAC/T-75-4 (5). 1-46.

Goldman, C.R. 1970. Antarctic freshwater ecosystems. *In* Antarctic ecology. Holgate, M.W. ed. Academic Press, London.

Göthberg, A. 1973. Trichopterernas flygaktivitet vid Ricklean. *Zoologisk Revy.* 35: 125-130.

Göthberg, A. and Karlström, U. 1975. Ecological research in running waters in northern Sweden. *Rapport från Rickleå Fältstation.* 64: 1-30.

Gressitt, J.L. 1970. Subantarctic entomology, particularly of South Georgia and Heard Island. *Pacific Insect Monogr.* 23: 1-374.

Gressitt, J.L. and Leech, R.E. 1961. Insect habitats in Antarctica. *Polar Record* 10: 501-505.

Grigoriev, N.F. 1959. "On the influence of water basin on geocryologic conditions of the coastal lowland of the Yana River mouth region of the Yakut SSR". Izv. Akad. Nauk SSSR (Ser. Geofiz.) 1959: 202-206.

Grimas, U. and Wiederholm, T. 1979. Biometry and biology of *Constempellina brevicosta* (Chironomidae) in a subarctic lake. *Holarct. Ecol.* 2: 119-124.

Grobbelaar, J.U. 1978. Mechanisms controlling the composition of freshwaters on the sub-antarctic island Marion. *Arch. Hydrobiol.* 83: 145-157.

Hobbie, J.E. 1973. Arctic limnology: a review. *Arctic Instit. N. Amer. Tech. Pap.* 25: 127-168.

Holmen, K. and Scotter, G.W. 1967. *Sphagnum* species of the Thelon River and Kaminuriak Lakes regions, N.W.T. *Bryologist.* 70: 432-437.

Hooper, F. 1947. Plankton collections from the Yukon and Mackenzie River Systems. *Trans. Amer. Microscop. Soc.* 66: 74-84.

Hynes, H.B.N. 1970. The ecology of running waters. University of Toronto Press, Toronto.

Hynes, H.B.N., Kaushik, N.K., Lock, M.A., Lush, D.L., Stocker, Z.S.J.
    Wallace, R.R. and Williams, D.D. 1974. Benthos and
    allochthonous organic matter in streams. *J. Fish. Res. Bd
    Canada*. 31: 545-553.
Illies, J. 1956. Seeausfluss-Biozonen lappländischer Waldbäche.
    Ent. Tidskr. 77: 138-153.
Ives, J.D. and Barry, R.G. (eds.) 1974. Arctic and alpine
    environments. Methuen, London.
Johansen, F. 1911. Freshwater life in northeast Greenland.
    Meddr. Grønland. 45: 321-337.
Johnson, T. 1978. Aquatic mosses and stream metabolism in a
    North Swedish river. *Verh. int. Verein. theor. angew.
    Limnol*. 20: 1471-1477.
Johnston, G.H., and Brown, R.J.E. 1964. Some observations on
    permafrost distribution at a lake in the Mackenzie Delta,
    N.W.T., Canada. *Arctic* 17: 163-175.
Kaiser, E. and Müller, K. 1971. Flugverhalten von Sialiden in
    hohen nordischen Breiten. *Berichte der Ökologischen
    Station Messaure*. 7: 1-10.
Kalff, J. 1968. Some physical and chemical characteristics of
    arctic freshwaters in Alaska and Northwestern Canada.
    J. Fish. Res. Bd Canada. 25: 2575-2587.
Karlström, U. 1978. Role of the organic layer on stones in
    detrital metabolism in streams. *Verh. int. Verein. theor.
    angew. Liminol*. 20: 1463-1470.
Karlström, U, and Backlund, S. 1977. Relationship between algal
    cell number, chlorophylla and fine particulate organic
    matter in a river in Northern Sweden. *Arch. Hydrobiol*.
    80: 192-199.
Kogl, D.R. 1965. Springs and ground water as factors affecting
    survival of chum salmon in a sub-arctic stream. M. Sc.
    Thesis, University of Alaska.
Kureck, A. 1966. Schlüpfrhythmus von *Diamesa arctica*
    (Diptera Chironomidae) auf Spitzbergen. *Oikos* 17: 276-277.
Kureck, A. 1969. Tagesrythmen lappländischer Simuliiden
    (Diptera). *Oecologia* 2: 385-410.
Levanidov, V. Ya. 1977. The benthos biomass of some streams on
    the Chukotka Peninsula. *Hydrobiol. J*. 13: 46-52.
Lewis, C.R. 1962. Ice mound on Sadlerochit River, Alaska.
    Arctic 15: 145-150.
Livingstone, D.A. 1963. Alaska, Yukon, Northwest Territories
    and Greenland. *In* Limnology in North America. Frey, D.G.
    ed. University of Wisconsin Press, Madison.
Loftus, W.F. and Lenon, H.L. 1977. Food habits of the salmon
    smolts *Oncorhynchus tshawytscha* and *O. keta* from the Salcha
    River, Alaska. *Trans. Am. Fish. Soc*. 106: 235-240.
Mackay, D.K. and Løken, O.H. 1974. Arctic Hydrology. *In*
    Arctic and alpine environments. Ives, J.D. and Barry, R.G.
    eds. Methuen, London.

Madder, M.C.A., Rosenberg, D.M. and Wiens, A.P. 1977. Larval
    cocoons in *Eukiefferiella claripennis* (Diptera; Chironomidae).
    Can. Ent. 108: 891-892.
Magnin, E. 1978. Recherches en écologie des eaux douces
    effectuées sur le territoire de la Baie James de 1973-1975
    (résumé). *Verh. int. Verein. theor. angew. Limnol.* 20: 232.
McAlpine, J.F. 1965. Insects and related invertebrates of Ellef
    Ringnes Island. *Arctic* 18: 73-103.
McCann, S.B. and Cogley, J.G. 1972. Hydrological observations on
    a small arctic catchment, Devon Island. *Can. J. Earth Sci.*
    9: 361-365.
McCann, S.B. and Cogley, J.G. 1977. Floods associated with
    glacier margin drainage in Ellesmere Island, N.W.T.
    Proc. Can. Hydrol. Symp. 77: 13-23.
McCart, P. and Craig, P. 1973. Life history of two isolated
    populations of arctic char *(Salvelinus alpinus)* in spring-fed
    tributaries of the Canning River, Alaska. *J. Fish. Res. Bd
    Canada.* 30: 1215-1220.
McKinnon, G.A., Sutherland, B.G., and Robinson, P.R. 1978.
    Preliminary data on the aquatic resources of three
    Mackenzie River tributaries to be crossed during highway
    construction. *Canada Department of Environment, Fisheries
    and Marine Service, Man. Rep.* 1481: 1-32.
Melin, R. 1947. Undersökningor vid Sveriges meteorologiska och
    hydrologiska institut över vattendragens isförhållanden.
    Svenska Vattenkraftföreningens Publikat. 393: 1-150.
Mendl, H. and Müller, K. 1978. The colonization cycle of
    Amphinemura standfussi Ris (Ins.: Plecoptera) in the Abisko
    area. *Hydrobiologia* 60: 101-111.
Moore, J.W. 1974a. Benthic algae of Southern Baffin Island. I.
    Epipelic communities in rivers. *J. Phycol.* 10: 50-57.
Moore, J.W. 1974b. Benthic algae of Southern Baffin Island.
    III. Epilithic and epiphytic communities. *J. Phycol.*
    10: 456-462.
Moore, J.W. 1977a. Ecology of algae in a subarctic stream.
    Can. J. Bot. 55: 1838-1847.
Moore, J.W. 1977b. Some factors influencing the density of
    subarctic populations of *Bosmina longirostris, Holopedium
    gibberum, Codonella cratera* and *Ceratium hirundinella.*
    Hydrobiologia. 56: 199-207.
Moore, J.W. 1977c. Some factors affecting algal consumption in
    subarctic Ephemeroptera, Plecoptera, and Simuliidae.
    Ocecologia 27: 261-273.
Moore, J.W. 1978a. Some factors influencing the density and
    birth rate of three subarctic rotifer populations. *Arch.
    Hydrobiol.* 83: 251-271.
Moore, J.W. 1978b. Distribution and abundance of phytoplankton
    in 153 lakes, rivers, and pools in the Northwest
    Territories. *Can. J. Bot.* 56: 1765-1773.

Moore, J.W.   1979a.   Seasonal succession of phytoplankton in a
    large subarctic river. *Hydrobiologia*   67:   107-112.
Moore, J.W.   1979b.   Distribution and abundance of attached,
    littoral algae in 21 lakes and streams in the Northwest
    Territories. *Can. J. Bot.*   57:   568-577.
Müller, K.   1954a.   Investigations on the organic drift in North
    Swedish Streams. *Rep. Instit. Freshwat. Res. Drottningholm*
    35:   133-148.
Müller, K.   1954b.   Faunistisch-ökologische Untersuchungen in
    Nordschwedische Waldbäche. *Oikos*   5:   77-93.
Müller, K.   1955.   Produktionsbiologische Untersuchungen in
    Nordschwedischen Fliessgewässern 3.   Die Bedeutung der
    Seen and Stillwasserzonen für die Produktion in
    Fliessgewässern. *Rep. Instit. Freshwat. Res. Drottningholm*
    36:   148-162.
Müller, K.   1962.   Iakttagelser zörande laxöringens vandringar i
    en lappländsk skogsbäck. *Fauna Flora*   1962:   200-206.
Müller, K.   1968.   Freilaufende circadiane Periodik von
    Ellritzen am Polarkreis. *Naturwissenschaften*   55:   140.
Müller, K.   ed.   1970.   Phänomene der Tages-und Jahresperiodik in
    hohen nordischen Breiten. *Oikos, Suppl.*   13:   1-142.
Müller, K.   1973a.   Seasonal phase shift and the duration of
    activity time in the burbot *Lota lota* (L.)   *J. Comp. Physiol.*
    84:   357-359.
Müller, K.   1973b.   Circadian rhythms of locomotor activity in
    aquatic organisms in the subarctic summer. *Aguilo, Ser.*
    Zoologica   14:   1-18.
Müller, K.   1976.   Chronobiological studies on whitefish
    (*Coregonus lavaretus* L.) at the arctic circle. *Arch. Fisch.*
    Wiss.   27:   121-132.
Müller, K. and Benedetto, L.A.   1970.   Die lokomotorische
    Aktivitat der Larven von *Dinocras cephalotes* Curtis
    (Plecoptera). *Oikos, Suppl.*   13:   69-74.
Müller, K. and Thomas, E.   1972.   Bäckslandornas rytmik i
    Messaureområdet. *Fauna Flora*   67:   191-195.
Müller, K., Müller-Haeckel, A., Thomas, E. and Göthberg, A.
    1970.   Der Kaltisjokk. *Öst. Fisch.*   23:   76-135.
Müller, K., Mendl, H., Dahl, M. and Müller-Haeckel, A.   1976.
    Biotopwahl and Kolonization Cycle von Plecopteren in
    einem subarktischen Gewässersystem. *Ent. Tidskr.*   97:   1-6.
Müller-Haeckel, A.   1967.   Tages-und Jahresperiodik von
    Ceratoneis arcus Kütz (Diatomeae). *Oikos*   18:   351-356.
Müller-Haeckel, A.   1970a.   Messung der tagesperiodik
    Neukolonization von Algenzellen in Fliessgewässern.
    Oikos, Suppl.   13:   14-20.
Müller-Haeckel, A.   1970b.   Tages-und Jahresperiodik einzelliger
    Algen in fliesseuden Gewässern. *Öst. Fisch.*   23:   97-101.
Müller-Haeckel, A.   1973.   Experiments zum Bewegungsverhalten
    von einzelligen Fliesswasseralgen. *Hydrobiologia.*   41:   221-
    246.

Müller-Haeckel, A. and Håkansson, H.   1978.   The diatomflora of a
    small stream near Abisko (Swedish Lapland) and its annual
    periodicity, judged by drift and colonization.   *Arch.*
    Hydrobiol.   84: 199-217.

Nava, J.A. and Morrison, P.   1974.   A note on hot springs in the
    interior of Alaska.   *Arctic*   27: 241-243.

Oliver, D.R.   1968.   Adaptations of arctic Chironomidae.
    Ann. Zool. Fennici.   5: 111-118.

Oliver, D.R. and Corbet, P.S.   1966.   Operation Hazen: aquatic
    habitats in a high arctic locality: the Hazen camp study
    area, Ellesmere Island, N.W.T. Defence Research Board of
    Canada, Report D. Phys. R(G) Hazen.   26: 1-115.

Oliver, D.R., Corbet, P.S. and Downes, J.A.   1964.   Studies on
    arctic insects: the Lake Hazen project.   *Can. Ent.*   96:
    138-139.

Parker, B.C. and Simmons, G.M.   1978.   Ecosystem comparison of
    oasis lakes and soils.   *Antarctic J. United States*
    13: 168-169.

Peterson, B.V.   1977.   The blackflies of Iceland (Diptera:
    Simuliidae).   *Can. Ent.*   109: 449-472.

Pissart, A.   1967.   Les modalités de l'écoulement de l'eau sur
    l'Ile Prince Patrick.   *Biuletyn Peryglacjalny*   16: 217-224.

Polunin, N.   1948.   Botany of the Canadian Arctic.   III.
    Vegetation and ecology.   *Nat. Museum Canada Bull.*   104: 1-304.

Popova, E.I.   1957.   Comparative hydrobiological characterization
    of the Kolva and Kos'Yu Rivers (Basin of the Pechora River).
    Trans. 6th Conf. on the Biology of Inland Waters   190-197.

Post, A. and Mayo, L.R.   1971.   Glacier dammed lakes and outburst
    floods in Alaska.   *United States Geol. Sur. Hydrol. Invest.*
    Atlas, HA-455.   10 pp.

Power, G.   1969.   The salmon of Ungava Bay.   *Arctic Inst. of N.*
    Amer. Tech. Pap.   22: 1-72.

Rawson, D.S.   1953.   Limnology in the North American arctic and
    subarctic.   *Arctic*   6: 198-204.

Regier, H.A.   1976.   Science for the scattered fisheries of the
    Canadian Interior.   *J. Fish. Res. Bd Canada*   33: 1213-1232.

Ridder, W.P. and Peckam, R.D.   1978.   A study of a typical spring-
    fed stream of Interior Alaska.   In Lake and Stream
    Investigations.   *Ann. Perf. Rep. Alaska Fish and Game*
    20 (G-III-G): 25-63.

Roeder, D.R., Crum, G.H., Rosenberg, D.M. and Snow, N.B.   1975.
    Effects of Norman Wells crude oil on periphyton in selected
    lakes and rivers in the Northwest Territories.   *Canada*
    Department of the Environment, Fisheries and Marine Service,
    Tech. Rep.   552: 1-31.

Rosenberg, D.M. and Snow, N.B.   1975a.   Effect of crude oil on
    zoobenthos colonization of artificial substrates in subarctic
    ecosystems.   *Verh. int. Verein. theor. angew Limnol.*
    19: 2172-2177.

Rosenberg, D.M. and Snow, N.B. 1975b. Ecological studies of aquatic organisms in the Mackenzie and Porcupine River drainages in relation to sedimentation. *Canada Department of the Environment, Fisheries and Marine Service, Tech. Rep.* 547: 1-86.

Rosenberg, D.M. and Wiens, A.P. 1975. Experimental sediment addition studies on the Harris River, N.W.T., Canada: the effect on macroinvertebrate drift. *Verh. int. Verein. theor. angew. Limnol.* 19: 1568-1574.

Rosenberg, D.M. and Wiens, A.P. 1976. Community and species responses of Chironomidae (Diptera) to contamination of freshwaters by crude oil and petroleum products, with special reference to the Trail River, Northwest Territories. *J. Fish. Res. Bd Canada* 33: 1955-1963.

Rosenberg, D.M. and Wiens, A.P. 1978. Effects of sediment addition on macrobenthic invertebrates in a northern Canadian river. *Wat. Res.* 12: 752-763.

Rosenberg, D.M., Wiens, A.P. and Saether, O.A. 1977a. Responses to crude oil contamination by *Cricotopus (C.) bicinctus* and *C. (C.) mackenziensis* (Diptera: Chironomidae) in the Fort Simpson Area, Northwest Territories. *J. Fish Bd Canada* 34: 254-361.

Rosenberg, D.M., Wiens, A.P. and Saether, O.A. 1977b. Life histories of *Cricotopus (Cricotopus) bicinctus* and *C. (C.) mackenziensis* (Diptera: Chironomidae) in the Fort Simpson area, Northwest Territories. *J. Fish. Res. Bd Canada* 34: 247-253.

Saether, O.A. 1968. Chironomids of the Finse area, Norway, with special reference to their distribution in a glacier brook. *Arch. Hydrobiol.* 64: 426-483.

Salt, R.W. 1961. Principles of insect cold-hardiness. *Ann. Rev. Ent.* 6: 55-74.

Samsel, G.L. and Parker, B.C. 1972. Limnological investigations in the area of Anvers Island, Antarctica. *Hydrobiologia* 40: 501-511.

Schindler, D.W., Welch, H.E., Kalff, J., Brunskill, G.J. and Kritsch, N. 1974. Physical and chemical limnology of Char Lake, Cornwallis Island (75°N Lat.). *J. Fish. Res. Bd Canada* 31: 585-607.

Schmidt, H.W. and Müller, K. 1967. Zur Tages-und Jahresperiodik der Gattung *Lebertia* (Hydrachnellae, Acari). *Oikos* 18: 357-359.

Scholander, P.F., Flagg, W., Hock, R.J., Irving, L. 1953. Studies on the physiology of frozen plants and animals in the arctic. *J. cell. comp. Physiol.* 42, suppl.: 1-56.

Sheridan, W.L. and McNeil, W.J. 1968. Some effects of logging on two salmon streams in Alaska. *J. Forest.* 66: 128-133.

Slack, K.V., Nauman, J.W. and Tilley, L.J. 1977. Benthic invertebrates in an arctic mountain stream, Brooks Range, Alaska. *J. Res. U.S. Geol. Surv.* 5: 519-527.

Slack, K.V., Nauman, J.W. and Tilley, L.J.   1979.   Benthic
    invertebrates in a north-flowing stream, Brooks Range,
    Alaska. *Wat. Res. Bull.* 15: 108-135.
Slatt, R.M.   1970.   Geochemistry of meltwater streams from nine
    Alaskan glaciers. *Bull. Geol. Soc. Amer.* 83: 1125-1132.
Snow, N.B. and Chang, P.   1975.   Aspects of zoobenthos ecology of
    the Mackenzie Delta, N.W.T. *Verh. int. Verein. theor. angew.
    Limnol.* 19: 1562-1567.
Snow, N.B. and Rosenberg, D.M.   1975.   The effects of crude oil
    on the colonization of artificial substrates by zoobenthos
    organisms. *Canada Department of the Environment, Fisheries
    and Marine Service, Tech. Rep.* 551: 1-35.
Snow, N.B., Rosenberg, D.M. and Moenig, J.   1975.   The effects of
    Norman Wells crude oil on the zoobenthos of a northern
    Yukon stream one year after experimental spill. *Canada
    Department of the Environment, Fisheries and Marine
    Service, Tech. Rep.* 550: 1-8.
Södergren, S.   1963.   Undersökningar av driftfaunan i Ricklean.
    Laxforskningsinstitut Meddelande. 5: 1-25.
Steffan, A.W.   1971.   Chironomid (Diptera) biococenoses in
    Scandinavian glacier brooks. *Can. Ent.* 103: 477-486.
Steffan, A.W.   1972.   Zur Produktionsökologie von Gletscherbächer
    in Alaska und Lappland. *Verh. dk. Zool. Ges.* 65: 73-78.
Steinböck, O.   1934.   Die Tierwelt der Gletschergewässer.
    Z. Deut. Öst. Alpenvereins. 65: 263-265.
Steine, I.   1972.   The number and size of drifting nymphs of
    Ephemeroptera and Simuliidae by day and night in river
    Stranda, Western Norway. *Norsk ent. Tidskr.* 19: 127-131.
Stocker, Z.S.J. and Hynes, H.B.N.   1976.   Studies on the
    tributaries of Char Lake, Cornwallis Island, Canada.
    Hydrobiologia 49: 97-102.
Stone, K.H.   1963.   The annual emptying of Lake George, Alaska.
    Arctic 16: 27-40.
Svensson, P.O.   1966.   Growth of nymphs of stream living stone-
    flies (Plecoptera) in northern Sweden. *Oikos* 17: 197-206.
Syrjämäki, J.   1969.   Diel patterns of swarming and other
    activities of two arctic Dipterans (Chironomidae and
    Trichoceridae) on Spitsbergen. *Oikos* 19: 250-258.
Thienemann, A.   1941.   Lappländische Chironomiden und ihre
    Wohngewässer. *Arch. Hydrobiol. Suppl.* 17: 1-253.
Thomas, E.   1970.   Die Oberflächendrift eines lapplandischen
    Fliessgewässers. *Oikos. Suppl.* 13: 45-64.
Tobias, W. and Thomas, E.   1967.   Die Oberflachendrift als
    Indikator periodischer Aktivitätsabläufe bei Insekten.
    Ent. Zeitsch. 77: 153-163.
Tuxen, S.L.   1944.   The hot springs of Iceland, their animal
    communities and their zoogeographical significance.
    Zool. Iceland 1(II): 1-206.

Twinn, C.R. 1952. A review of studies of bloodsucking flies in
    northern Canada. *Can. Ent.* 84: 22-28.
Ulfstrand, S. 1967. Microdistribution of benthic species
    (Ephemeroptera, Plecoptera, Trichoptera, Diptera:
    Simuliidae) in Lapland streams. *Oikos* 18: 293-310.
Ulfstrand, S. 1968a. Life cycles of benthic insects in Lapland
    streams (Ephemeroptera, Plecoptera, Trichoptera, Diptera:
    Simuliidae). *Oikos* 19: 167-190.
Ulfstrand, S. 1968b. Benthic animal communities in Lapland
    streams. *Oikos, Suppl.* 10: 1-120.
Vannote, R.L., Minshall, G.W., Cummins, K.W., Sedell, J.R. and
    Cushing, C.E. 1980. The river continuum concept.
    Can. J. Fish. Aqu. Sci. 37: 130-137.
Watson, D.G., Hanson, W.C., Davis, J.J. and Cushing, C.E. 1966.
    Limnology of tundra ponds and Ogotoruk Creek. *In* Environment
    of the Cape Thompson Region, Alaska. Wilimovsky, N.J. and
    Wolfe, J.N. eds. *U.S. Atomic Energy Comm.* PNE-481:
    415-435.
Whitton, B.A. ed. 1975. River ecology. University of
    California Press, Berkeley.
Woo, M.K. and Marsh, P. 1977. Effect of vegetation on
    limestone solution in a small High Arctic basin.
    Can. J. Earth Sci. 14: 571-581.
Wynne-Edwards, V.C. 1952. Freshwater vertebrates of the arctic
    and subarctic. *Bull. Fish. Res. Bd Canada* 94: 1-28.
Young, G.J. 1977. Glacier outburst floods. *Proc. Canadian
    Hydrol. Symp.* 77: 1-13.
Yukhneva, V.S. 1971. Chironomid larvae in the lower reaches of
    the Ob-Irtysh basin. *Hydrobiol. J.* 7: 28-31.
Zhadin, V.I. and Gerd, S.V. 1961. Fauna and flora of the rivers,
    lakes and reservoirs of the USSR. Israel Program for
    Scientific Translations, Jerusalem (1963).
Zvereva, O.S. 1957. Some features of the distribution of
    aquatic fauna and conditions of biological production in
    water bodies of the northeastern European USSR.
    Trans. 6th Conf. on the Biology of Inland Waters. 182-189.

RUNNING WATER ECOLOGY IN AFRICA

J.B.E. Awachie

Hydrobiology/Fishery Research Unit
Department of Zoology
University of Nigeria
Nsukka, Nigera

INTRODUCTION

The environment of running water has provided the focus of human development activities from ancient times. In Africa, as elsewhere, notable ancient civilizations were associated with the major river systems such as the Niger and the Nile. This is probably related to the fact that river basins provided easy access to essential requirements of life viz: cheap and reliable source of water, food from both aquatic and terrestrial components, as well as cheap water transport system.

As practically all the large and sizeable bodies of running water on the continent are located in sub-Saharan Africa (Fig. 1), it is not surprising that most data presented here relate largely to the humid tropical zone. In spite of the foregoing, and because lotic studies have, as elsewhere, received much less attention than lentic environments in Africa (apparently due to the inherent difficulties in investigating them) the present state of our knowledge on the ecology of rivers and streams is very poor compared to that of the Great Lakes of East Africa and Lake Chad in central Africa. Indeed, early and usually fragmentary data on African running water came from reports of collections of occasional expeditions from Europe by diverse interest groups (Bourne 1893; Ekman, 1903; Sars, 1909; Keifer, 1949 *inter alia*). These reports were often sandwiched in lacustrine studies.

A major impediment to the elucidation of lotic community ecology has been the generally poor knowledge of the taxonomy of most aquatic groups especially benthic invertebrates, in an age when systematics has become unfashionable. Considerable advances

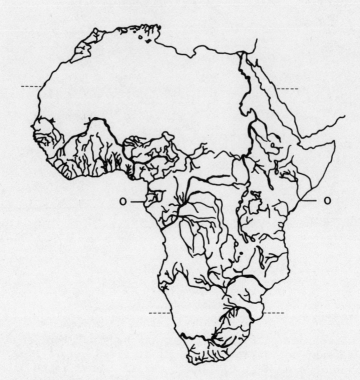

Figure 1.   Distribution of the running water systems of Africa
            (note that practically all are sub-Saharan).

achieved over the last 20-25 years have been facilitated by the
contributions of individuals like Hynes, Rzoska, and Worthington,
and international study programs such as the International Biological
Program (IBP) and Man and Biosphere (MAB). Added to this is the
belated but growing appreciation that baseline ecological data
are necessary to optimize the benefits expected to accrue from
river impoundments which have been springing up rapidly all over
Africa, especially in the humid tropical belt.

     Recent data (MacArthur, 1969; Hynes, 1970) confirm that
tropical lotic communities obey the general ecological rule in
being more diverse, with large numbers of species and more complex
inter-relationships than temperate communities. In addition, the
tropical African environment, especially in the humid areas, is
less seasonal and thus more productive (MacArthur, 1969; Lowe-
McConnell, 1977). Even though more difficult to prove, lowland
tropics are, in general, more productive.

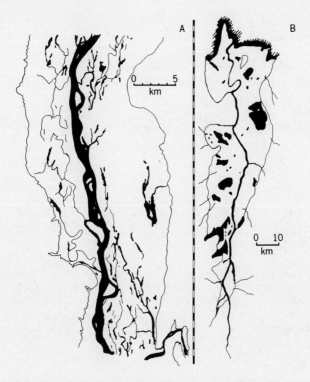

Figure 2. Characteristic features of floodplain river systems.
          Two basic varieties are typified by the Niger (A) and
          Luapula (B) rivers, at low water mark. Note main river
          channels, permanent pools, swamps and lagoons (black)
          (adapted from Welcomme, 1976).

MORPHOLOGICAL STUDIES

     Available biologically oriented studies on the morphology of
African running waters have been recently reviewed by Welcomme
(1974, 1979). In general, they vary from those which do not flood
their banks (without floodplains) to those which seasonally flood
large areas of their banks and internal deltas (with extensive
floodplains). The latter, floodplain rivers and streams, provide
more varied habitats and are, therefore, more productive.
Because of their importance in fishery and agricultural production,
they have been the subject of more investigations especially
those aspects that influence biological productivity (Gosse, 1963;
Leopold et al., 1964; Daget and Iltis, 1965; Rzoska, 1974, *inter
alia*).

     The basic features of floodplain rivers are illustrated in
Figures 2 and 3. They comprise the main river channel, floodplain

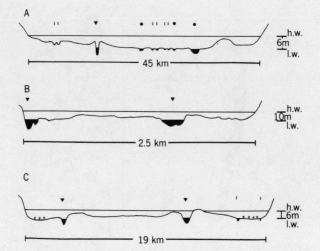

Figure 3.   Cross section of three river floodplain systems to show
            differences in basin morphology and amplitude of floods
            (A-Kafue; B-Niger at Onitsha; C-Oueme at Zinvie).
            ▲ Main channel; ● Lagoon; | Streams/secondary channels;
            ** Swamps (adapted from Welcomme, 1975).

pools, lagoons and swamps, channels and creeks, and levees
(raised banks bordering the main river channel). Levees are often
important in directing the path of flow of floodwaters, as well
as in controlling their rate of recession and, hence, the drying
of the plain.

    Two broad types of the main river channel are distinguishable
viz: simple and the braided. The latter are common along the
Niger, Zaire (Congo) and Zambezi rivers and have many complex
anastomosing water courses which result in the isolation of numerous
sand or vegetated islands. The floodplain, though somewhat intermediate in its flow characteristics between lotic and lentic
waters, in having running water during the flood and being almost
lacustrine at the height of the dry phase, is considered an
essential part of the floodriver system. Comparative data on the
fluctuations of the area under water during flood and low water
seasons are given in Tables 1 and 2.

Table 1. Hydrological characteristics of some African
          floodplain river systems (adapted from Welcomme, 1974).
          * Fringing floodplains
          ▲ Part of Larger system

Floodplain	A_1 Area at peak flood (ha)	A_2 Area during low water (ha)	$\dfrac{A_2}{A_1}$ x 100
Barotse (Zambesi)	512 000	32 915	6
Benue* (Nigeria)	310 000	129 000	42
Kafue flats	434 000	145 560	34
Massilli*	1 500	200	13
Niger (Central Delta)	2 000 000	300 000	15
Niger* (Niger)	90 704	27 000	30
Niger* (Dahomey)	27 440	3 200	12
Niger* (Nigeria)	480 000	180 000	38
Okavango	1 600 000	312 000	20
Ogun*	4 250	2 500	59
Oshun*	3 740	2 000	53
Ouémé	100 000	5 170	5
Pongolo*	10 416	2 927	28
Senegal*	1 295 000	78 700	6
Elephant and Ndinde Marshes (Shire) ▲	67 300	46 049	68
Sudd (Nile)	9 200 000	1 000 000	11
Volta* (White Volta) ▲	85 324	10 221	12
Yaérés (Logone) ▲	460 000	2 450	0.5

Table 2. Differences in the dry season composition of some African floodplains (from Welcomme, 1974).
- indicates negligible; numbers in brackets represent % of the total

Floodplain	River plus channels	Swamp	Lagoon/lake	Total	Dry area (ha)
Kafue	5 380 (4)	130 000 (89)	10 180 (7)	145 560	288 440
Shire	724 (2)	15 500 (84)	6 000 (14)	46 049	21 251
Pongolo	427 (15)	-	2 500 (85)	2 927	7 489
Senegal	28 100 (36)	-	50 600 (64)	78 700	1 216 300
Ouémé	1 402 (27)	-	3 768 (73)	5 170	94 830

HYDROLOGICAL FEATURES

Depending on their lengths and direction of flow, practically
all lotic systems in tropical Africa experience annual or biannual
flow regimes due to the seasonal pattern of rainfall. Thus while
the relatively shorter Oueme and Benue rivers exhibit single annual
flood during the wet season, the much longer Niger experiences two
floods in its lower reaches viz: the small siltless 'black flood'
which emanates from earlier rains upstream near its source, and
the larger turbid 'white flood' which results from local rainfall
in Nigeria (Fig. 4). The two flood peaks pass through the Onitsha
area (Nigeria) in January/February and August/September
respectively (Awachie, 1975).

The details of the hydrology of the Nile system have been
reviewed by Rzoska (1976). Comparative data on African rivers and
streams indicate two distinctive sources of flood which differ in
origin and flow direction: (i) local rainfall – here the water
arises from the surrounding plain and forests, and flows into the
river channel. It has been recorded for the Gambia and Kafue
rivers (Svensson, 1933; Carey, 1971) and is probably common to all
systems to a greater or lesser extent (Welcomme, 1974). It is

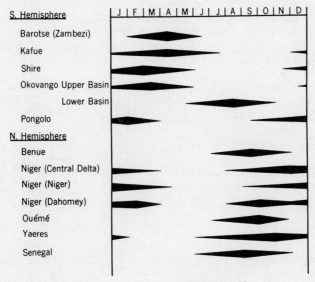

Figure 4. The flood characteristics of major floodplain river
           systems of Africa (adapted from Welcomme, 1975).

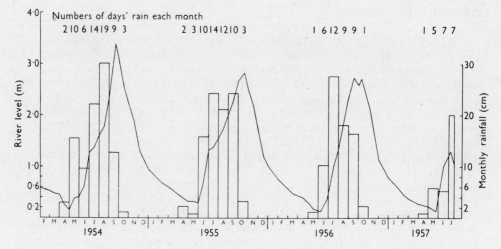

Figure 5.   The relationship between the variation in monthly
            rainfall pattern and flood curve fluctuations of River
            Sokoto at Birnin Kebbi (from Holden & Green, 1960).

responsible for the high incidence of terrestrial fauna in the
stomachs of many tropical omnivorous and opportunistic feeders,
e.g. *Clarias* spp. in the Lower Niger-Anambra system during early
rainy season in Nigeria (Awachie and Ezenwaji, 1980). (ii) overspill
from the river channel, which flows from the river to the lower
fringing plain and occurs when a flood crest travelling down the
river channel overflows the banks. It is typical of the floodplain
river system and provides for the essential links in the biology
and ecology of floodriver communities. The typical flood characteristics of major African floodplains are summarised in Figures 4 and
5. They indicate that, in general, each phase of the annual water
cycle lasts about six months. The periods become increasingly
unequal further north or south of the equator.

A third source of flood, tidal action, is mentioned for
completeness. It is limited to estuaries and coastal floodplains,
and results in intricate longitudinal zonation of the biota due to
the combined action of varying gradients of tidal flow, current
velocity and salinity (Olaniyan, 1957, 1961, 1969; Volker, 1966
Ikusemiji, 1976, 1969).

In the so-called 'reservoir' rivers, as distinct from 'flood'
rivers, internal factors which may be structural or geographical,
operate to reduce the variability of the floodwater flow, thus
suppressing the flood phase. Among such factors are the presence of
extensive swamps which store water, and the arrival at different
periods of the flow from many tributaries which arise in
different latitudes. The condition of Zaire river below

Kisangani approximates closely to this type (Welcomme 1979).

A further hydrologic feature of relevance to running water
communities in the humid tropics, is the problem of water balance
and the allied issue of discharge rates from each lotic system.
These inevitably influence the bionomics of biotic communities
but will not be discussed further here for want of space. Further
information on the subject may be obtained from the recent studies
of Bryson (1974), Van der Leeden (1975), and Welcomme (1979).

PHYSICO-CHEMICAL STUDIES

The physical and chemical parameters dealt with here are
those which have considerable bearing on the life of lotic
organisms in tropical Africa. They include temperature, current,
dissolved oxygen, hydrogen ion concentration, and conductivity
since salts in freshwater are almost fully dissociated into ions
and as Beadle (1974) pointed out, it is the balance of ions rather
than their weight in a given quantity of water that is most
important biologically. Current or water movement is an essential
and pervading characteristic of running waters. It operates to
even out physical and chemical changes and will not be given a
separate treatment here.

Available data from various systems show that the dynamics
of physical and chemical parameters follow rather closely on the
flood cycle. In floodriver systems, differences between the river
channel and floodplain conditions only arise during the dry season,
with their separation.

Temperature: Generally, the surface water temperature follows
ambient air temperature but is influenced by insolation, substrate
composition, turbidity, vegetation cover, ground water and rain-
water inflows (Egborge, 1972, 1979). For most tropical areas, a
pattern in which dry season temperatures are higher than those of
the wet season, is the rule. Around the equatorial belt, however,
running water temperature is usually stable diurnally and seasonally,
while in higher latitudes, rather low figures are obtained. Thus
long rivers, like the Nile, which traverse a wide belt of
latitudes, exhibit longitudinal or North-South temperature
gradients: 28°C in southern Sudan to 17-19°C in Khartoum - Cairo
(Talling and Rzoska, 1967; Ramadan, 1972; Talling, 1976).

In warm springs (Egborge and Fagade, 1970) and at the fringes
of the Sahara desert in West Africa, running water temperatures
of up to 33°C and above have been recorded.

Dissolved Oxygen: Dissolved oxygen is usually not limiting in
running water except in situations where the depletion of oxygen

is caused by intensive biological activity as in areas with a mat
of vegetation like *Cyperus papyrus, Echinochloa* and *Eichhornia*
of the Nile Sudd, and *Pistia, Ceratophyllum*, planktonic and
filamentous algae blooms in the floodplain and pools along the
Niger and Volta river systems (Talling, 1957; Holden and Green,
1960; Rzoska, 1974).

Levels of dissociated oxygen fluctuate seasonally due to the
aerating action of the wind and turbulence (Egborge, 1971).
During the dry season, pools and shallow ponds associated with
running water may become grossly deoxygenated causing mass mortality
of the biota. At Faku, below the Kainji dam on the Niger, the
lowering of oxygen related to effluents from upstream hydroelectric
works, has been responsible for changes in the ichthyofauna of the
area with *Lates niloticus* giving way to siluroid species which are
adapted to living in low oxygen conditions (Adeniji, 1979). In
this connection, it may be added that the concentration of dissolved
carbon dioxide is inversely related to that of oxygen (Talling,
1957, 1976).

Hydrogen ion concentration (pH): levels of pH vary widely between
different rivers and streams, and are greatly influenced by the
bicarbonate-carbonate alkalinity and the concentration of free
carbon dioxide (Talling, 1976). As observed by Egborge (1971),
pH is generally lower in the flood season and rises in the dry
season in river channels. Forest rivers tend to be more acid
(pH 4-7) than their grassland equivalent which are neutral or
slightly alkaline.

Other factors which produce variability in pH include the
nature of the soil. Typical tropical soils are lateritic and thus
acidic. The presence and extent of associated swamps and floodplain
pools which are progressively acidic during low water level, is
also relevant. However, Holden and Green (1960) have observed
that in open floodplain lagoon systems in open savanna country, the
concentration of calcium salts due to evaporation, produces
alkaline conditions. Diurnal fluctuations in pH, where they occur,
are associated with phytoplankton metabolism.

Conductivity and Salinity: Conductivity which measures the total
ionic composition of water and, therefore, its overall chemical
richness, is vital as it indicates the biogenic potential of any
body of water. Like other physical and chemical parameters
discussed above, it tends to be higher during the dry season for
most African river systems as shown by Egborge (1971) for Oshun
river, Carey (1971) for the Kafue, Talling and Talling (1965) for
the Nile, and Centre Technique Forestier Tropicale (1972) for the
Senegal. Holden and Green (1960) have explained that this does
not necessarily indicate that the total quantity of salts is less
during the flood but that rather it is a reflection of the more
diluted condition.

The above has implications for life activity rhythms of
communities in both the river channel and associated floodplains
and will be further dealt with later. Salinity and conductivity
are also related to discharge rates and this relationship depends
on the balance between precipitation and evaporation, as well as
the geochemistry of the catchment area and substratum (Nduku and
Robertson, 1977). For similar reasons, longitudinal variations in
conductivity may occur in long rivers like the Niger, Nile and
Zaire.

PLANKTON

There is concensus on the remarkable similarities in the
composition and distribution of African lotic plankton and the
dominant role of the seasonal pattern of rainfall and floods on
their abundance and dynamics (Holden and Green, 1960; Thomas, 1965;
Talling and Rzoska, 1967; Egborge, 1973). The importance of the
adventitious component is also recognized.

Phytoplankton is generally dominated by Bacillariophyceae
in the open river, while in associated backwaters, floodplains and
reservoirs, the Volvocales and Dinophyceae become preponderant
(Egborge 1974, 1979). Phytoplankton production is also strongly
correlated with conductivity and transparency, and inversely
related to water level and current velocity (Egborge, 1974).

Zooplankton production characteristics are similar to those
of phytoplankton, higher production having been noted in the
floodplain during low water (FAO/UN, 1969, 1970). In the Nile
Sudd, the mean number of individuals was 4460 m^{-3} at dry season
compared to 2070 m^{-3} during wet season (Monakov, 1969). The
seasonality of zooplankton as exemplified by Sakoto river is
illustrated in Figure 6.

Zooplankers may be grouped under three categories on the
basis of their geographical distribution viz: African,
circumtropical, and cosmopolitan (Rzoska, 1976). Among the
Crustacea, essentially African species are few and include
diaptomid species and *Daphnia barbata*. *Daphnia lumholtzi* extends
to the Middle East. Circumtropical elements include many
cladocerans - *Diaphanosoma excisum, Ceriodaphnia dubia, C. cornuta*
and *Moina dubia* while cosmopolitan forms contain the cladocerans
Bosmina longirostris, Simosa spp., *Chydorus sphaericus* and some
cyclopids. Other components of tropical African zooplankton include
a large number of species of the Rotifera, the most common of which
are *Keratella tropica* and *K. cochlearis*.

Sudden seasonal pulses in total zooplankton are contributed
largely by rotifers. Nauplii production, which derives from

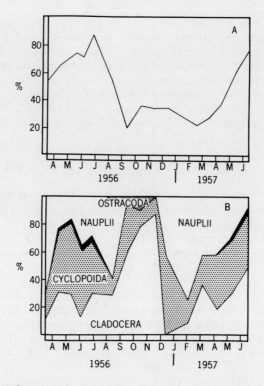

Figure 6.   A – Seasonal dynamics of the main components of
crustacean zooplankton in Sokoto river, Nigeria (from
Green, 1962).   B – Crustacean component as percentage of
total zooplankton in the river.

fringing backwaters and are washed into river channels with flood
rise, and elements of the Rhizopoda e.g. *Arcella* spp., *Difflugia*
spp. and the freshwater medusa *Limnocnida* which have wide
continental distribution, are also contributory components (Holden
and Green, 1960, Green, 1963).

The repopulation of many running water habitats, sequel to
the dry season, is facilitated by 'reservoir populations' from
the floodplain pools, diapausing eggs and larvae and cladoceran
ephyppia.   The occurrence of the latter in tropical Africa,
throughout the year, indicates that unlike in temperate waters,
climatic stress is not a major factor in their formation.   The
role of remarkably fast growth and reproduction by surviving
organisms is thought to be the dominant factor in the rapid
repopulation process.

BENTHOS

Unlike plankton, both qualitative and moreso, quantitative
studies on the benthos of African running water are few and
scattered in the literature. Williams and Hynes (1971) have
observed that poor taxonomic knowledge of African freshwater
faunas in general, allows few groups to be identified beyond the
generic level, thus limiting detailed analysis of benthic
communities. Also unlike plankton, river zoobenthos is locally
differentiated, with the banks and deeper sediments being the main
habitats. Drifting may, however, occur in some places (Rzoska,
1976).

At the global level, Hynes (1970) distinguished between
two broad substrate types; the hard and the soft, with distinctive
faunal characteristics. Despite the paucity of data on the
African scene, the investigations of Hynes (1952, 1953), Hynes,
Williams and Kershaw (1961), Omer-Cooper (1971) for Central and
East Africa, Manakov (1969) for the Nile system and Awachie
(unpublished observations) for the Lower Niger/Anambra system,
indicate that the dominant elements are the Tubificidae,
Chironomidae, Simuliidae, Ephemeroptera, Odonata, Trichoptera,
Coleoptera and various molluscs such as Bulinidae, Planorbidae,
Sphaeriidae and Unionidae.

The detailed characterization of the composition, abundance
and distribution of the benthos is related to fine variations in
the nature of the substrate – mud, sand, stone, and fringing,
submerged or emergent vegetation. In Malawi, Shepherd (1976)
recorded 2073 individuals m^{-2} in a mud bottom under *Nymphaea*, with
the major elements being oligochaetes, molluscs, Coleoptera and
Diptera. Whereas in *Ceratophyllum* beds 362 individuals m^{-2}
comprising hemipterans, molluscs and oligochaetes were taken. In
Zimbabwe, Mills (1976) observed that *Chaoborus* larvae dominated the
invertebrate drift from Mwenda river into Lake Kariba, while in
the middle Niger, Blanc et al. (1955) recorded the clear
differences between the molluscan communities of sandy and muddy
reaches.

Characteristic crustacean species associated with bank-edge
vegetation are found all over tropical Africa. In West Africa,
the freshwater shrimps *Macrobrachium, Caridina, Palaemon,
Eurynchoides holthuisi* and *E. edingtoni*, as well as the crab
Sudanautes africanus, a vector for human lung flukes are common.
East and Central African representatives include *Macrobrachium,
Caridina niloticus, P. niloticus* and the crab *Potamonautes* spp.
It is interesting to note that one of the species, *Potamonautes
niloticus*, is associated with early stages of *Simulium neavei*, a
vector of human river blindness.

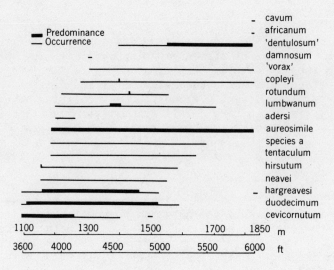

Figure 7. Altitudinal zonation of *Simulium* species (Diptera) on
the slopes of Mount Elgon, East Africa (from Williams
and Hynes, 1971).

Altitudinal zonation is exhibited by elements of highland
fauna of tropical Africa. This is evident from the studies of
Someren (1952) on Mount Kenya; Harrison (1965), Harrison and
Elsworth (1958) in South Africa; Crisp (1956) in Ghana;
Marlier (1954) in Eastern Congo; Hynes and Williams (1962) and
Williams and Hynes (1971) in Uganda and Mount Elgon respectively.
At Mount Elgon clear altitudinal zonation with upper and lower
limits, was shown by 17 species of *Simulium* (Fig. 7) and 15
species of elminthid beetles. Not all the benthic invertebrates
were, however, limited in their distribution. Thus the triclad
Dugesia, ephemeropterans *Caenis* and *Afronurus*, stonefly *Neoperla*
spio as well as some species of Simuliidae and Chironomidae occurred
in all zones of altitude. A somewhat similar picture is available
for the distribution of chironomids and the ephemeropteran
Povilla adusta in West Africa (Bidwell and Clarke, 1977).

The general finding for tropical Africa, is therefore, that
over a wide belt, there is a close similarity of the fauna of
swift water streams and rivers, and that on high mountains; there
is a clear evidence of altitudinal zonation caused probably by
temperature. The problem of faunal associations has received little
attention because of the usually poor detailed taxonomic knowledge
mentioned earlier. As, however, the taxonomy of the Simuliidae was

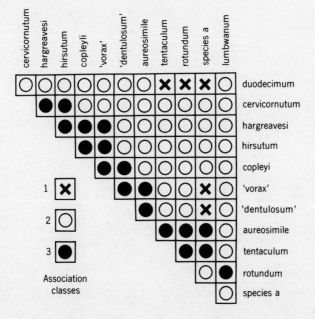

Figure 8.   Association classes of the simuliid fauna of Mount
            Elgon streams, East Africa. Pairs of species never
            found together (x); pairs of species sometimes found
            together (o); pairs of species closely associated and
            neither occurs at all often, without the other (●)
            (from Williams and Hynes, 1971).

known, it was possible to establish interesting 'species
association classes' of the commoner elements of the *Simulium*
(black fly) community on Mount Elgon (Fig. 8) (Williams and Hynes,
1971).

FISH STUDIES

     As with other features considered above, the biology of
running water fish in Africa is profoundly influenced by the
seasonality of flooding consequent on the annual or biannual
rainfall. Lowe-McConnell (1977) has observed that riverine fish
are extremely mobile moving long distances up and down the river.
It is not surprising, therefore, that most recent studies are on
floodriver fish, especially as they are more productive (Fig. 9).
Welcomme (1979) has provided a comprehensive review of these
fishes.

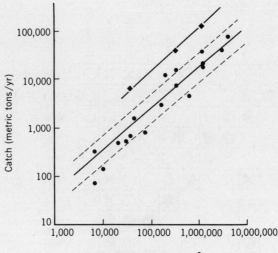

Figure 9.   Relationship between African floodplain size (area)
and biological production as illustrated by fish yield.

Composition, distribution and adaptations: all groups, and nearly
all genera and species of fish which live in freshwater are to be
found at times in rivers and streams, and it is fairly certain
that fishes moved into freshwater from the sea by way of the rivers,
and may still do so as a normal part of their life history (Hynes,
1970). In this connection it is noteworthy that between 1970-1980,
Awachie caught essentially marine and brackishwater species of both
cartilaginous and bony fish far upstream on the Niger and Benue
drainage complex. *Solea* and the sea catfish *Arius* as well as the
elasmobranchs *Trygon* and *Dasyatis* were taken at Onitsha on the Niger,
while the shark *Scoliodon* together with *Trygon* have been recorded
as far upstream as Numan, Lau and Yola on the Benue river, in July
1974. In some large and old rivers like the Zaire, obstructions
like rapids and slow flow, have prevented marine fishes from
penetrating far up the river.

Two main groups of fish may, therefore, be recognised, the
freshwater and the brackishwater species. The freshwater components
which dominate the African inland water ichthyofauna, including
those of running water, feature the most successful and widely
distributed Cichlidae and siluroid complex as well as such common
species as *Lates niloticus, Heterotis niloticus,* characoids,
mormyroids, *Polypterus,* the lung fish *Protopterus* and African
carps *Labeo* spp.

Another grouping, based mainly on behavioural responses
to environmental stress, differentiates between (i) 'whitefish'
which inhabit the savanna belt and avoid adverse conditions in
rivers and floodplains by migrating into and within the main river
channels, and (ii) 'black fish' which have adaptations to exist in
hot and often dry conditions with low oxygen levels during the dry
season. They usually live in forest zones. Notable examples are
the lung fish (*Protopterus*) which forms and lives in cocoons,
Polypterus, Heterotis and *Gymnarchus* which possess external gills
in the relevant phase of the life cycle. The possession of accessory
breeding structures by *Clarias, Heterobranchus* and *Parophiocephalus*
species and the production of dormant eggs by *Aphyosemion* and
Nothobranchus species, serve the same need. All the latter group
are typical of the African floodplain system.

Fish migration investigations from various regions of Africa
show that fish migrate for various reasons including feeding and
breeding (Matagne, 1950; Blache, 1964; Williams, 1971). These
migrations are often correlated with the flood regime (Welcomme,
1969; Durand, 1970, 1971). The following six phases in the
migration and distribution of fish are distinguishable, (i)
longitudinal migrations within main channels (ii) lateral
migration to the floodplain (iii) local movements within the
floodplain (iv) lateral migration from the floodplain towards the
main channel (v) longitudinal migration within the main channel,
usually downstream (vi) local movements within dry season
habitats e.g. river, lake, or the sea in the case of estuarine
species (Welcomme, 1979). Potamodromous species undertake
spectacular and intense migrations as exhibited by *Labeo* in Zaire
and Kenya (Matagne, 1950 , Cadwalladr, 1965) and *Clarias* species in
Nigeria (Awachie and Ezenwaji, 1980).

Feeding, growth and reproduction: activities associated with
feeding, growth, reproduction and mortality are seasonal and also
geared to the flood regime. Feeding habits of many tropical species
are specialised but flexible in order to optimize the exploitation
of foods which vary seasonally in both variety and quantity.
Nevertheless, broad trophic groups, ranging from planktivorous to
omnivorous, are distinguishable with the floodplain component
being dominated by piscivores and detritivores.

The flood phase is the period of intensive feeding and growth,
during which fat is stored for sustenance during dry season when
food is scarce. Increase in both weight and length occurs at this
season and may be reversed during the dry season when loss of
condition is prevalent (Daget, 1952, Durand and Loubens, 1970, for
the Chari; Reizer, 1971, for the Senegal; Daget, 1957, for the Niger;
Greenwood 1958, 1976 for the Nile and East Africa).

Growth in any one year is positively correlated with the
duration and intensity of the flood that year, resulting in good
and poor 'growth-year' classes (Dudley, 1974; Kapetsky, 1974).

The characteristics of the flood curve are also important
in determining the latter (Welcomme, 1976, 1979). It is also
noteworthy that a large proportion of the increase in fish
biomass is, as a result, contributed by the rapid growth of young-
of-the-year. This is particularly so because, as Lowe-McConnell
(1977) noted, floodriver fishes tend to have short life cycles;
many of them mature within one year and are ready to spawn by the
next flood season. This life pattern is typified by *Sarotherodon*
and *Tilapia* species, and parallel similar life cycle characteristics
recorded in the invertebrates.

For most river fishes, breeding commences with flood rise
although a few start just before this event. Like growth, breeding
success is closely related to the flood curve. Smooth and gradual
rise in water level, with high flood amplitude of long duration,
produces higher numbers of the young-of-the-year than floods with
the opposite features. The overall effect is that in good years,
young fishes emerge at a period of abundant food availability in
the form of large populations of micro-organisms, plankton, insects,
molluscs and aquatic vegetation. The latter, as well as submerged
terrestrial vegetation provide protection for young fish from their
predators, thus enhancing survival and reproductive success.

Most of the major factors causing fish mortality, whether
natural or by fishing, are density-dependent and also related to
the flood cycle. Mortality is thus generally low during low
water and rises with the flood except where fishing for migrating
breeders is undertaken (Awachie and Ezenwaji, 1980). Added to this,
numerous fish, both adult and young, are stranded on the floodplain
and shallow depressions where they are readily picked up by man,
otters, birds and other piscivorous animals. This results in a
massive loss of fish which could become extreme if the flood
recession curve is steep. This situation is often exploited by
local fishermen who erect temporary barriers to facilitate the
stranding of fish.

Estimates of standing crop in running water environments seem
too variable to be reliable. However, the results of Kapetsky's
(1974) studies on the production of cichlids in the Kafue system,
apparently the only one of its type so far in tropical Africa,
indicate that production and biomass of fish also fluctuate with
the flood regime. Thus good floods give high fish stock and vice
versa.

The above data relate primarily to fin-fish since ecological
and biological production information on river and stream shrimps,

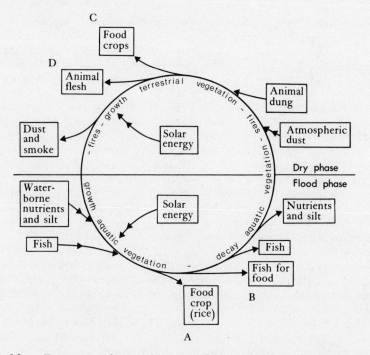

Figure 10. Energy and nutrient cycling in African running waters
            as typified by a floodriver system (adapted from
            Welcomme and Hagberg, 1976) (note the important
            component of allochthonous material).

crabs, molluscs and other vertebrates - amphibians, reptiles, birds,
and mammals is scanty.

MANAGEMENT STUDIES

     A rather diagnostic difference between the running water
environment and the other major inland ecosystems, the terrestrial
and lacustrine, is the basic issue of energy source and recycling.
Unlike the latter ecosystems, rivers and streams obtain their energy
not only from solar radiation but also from allochthonous organic
matter which is of great importance to them (Fig. 10). What is more,
any recycling of energy does not take place in one area of the lotic
environment but is continually wafted downstream (Hynes, 1970).
The situation for management studies is made even more difficult
by the sparsity of data on tropical African river and stream
systems and the current upsurge in the use of river basins as
modules for integrated rural development without adequate basic
information on them (Awachie, 1979, 1980). Added to the above is

the rapidity with which changes occur in the humid tropics.

However, the seasonality of the flood regime is a fairly
reliable reference point for the management of running water
ecosystems of Africa, even if empirically, since both biotic and
abiotic phenomena are correlated with it, as established above.
Informed manipulation of the flood cycle could, therefore, prove
a useful tool with which to manage the resources of running water
systems without disrupting their integrity. Indeed, the growing
problems of water quality, intractable erosion of deforested land
put under agricultural and human habitation together with the allied
issues of eutrophication and pollution, with their negative impacts
on fisheries production (Alabaster, 1980), may find solution or at
least remedies if the relationships of their incidence and flood
characteristics such as flow patterns and seasonal discharge rates
in a catchment area are known.

With particular reference to fisheries, management of the
hydrological regime has proved effective in area where controlled
release of flood water from an upstream dam, through channels, has
refilled desiccated floodplain pools and lakes downstream and
stimulated fish to migrate and breed, thus increasing overall fish
production of the system (Phelines et al., 1973). Added to this
is the finding that rational exploitation of river fishery is
based on the fact that wild stock concentrates at three points of
the annual flood cycle, flood rise, peak and flood fall. Uncontrol-
led exploitation of fish migrating early at flood rise, reduces
breeding fish and is harmful to the stock (Cadwalladr, 1965).
Similarly, some fisheries may be harmful, during flood fall, to
effective recruitment due to excessive removal of young fish
returning to the main river channels.

Fish diseases and management: the fish disease situation and its
bearing on the natural mortality, productivity and management of
fisheries was reviewed by Awachie et al. (1977). Flow rate and
flood amplitude were found to have direct bearing on the epizootio-
logy of parasites. Ectoparasitic forms comprising fungal, bacterial,
protozoan, crustacean, and monogenean parasites are more important
than endoparasites in the humid tropics. The bacterium *Aeromonas*,
fungus *Saprolegnia*, protozoans *Ichthyophthirius, Costia,* and
Trichodina, copepod *Ergasilius* and gyrodactylid gill flukes are
pathologically important elements that have wide distribution in
Africa (Fryer, 1968; Khalil, 1971; El Bollock and El Sarnagawi,
1975; Awachie, 1978). With the possible exception of the eye-socket
worm *Nematobothrium* and the metacercariae of *Clinostomum* which
make infected siluroid fish aesthetically unacceptable, endoparasites have not been found to cause high fish mortalities
(Awachie, 1965, 1972, 1978).

Fish in river and stream channels have very low infestation
rates. In the slow flowing to lentic conditions of the floodplain pools, infestation rates and mortalities could be high.
During the dry season, the hot backwaters and floodplain systems
provide ideal situations for rapid increase in parasite population,
consequent severe epizootics, loss of condition of hosts, and mass
mortalities due to ectoparasites.

In the absence of detailed biological knowledge of most
causative organisms on the continent, effective management of
parasitosis in lotic systems is difficult.

Public Health aspects: in most of tropical Africa, untreated
surface water from rivers and their floodplains is the main source
of drinking water for rural riverine people. Consequently, water-
borne diseases such as cholera and typhoid exhibit sporadic out-
breaks, while health hazards from dangerous biocides, emanating
from agricultural practices and disease control programs, are high.

Some elements of lotic water faunas serve as vectors for
important tropical diseases. These include *Similium* species for
river blindness, the snails *Bulinus* and *Biomphalaria* for
bilharziasis, and the mosquito *Anopheles* for malaria. In addition,
rich fringing vegetation along rivers and streams provides cover
and resting places of tse-tse fly, *Glossina*, which vectors both
human and animal forms of sleeping sickness.

Malaria is holoendemic in tropical Africa with high morbidity
and mortality especially among rural people. Waddy (1975, 1979)
and Walsh (1970) reviewed the health hazards posed by malaria,
river blindness and bilharziasis in the major river basins of
Africa. A rather interesting finding from the many investigations
on the issue of river blindness in lotic environments is that
contrary to predictions, neither the development of man-made lakes
nor the use of insecticides have succeeded in restricting the
distribution of the disease and its vector *Similium* in African
river basins for any reasonable length of time. And the recent
discovery of the association of immature stages of the *Similium
neavei* complex with the river crab *P. niloticus* in East African
streams (Williams, 1968; Williams and Hynes, 1971) has introduced
new dimensions to the problem of *Similium* management and control in
Africa.

FUTURE STUDIES

The poor state of knowledge on the running water ecosystem
in Africa and some of the contributory factors have been briefly
indicated. In spite of this, current development efforts, aimed
at bridging the gap between urban and rural peoples of the continent,
are being located in river basins as a matter of deliberate policy.

More often than not, development involves the construction of dams
and other water control systems. This has led to the disruption
of the natural production system of running waters without prior
development of the necessary predictive capability to abate stressed
resources and hence protect environmental integrity. It is held
that adequate pre-development/alteration studies be undertaken,
in future, to provide for informed planning and management of
affected aquatic and terrestrial resources.

Various developmental options are to be considered. Based
on African experience, so far, an ecosystems approach which will
ensure that all relevant factors - physical, chemical, biological
and human, are critically evaluated. In this connection, the multi-
purpose use concept seems to provide the most equitable option on
which integrated management of aquatic and terrestrial production
sub-systems of the lotic environment can be based. Thus a
concept which links the management of fisheries, forestry and
agriculture to those of agro-industrial and hydro-electric
developments will facilitate desirable recycling of nutrients,
optimize the production of the lotic system, minimize pollution,
eutrophication, biocides and toxic substances as well as health
hazards consequent on changes in the hydrobiological and socio-
cultural milieu e.g. bilharziasis and aquatic vectors of disease.

In the case of fisheries, a vital resource to riverine people,
which often receives severe multiple negative impacts in river
developments, the Mekong model where comprehensive studies of
integrated development of the river basin resources were conducted
before the alteration of the basin, has shown that it is possible,
using predictive models, to offset fishery losses amounting to US
$1.05 million by providing compensatory measures, thus optimising
production from all sectors in the Mekong valley (Pantulu, 1979).

Environmental conservation: recent experience with major dams,
irrigation, and navigation works in Africa has highlighted the
scale of alteration which can occur and, therefore, the need for
increased efforts to conserve the environment of running water
which is often involved. This aspect of river and stream ecology
should receive more attention in future.

Negative impacts of integrated basin development could be
severe at the current rate of economic development on the
continent. It may also lead to complete and an unnecessary change
of the ecology of the target areas of development. It is therefore
envisaged that conservation measures would be inscribed in the
policies governing strategic planning, management and development
of running water ecosystems and other aquatic and terrestrial
systems for that matter. All this will not come to fruition unless
conscious and sustained efforts are made to provide, without

further delay, essential basic biological data on such subjects
as the taxonomy, community ecology and productivity dynamics of
rivers and streams in Africa. It is on such data that meaningful
applied and development-oriented studies are to be based in order
to optimise the utilization of the rich natural resources of
Africa for the greatest welfare of its peoples.

REFERENCES

Adeniji, H.A. 1979. Waterfalls and dissolved oxygen concentration,
    possible effects on the fishery of the proposed Jebba Lake.
    *Proc. Internat. Conf. Kainji lake and River Basins
    Development in Africa.* 1: 178-183.
Alabaster, J. 1980. Present and potential effects of pollution
    on fisheries in East Africa. *FAO/CIFA Seminar on River Basin
    Management and Development, Malawi, Dec. 1980,* (Proceedings
    1980.
Awachie, J.B.E. 1965. Preliminary notes on the parasites of fish
    in the area of the Kainji reservoir In "The First Scientific
    Report of the Kainji Biological Research Team." White, E.
    ed. Liverpool Univ. Press.
Awachie, J.B.E. 1972. On a Didymozoid trematode parasitic in the
    eyesockets of African carps (*Labeo* spp) from the River Niger,
    Nigeria. *Acta Parasit. Polon.* 20: 449-500.
Awachie, J.B.E. 1975. Fish culture possibilities on the flood-
    plains (ofadamas) of the Niger-Benue drainage system *FAO/CIFA
    Symp. on Aquaculture in Africa.* 256-281.
Awachie, J.B.E. 1978. On the dynamics of the parasites of fish
    in the Kainji man-made lake in Nigeria. *Proc. 4th Internat.
    Conf. Parasitology, Section* H: 20.
Awachie, J.B.E. 1979. On fishing and fisheries management in
    large tropical rivers with particular reference to Nigeria.
    FAO Fish. Tech. Pap. 1974: 37-48.
Awachie, J.B.E. 1980. Some general considerations on River Basins
    in Africa and their management and development in relation to
    fisheries. *Keynote Address. FAO/CIFA Seminar on River Basin
    Management and Development, Malawi,* Dec. 1980. CIFA/80/3:
    1-22.
Awachie, J.B.E. and Ezenwaji, G.M.H. 1980. The importance of
    Clarias species in the freshwater fisheries development of
    Anambra river basin, Nigeria. *FAO/CIFA Seminar on River
    Basin Management and Development, Malawi,* Dec. 1980
    (Proceedings in press).
Awachie, J.B.E., Ilozumba, P.C.O. and Azugo, W.I. 1977. Fish
    parasites in the ecology, management and productivity of
    river and floodplain fisheries in Africa. *FAO/CIFA
    Symposium on River and floodplain fisheries in Africa.*
    CIFA/T.5: 296-312.

Beadle, L.C. 1974. The inland waters of tropical Africa: an
    introduction to tropical limnology. Longman, London.
Bidwell, A. and Clarke, N.V. 1974. The invertebrate fauna of
    Lake Kainji, Nigeria. *Nig. Field.* 42: 104-110.
Blache, J. 1964. Les poissons du bassin du Tchad et du Bassin
    adjacent Mayo Kebbi. *Mém. ORSTOM.* 4: 1-483.
Blanc, M., Daget, J. and D'Aubenton, F. 1955. Recherches
    hydrobiologiques dans le bassin du Moyen-Niger. *Bull. Inst.
    Fr. Afr. Noire A. Sci. Nat.* 17: 619-746.
Bourne, G.C. 1893. On two new species of copepoda from
    Zanzibar. *Proc. Zool. Soc. Lond.* I: 164-166.
Bryson, R.A. 1974. A prospective climatic change. *Science*
    184: 753-760.
Cadwalladr, D.A. 1965. The decline in the *Labeo victorianus*
    Blgr. (Pisces: Cyprinidae) fishery of Lake Victoria and an
    associated deterioration in some indigenous fishing methods
    in the Nzoia river, Kenya. *East Afr. Agric. For. J.*
    30: 249-256.
Carey, T.G. 1971. Hydrobiological survey of the Kafue floodplain.
    Fish. Res. Bull; Zambia 5: 245-295.
Central Forestier Technique Tropicale. 1972. Incidences sur la
    pêche de l'aménagement hydro-agricole du bassin du Senegal.
    II. Influence des travaux d'aménagement sur les ressources
    piscicoles. *Nogent-sur-Marne.* CTFT: 1-119.
Crisp, G. 1956. An ephemeral fauna of torrents in the Northern
    Territories of the Gold Coast, with special reference to the
    enemies of *Simulium.* *Ann. trop. Med. Parasit.* 50: 260-267.
Daget, J. 1952. Memoires sur la biologie des poissons du Niger.
    1. Biologie et croissance des éspèces du genre *Alestes.*
    Bull. Inst. Fr. Afr. Noire 14: 191-225.
Daget, J. 1957. Données récentes sur la biologie de poissons
    dans le delta central du Niger. *Hydrobiologia* 9: 321-347.
Daget, J. and Iltis, A. 1965. Poissons de Côte d'Ivoire eaux
    douces et saumâtres. *Mem. IFAN.* 74: 1-385.
Dudley, R.G. 1974. Growth of *Tilapia* of the Kafue floodplain,
    Zambia: predicted effects of the Kafue Gorge Dam. *Trans. Am.
    Fish. Soc.* 103: 281-291.
Durand, J.R. 1970. Les peuplements ichtyologiques de l'El Beid
    Première note. Presentation du milieu et resultats
    généraux. *Cah. ORSTOM (Hydrobiol.)* 4: 3-36.
Durand, J.R. 1971. Les peuplements ichthyologiques de l'El Beid.
    2e note. Variations inter et intraspecifiques. *Cah.
    ORSTOM (Hydrobiol.)* 5: 147-159.
Durand, J.R. and Loubens, G. 1970. Variations du coefficient de
    condition chez les *Alestes baremose* (Pisc. Charac.) du bas
    Chari et du Lac Tchad. *Cah. ORSTOM (Hydrobiol.)* 4: 61-81
Egborge, A.B.M. 1971. The chemical hydrology of the River Oshun,
    Western State, Nigeria. *Freshwat. Biol.* 1: 257-272.

Egborge, A.B.M. 1972. The physical hydrology of the River Oshun, Western State, Nigeria. *Arch. Hydrobiol.* 70: 72-81.

Egborge, A.B.M. 1973. A preliminary check-list of the phytoplankton of the Oshun River, Nigeria. *Freshwat. Biol.* 3: 569-572.

Egborge, A.B.M. 1974. The seasonal variation and distribution of phytoplankton in the River Oshun, Nigeria. *Freshwat. Biol.* 4: 177-191.

Egborge, A.B.M. 1979. Observations on the vertical distribution of the zooplankton in Lake Asejire, Nigeria. *Proc. Internat. Conf. on Kainji Lake and River Basins Development in Africa* 2: 203-218.

Egborge, A.B.M. and Fagade, S.O. 1979. Notes on the hydrobiology of the Vikki warm springs, Yankari Game Reserve, Nigeria. *Pol. Arch. Hydrobiol.* 26: 313-322.

Ekman, S. 1903. Cladoceren und freischwimmende Copepoden aus Agypten und Sudan. Results of the Swedish Zoological Expedition to Egypt and White Nile 1901. *Uppsala Nr.* 26: 1-18.

El Bolock, A. and El Sarnagawi, D. 1975. Some diseases recorded on cultivated fishes in Egypt. *FAO/CIFA Symp. on Aquaculture in Africa* FAO, Rome. 4 pp.

FAO/UN. 1969. Report to the Government of Zambia on fishery development in the Central Barotse floodplain. Second phase. Based on the work of D. Duerre. *Rep. FAO/UNDP(TA).* 2638: 1-80.

FAO/UN. 1970. Report to the Government of Nigeria on fishery investigations on the Niger and Benue rivers in the northern region and development of a program of riverine fishery management and training. Based on the work of M.P. Motwani. *Rep. FAO/UNDP(TA).* 2771: 1-196.

Fryer, G. 1968. The parasitic Crustacea of African freshwater fishes; their biology and distribution. *J. Zool. Lond.* 156: 45-95.

Gosse, J.P. 1963. Le milieu aquatic et l'écologie des poissons dans la région de Yangambi. *Ann. Mus. R. Afr. Cont.* 116: 113-271.

Green, J. 1963. Zooplankton of the River Sokoto, the Rhizopoda Testacea. *Proc. Zool. Soc. Lond.* 141: 497-514.

Greenwood, P.H. 1958. Reproduction in the East African lungfish *Protopterus aethiopicus* Heckel. *Proc. Zool. Soc. Lond.* 130: 547-567.

Greenwood, P.H. 1976. Fish fauna of the Nile. *In*: "The Nile: biology of an ancient river." Junk, The Hague.

Harrison, A.D. 1965. River zonation in Southern Africa. *Arch. Hydrobiol.* 61: 380-386.

Harrison, A.D. and Elsworth, J.F. 1958. Hydrobiological studies on the Great Berg River, Western Cape Province. Part I. General description, chemical studies and main features of the flora and fauna. *Trans R. Soc. S. Afr.* 35: 281-385.

Holden, M.J. and Green, J. 1960. The hydrology and plankton of
    the river Sokoto. *J. Anim. Ecol.* 29: 65–84.
Hynes, H.B.N. 1952. The Neoperlinae of the Ethiopian region
    (Plecoptera, Perlidae). *Trans. R. Ent. Soc. Lond.* 103:
    85–108.
Hynes, H.B.N. 1953. The nymph of *Neoperla spio* (Newman)
    (Plecoptera: Perlidae). *Proc. R. Ent. Soc. Lond.* 28:
    93–99.
Hynes, H.B.N. 1970. The Ecology of Running Waters. Liverpool
    University Press, Liverpool.
Hynes, H.B.N. and Williams, T.R. 1962. The effect of DDT on the
    fauna of a central African stream. *Ann. trop. Med.*
    Parasit. 56: 78–91.
Hynes, H.B.N., Williams, T.R. and Kershaw, W.E. 1961. Fresh-
    water crabs and *Simulium neavei* in East Africa. I.
    Preliminary observations made on the slopes of Mount Elgon
    in December, 1960 and January, 1961. *Ann. trop. Med.*
    Parasit. 55: 197–201.
Ikusemiju, K. 1976. Distribution, reproduction and growth of
    the catfish, *Chrysichthys walkeri* (Guntheri) in the Lekki
    lagoon, Nigeria. *J. Fish. Biol.* 8: 453–458.
Ikusemiji, K. 1979. Possible impacts of river basin development
    on the lagoons of South-west Nigeria. *Proc. Internat. Conf.*
    on Kainji Lake and River Basins Development in Africa. 2:
    371–374.
Khalil, L.F. 1971. Checklist of the Helminth parasites of
    African freshwater fishes. Commonwealth Helminthological
    Bureau, St. Albans.
Kapetsky, J.M. 1974. The Kafue river floodplain: an example of
    pre-impoundment potential for fish production. *In* "Lake
    Kariba: a man-made tropical ecosystem in Central Africa."
    Balon, E.K. and Coche, A.G. eds. W. Junk, The Hague.
Keiffer, F. 1949. The Armstrong College Zoological Expedition
    to Siwa Oasis (Libyan desert) 1935. Freilebende Rudder-
    fusskrebse (Crustacea, Copepoda). *Proc. Egyptian Acad. Sci.*
    Cairo 4: 62–112.
Leopold, L.B., Wolman, M.B. and Miller, J.P. 1964. Fluvial
    processes in geomorphology. W.H. Freeman, San Francisco.
Lowe-McConnell, R.H. 1977. Ecology of fishes in tropical
    waters. Edward Arnold, London.
Macarthur, R.H. 1965. Patterns of species diversity. *Biol. Rev.*
    40: 510–533.
Marlier, G. 1954. Recherches hydrobiologiques dans les revieres
    du Congo Oriental II. Etude ecologique. *Hydrobiologia*
    6: 225–264.
Matagne, F. 1950. Premières notes au sujet de la migration de
    pumbu (*Labeo* sp.). Bief du Luapula-Moera. *Bull. Agric.*
    Congo Belge. 41: 794–834.

Mills, M.L.   1976.   The *Chaoborus* (Diptera: Chaoboridae) component
     of the invertebrate drift of the Mwenda river Lake Kariba,
     Central Africa.   *Hydrobiologia*   48: 247-250.
Monakov, A.V.   1969.   The zooplankton and zoobenthos of the White
     Nile and adjoining waters in the republic of the Sudan.
     Hydrobiologia   33: 161-185.
Nduku, W.K. and Robarts, R.D.   1977.   The effect of catchment
     geochemistry and geomorphology on the productivity of a
     tropical African montane lake (Little Connemara Dam No. 3,
     Rhodesia).   *Freshwat. Biol.*   7: 19-30.
Olaniyan, C.I.O.   1961.   Observations on the salinity and
     stratifications of the tidal current in Lagos Harbour,
     Nigeria, with a description of a simple method for the
     collection of water samples at various depths for salinity
     determinations.   *J. West Afric. Sci. Ass.*   7: 49-58.
Olaniyan, C.I.O.   1969.   The seasonal variation in the hydrology
     and total plankton of the lagoons of southwest Nigeria.
     Nig. J. Sci.   3: 101-119.
Omer-Cooper, J.   1971.   Taxonomic studies on some African
     Hyphydrus (Coleoptera: Dytiscidae).   *J. ent. Soc. sth. Afr.*
     34: 277-288.
Pantulu, V.R.   1979.   Fishery problems associated with multiple
     uses of large rivers.   *FAO Fish. Tech. Pap.*   194: 48-53.
Phelines, R.F., Coke, M. and Nicol, S.M.   1973.   Some biological
     consequences of the damming of the Pongolo river.   *Proc
     IIème Congrès des Grands Barrages*   175-190.
Ramadan, F.M.   1972.   Characterization of Nile waters prior to the
     High Dam.   *Z. Wasser Abwasser Forsch.*   5: 21-24.
Reizer, C.   1971.   Contribution a l'etude hydrobiologique du Bas-
     Sénégal.   Premières recommandations d'aménagement halieutique.
     Nogent-sur-Marne, C.T.F.F.: 1-142.
Rzoska, J.   1974.   The Upper Nile Swamps, a tropical wetland study.
     Freshwat. Biol.   4: 1-30.
Rzoska, J. ed.   1976.   The Nile: biology of an ancient river.
     Junk, The Hague.
Sars, G.O.   1909.   Zoological results of the third Tanganika
     Expedition conducted by Dr. W.A. Cunnington 1904-1905.
     Report on the Copepoda.   *Proc. Zool. Soc. Lond.*   31: 31-77.
Shepherd, C.J. ed.   1976.   Investigation into fish productivity
     in a shallow freshwater lagoon in Malawi 1975/76.   Ministry
     of Overseas Development, London.
Someren, V.D.   1952.   The biology of trout in Kenya colony.
     Government Printer, Nairobi.
Svenson, G.S.O.   1933.   Freshwater fishes from the Gamia River
     (British West Africa).   Results of the Swedish expedition,
     1931.   *K. Svensk. Wetenskskapsakad. Handl.*   12: 1-13.
Talling, J.F.   1957.   The longitudinal succession of water
     characteristics in the White Nile.   *Hydrobiologia*
     9: 73-89.

Talling, J.F. 1976. Water Characteristics. *In*: The Nile: biology
    of an ancient river. Rzoska, J. ed. Junk, The Hague.
Talling, J.F. and Talling, I.B. 1965. The chemical composition
    of African lake waters. *Int. Revue ges. Hydrobiol.* 50:
    421-463.
Talling, J.F. and Rzoska, J. 1967. The development of plankton
    in relation to the hydrological regime in the Blue Nile.
    J. Ecol. 55: 637-662.
Thomas, M.P. 1965. Zooplankton studies. In: *The First Scientific
    report of the Kainji biological Research Team* White, E. ed.
    Liverpool. p. 17-20.
Van der Leeden, F. 1975. Water Resources of the World, Selected
    Statistics. Water Information Centre, New York.
Volker, A. 1966. Surface hydrology of deltaic area. *In*:
    Scientific problems of the Humid Tropical Zone deltas and
    their implications. *Proc. Dacca Symp.* UNESCO, Paris.
Waddy, B.B. 1975. Research into the health problems of man-made
    lakes with special reference to Africa. *Trans. Roy. Soc.
    Trop. Med. Hyg.* 69: 39-50.
Waddy, B.B. 1979. Health hazards associated with man-made lakes
    and approaches to their control, (with special reference to
    Kainji lake, Nigeria). *Proc. Internat. Conf. on Lake Kainji:*
    562-569.
Walsh, J.F. 1970. Evidence of reduced susceptibility to DDT in
    controlling *Simulium damnosum* (Diptera: Simuliidae) on the
    River Niger. *Bull. World Hlth Org.* 43: 316-318.
Welcomme, R.L. 1969. The biology and ecology of the fishes of a
    small tropical stream. *J. Zool. Lond.* 158: 485-529.
Welcomme, R.L. 1975. The fisheries ecology of African floodplains
    CIFA Tech. Pap. 3: 1-55.
Welcomme, R.L. 1976. Some general and theoretical considerations
    on the fish yield of African rivers. *J. Fish. Biol.* 8:
    351-364.
Welcomme, R.L. 1979. Fisheries ecology of floodplain rivers.
    Longman, London.
Williams, R. 1971. Fish ecology of the Kafue River and floodplain
    environment. *Fish. Res. Bull. Zambia* 5: 305-330.
Williams, R. 1971. Fish ecology of the Kafue River and floodplain
    environment. *Rhodesia Agric. J.* 57: 86-92.
Williams, T.R. and Hynes, H.B.N. 1971. A survey of the fauna of
    streams on Mount Elgon, East Africa, with special reference
    to the Simuliidae (Diptera). *Freshwat. Biol.* 1: 227-248.

RUNNING WATER ECOLOGY IN AUSTRALIA

W.D. Williams

Department of Zoology
University of Adelaide
Australia

INTRODUCTION

Less than a decade ago it was possible to state that "for the amount of work that has been published on the ecology of Australian rivers and streams, limnologists outside Australia might well be forgiven for thinking that no running waters exist in Australia at all" (Bayly and Williams, 1973). That statement was evidently so near the truth that no Australian limnologist was prepared to gainsay it. It can no longer stand, however, for even in the few years which have intervened a good deal of work on Australian running waters has been reported.

The present chapter does not attempt a comprehensive review of this work. Instead, I attempt to outline certain special features of Australian running waters by referring to work which illustrates these features. Since most of this is recent (and sometimes obscurely reported), the chapter will also serve a bibliographic function as well as update the partial review of Australian running waters by Bayly and Williams (1973).

It is appropriate to mention that amongst this recent work is a number of papers by H.B.N. Hynes derived from his own investigations during two antipodean sojourns (Hynes, 1974a,b, 1976a,b, 1978, 1980; Hynes and Hynes, 1975). Thus, trained in Britain, domiciled in Canada, Noel Hynes has nevertheless contributed significantly and directly to lotic studies half a world away! It is a privilege and pleasure to be associated with a book which honours a former teacher, a present friend and colleague, and a *quondam* Australian limnologist.

Table 1.   Rainfall and run-off for inhabited continents.
           From Department of National Resources, Australian
           Water Resources Council (1976).

Continent	Area (km^2 x 10^3)	Average annual rainfall (mm)	Evaporation and transpiration as a percentage of rainfall	Run-off as a percentage of rainfall
Africa	30210	660	76	24
North America	24260	660	60	40
South America	17790	1350	64	36
Asia	44030	610	64	36
Europe	9710	580	60	40
Australia	7690	420	87	13

PHYSICAL NATURE

     Two major determinants of river and stream type are climate and
topography. Both determinants are distinctive in Australia.
Climatically, Australia is a very arid continent (Table 1), with
high evaporation rates and low amounts of precipitation. These
combine to produce exceedingly low run-off values; the total annual
run-off in Australia, about 346 km^3, is only slightly in excess of
the volume discharged annually by one European river, the Danube.
Topographically, Australia is a flat continent. Only the Great
Dividing Range in the east and south-east and the Tasmanian highlands include land of significant elevation.

     The nature of Australian drainage basins (Fig. 1) relects
these features. Most of the western half of the continent is
arheically drained and such flowing waters as occur are episodic.
There is an immense endorheic basin in the centre of the
continent wherein flowing waters are episodic. And in the largely
exorheic part of the continent, permanent and fast-flowing streams
are more or less confined to the coast or highlands, with most
inland waters slow-flowing and frequently intermittent. In all
types of drainage basin, stream declivities are mostly slight. The
Murray-Darling system, the largest river system in Australia,
exemplifies inland flowing waters in exorheic basins and is dealt
with separately later.

     Whatever the nature of the drainage basin, a notable feature
of most permanent rivers is the wide variation in seasonal and

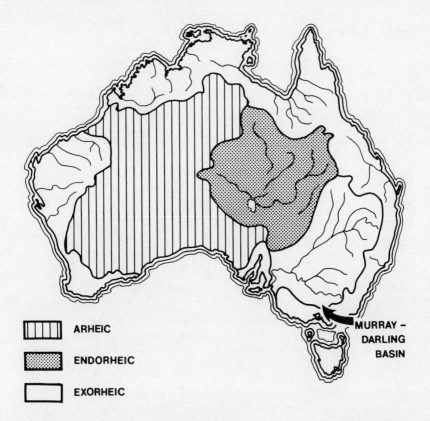

Figure 1. Main types of drainage basin in Australia.

annual discharge values (streamflow). Patterns of annual variation
are illustrated in Figure 2, which also shows that annual
variations are by no means synchronous throughout the continent.
Seasonal variation is especially marked in northern monsoonal areas
and north-eastern areas of high summer rainfall.

    Additional information on the hydrological characteristics of
Australian streams is provided by the Department of National
Resources and Australian Water Resources Council (1976, 1978a,b).
These list many further references.

    Little quantitative information is available on turbidities.
Characteristically, however, most lowland running waters appear
slightly to markedly turbid, and the major lowland rivers within
the Murray-Darling system rarely have Secchi disc readings over
2 m, with values frequently less than 1 m (Lake, 1967). The work
of Woodyer (1978) indicates that for the Darling River the high

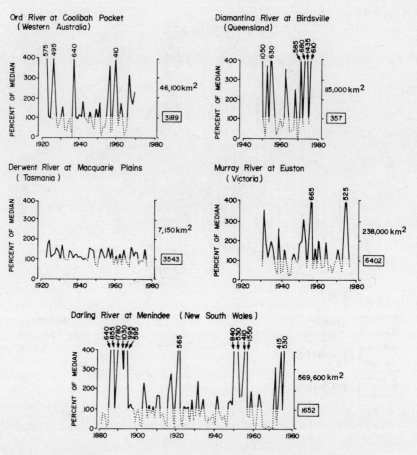

Figure 2. Variability of annual streamflow: year to year
           variations as percent of median streamflow of five rivers.
           Unenclosed figure at right of each graph indicates
           catchment area (km^2); figure at right enclosed in
           rectangle is median annual volume (m^3 x 10^6).
           Rearranged from Australian Water Resources Council,
           Department of National Development (1978).

turbidity is due more to the fineness of suspended sediments than
to their actual concentration. At least in part, the high
turbidities reflect catchment erosion for stream turbidities are
closely linked to catchment events (Williams, 1980b). Rates of
catchment denudation in parts of eastern Australia – including
discussion of both dissolved and particulate matter – have been
considered by Douglas (1973; see also 1966). Buckney's (1979)
careful study of the Onkaparinga River, South Australia, is also
of interest; *inter alia*, he demonstrated a gross correlation
between turbidity and discharge in this river.

CHEMICAL NATURE

Most running waters are said to be of the 'bicarbonate' type
with calcium and magnesium the dominant cations; running waters
with a different major ion dominance are regarded as unusual and
found particularly in arid areas (Hynes, 1970). Australia provides
many exceptions to this statement. In Tasmania, running waters
are more likely to be dominated by sodium than by the divalent
cations, and the same applies in Victoria, and south-western
Western Australia; elsewhere, however, bicarbonate waters occur
more frequently and the large inland rivers of the south-eastern
mainland are mainly of this type. It may be, as suggested by
Douglas (1968), that the supply of sodium chloride from precipitation
is more significant in the chemistry of Australian streams than
elsewhere.

Information on the major ionic composition of Australian streams
is to be found in a number of publications (recent ones include:
Buckney and Tyler, 1973; Williams and Buckney, 1976; Buckney, 1977;
Johnson and Muir, 1977; Walker and Hillman, 1977; Buckney, 1979;
Norris et al., 1980). In addition, much unpublished information is
available as a result of water quality monitoring by government
agencies. Examination of this material beyond immediate needs has
generally been neglected, as pointed out by Ross and Connell
(1979). Further examination would be of interest on many grounds;
if considered with area and run-off data, for example, it could be
used to derive a more up-to-date and credible mean composition of
Australian river water than that derived by Livingstone (1963) and
widely quoted (e.g. Benoit, 1969; Wetzel, 1975).

Examination of the relationships between climate, hydrology,
geology and major ion chemistry in individual waters has also been
neglected, although studies undertaken thus far underscore the
importance of such relationships (e.g. Douglas, 1968; Johnson and
Muir, 1977; Buckney, 1974, 1977). Few Australian authors (e.g.
Buckney and Tyler, 1973; Graham et al., 1978) have related their
data on river chemistry to Gibbs' (1970) worldwide scheme to
describe surface water chemistry.

The salinity (total dissolved salts) of Australian running
waters is another chemical characteristic of interest. Although
most Australian streams are fresh, with salinities $< 1,000$ mg 1^{-1},
a few have more elevated salinities. For example, the weighted
average salinity of the Hopkins River, a small river in western
Victoria, is $1,445$ mg 1^{-1}, but its salinity sometimes exceeds
$6,000$ mg 1^{-1}. Even higher values have been recorded in southwestern Western Australia; there, salinities $> 30,000$ mg 1^{-1} have
been recorded in the Pallinup River (Department of National Resources,
Australian Water Resources Council, 1976). No recently derived value

Table 2.   Australian river salinities. Distribution of 151
           selected rivers within various classes* of weighted
           average salinity. Data collated from Department of
           National Development, Australian Water Resources Council
           (1976).

Drainage Basin [+]	Weighted mean salinity (mg 1^{-1})		
	<59	59–120	>120
North-East Coast	14	14	7
South-East Coast	5	18	15
Tasmania	4	6	2
Murray-Darling	5	4	12
South Australian Gulf	0	0	8
South-West Coast	0	0	13
Indian Ocean	1	1	3
Timor Sea	0	2	1
Gulf of Carpentaria	7	2	3
Lake Eyre	1	1	1
Bulloo-Bancannia	1	0	0
Western Plateau	0	0	0
Totals	38	48	65

* Class values selected bearing in mind Livingstone's (1963)
value for the mean salinity of Australian rivers (i.e. 59 mg 1^{-1})
and his value for mean world river salinity (i.e. 120 mg 1^{-1}).

+ According to Department of National Resources, Australian
Water Resources Council (1976).

for the average salinity of all Australian rivers is available,
but were there one it would undoubtedly be greater than
Livingstone's (1963), viz. 59 mg 1^{-1}. On the basis of this
salinity, Australia has the most dilute rivers of any continent.
Examination of the weighted average salinity of 151 selected
rivers in all major drainage basins in Australia provides no
support for this (Table 2).

Whatever the mean river salinity in Australia, actual values
for several rivers have been unnaturally and significantly
increased by human activities in the recent past; irrigation, the
removal of catchment vegetation and agricultural mismanagement are
the chief activities implicated. Increased river salinity of this
type is now regarded as a major pollutional phenomenon in several
regions and is of considerable economic impact (Senate Select
Committee, 1970). Its ecological repercussions have yet to be
determined fully, but some unusual chemical events are already
known; for example, Morrissy (1974, 1979) has reported a downstream
decrease in salinity and the presence of haloclines in river pools
in a Western Australian river. A biological event of interest is
the apparently recent spread of an estuarine copepod, *Sulcanus
conflictus*, to saline pools in the Swan-Avon river system of
Western Australia (Rippingale, 1981). The more general effects of
rising salinities in the Avon River have been discussed by
Kendrick (1978). Literature about the increasing levels of
salinity in Australian rivers is voluminous but mostly of
peripheral interest to ecologists. An introduction can be gained
from the publications of Talsma and Philip (1971), Douglas (1972),
and Rixon (in press).

Finally, mention should be accorded plant nutrients, as early
observations suggested that nitrogen and phosphorus levels in
Australian waters were relatively high (Williams and Wan, 1972).
The observations remain to be confirmed, for despite escalating
interest in eutrophication, surprisingly little is known about
nutrients in Australian inland waters in general and rivers in
particular (Buckney, 1980). Wood (1975) collected the available
information for Australia up to 1975. Since then significant
contributions with respect to rivers have been made by Croome et al.
(1976), Walker and Hillman (1977), Whelan (1977), Campbell (1978),
Buckney (1979), and Holmes et al. (1980). An important point
which emerges is that turbidity, discharge and nutrient levels are
closely related. One of the salient results of Buckney's (1969)
study was that suspended material in the Onkaparinga River seems
to carry most of the nutrient load, as also was found by Walker
and Hillman (1977) in the River Murray near Albury and Wodonga.

FAUNAL COMPOSITION

The faunal composition of Australian inland waters as a whole
is distinctive in many ways, and this is particularly obvious when
the fauna of running waters is considered. In any event, it is
reasonable to suggest that because of the past (as well as
present) paucity of permanent natural freshwater lentic habitats
on the Australian mainland, the total Australian freshwater fauna
is one principally derived from that of rivers and associated waterbodies. The general features of the distinctive nature of the

Australian freshwater fauna have been discussed recently (Williams,
1980a), so that only salient points need noting here. (An up-to-
date taxonomic bibliography for all groups of Australian aquatic
invertebrates has been given by Williams (1980c).)

     For all but the microscopic fauna, endemicity at a variety of
taxonomic levels is well-developed. For many groups it is almost
complete at the species level and high at the generic level.
Additionally, there are various groups with families either
endemic to Australia or found only or mainly in the southern
hemisphere; of particular note in running waters are the freshwater
mussels, syncarids, phreaticoid isopods, crayfish, stoneflies, fish,
tortoises, and the platypus. Nearly all Australian freshwater
mussels belong to the Hyriidae, found only in Australasia and
South America. Within the Syncarida, all four families of the
primitive Anaspidacea are known from Australia, with only the
stygocarids known also from New Zealand and South America.
Anaspides tasmaniae (Anaspididae), the Tasmanian mountain shrimp,
is still relatively common in Tasmanian streams. Phreatoicid
isopods occur sporadically in a variety of inland waters throughout
Australia, and are common in many south-eastern Australian running
waters (particularly Tasmanian). The same may broadly be said of
crayfish, all of which belong to the southern hemispherical
Parastacidae. Of the stoneflies, about a quarter of the species
and a third of the genera are in the primitive family Eustheniidae,
found only in Australasia and South America. The most diverse
freshwater fish family, the Galaxiidae, is confined to the southern
hemisphere. Almost all freshwater turtles belong to the order
Pleurodira, now surviving only in Australia, New Guinea and South
America. And finally, the platypus, *Ornithorhynchus anatinus*
(Ornithorhynchidae), as need hardly be said, is the only aquatic
monotreme known and is confined to Australia.

     In addition to high endemicity, further distinctiveness is
provided by the absence of various animals often common in
running waters outside Australia, by an adaptive radiation within
some groups not markedly diverse on other continents, and by the
occurrence of special features within certain groups. Among notable
absentees are the mussel families Mutelidae, Margaritiferidae and
Unionidae, asellid isopods (such asellotes as occur are janirids),
phrynganeid caddisflies, cyprinid, salmonid and true percid fish
(all three fish families now introduced), ranid frogs (only one
northern species occurs), and Amphibia other than frogs. Among
groups which display an adaptive radiation in Australia are the
Leptophlebiidae (mayflies), Leptoceridae (caddisflies) and hylid
and leptodactylid frogs. Of groups with special features,
particular note may be accorded the fish. The Australian freshwater
fish fauna has a remarkably low species diversity, and has only
three species considered to be of "primary" freshwater origin

(i.e. marine affinities are very high), two of which are osteoglossids (*Scleropages* spp.) and one a lungfish (*Neoceratodus forsteri*).

In summary, then, high levels of endemicity, notable absences, considerable diversity in certain groups and the occurrence of special features in others combine to produce a fluviatile fauna unmistakably Australian in composition. Such a statement could well have been advanced many years ago, but could then only have been based on a synthesis of knowledge of individual components of the fauna. It is only recently that comprehensive surveys of individual streams have been produced to provide a more satisfactory basis for such a statement – at least, that is, so far as the macro-invertebrates are concerned. Note is made of the work of Jolly and Chapman (1966), Morrissy (1967), Thorp and Lake (1973), Aldenhoven (1975), Macmillan (1975), Blackburn (1976), McIvor (1976, Cadwallader and Eden (1977), Metzeling (1977), Robinson (1977), Smith et al. (1977, 1978), Suter and Williams (1977), Walker and Hillman (1977), Bishop and Tilzey (1978), Campbell (1978), Jackson (1978), Norris (1978, 1979), Yule (1978), Bishop (1979), Blackburn and Petr (1979), Fletcher (1979), Suter (1979), Blyth (1980) and Marchant (1980). Several of these authors were primarily concerned with the impact of pollutants or other disturbances on streams, but also provided good accounts of the invertebrate fauna of unpolluted or undisturbed sections of their study sites. One of the most comprehensive accounts is that by Suter (1979). This deals with the Latrobe River and its tributaries in eastern Victoria, and has been used to provide Table 3. The table will provide some flavour of how different is the composition of the macroinvertebrate fauna of an Australian river compared with that of rivers on other continents. It should be borne in mind, of course, that considerable regional differences also occur within Australia.

Comparable work on the lotic vertebrate fauna is lacking; recent accounts of vertebrates in individual Australian streams deal with restricted groups, e.g. frogs (Tyler and Crook, 1980), fresh-water turtles (Chessman, 1978) and fish (Lewis and Ellway, 1971; Pollard, 1974; Cadwallader, 1979; Bishop and Bell, 1978; Bishop, 1979; Jackson and Williams, 1980; Pollard et al. 1980).

THE MURRAY-DARLING RIVER SYSTEM

There are several reasons why the Murray-Darling River merits separate discussion. Although the fourth longest river system in the world and draining an immense catchment (one seventh of the continent and about one million km^2), it has an exceedingly small annual average discharge (22 m^3 x 10^9) vis-a-vis other major rivers of the world (cf. Amazon, 6,307; Mississippi, 593; Danube, 282) and exhibits very slight declivity over most of its course. It is, therefore, by any hydrological measure of rivers a distinctive one.

Table 3. Species list of benthic macroinvertebrates of the Latrobe River and tributaries. Modified from Suter (1979: Appendix 1).

PORIFERA: Spongillidae: 1 species
PLATYHELMINTHES: Tricladida : Dugesiidae: *Cura pinguis*
NEMATOMORPHA: *Gordius* sp.
ANNELIDA: Oligochaeta: Aeolosomatidae: *Aeolosoma* 4 spp; Haplotaxidae: *Haplotaxis* 3 spp.; Lumbriculidae: *Lumbriculus variegatus*; Phreodrilidae: *Phreodrilus* (nr) *nothofagi*, *P.* (*Phreodrilus*) *brianchiatus*; + *Phreodrilus* 2 spp.; Naididae: *Chaetogaster* sp., *Pristina osborni*, *P. proboscidea*, *P. longiseta*, *P. idrensis*, + *Pristina* 2 spp., *Nais communis*, *N. raviensis*, *N. elinguis*, *Slavina appendiculata*, *Dero* (*Aulophorus*) *furcatus*; Tubificidae: *Branchiura sowerbyi*, *Telmanodrilus multiprostratus*, *Tubifex tubifex*, *Limnodrilus hoffmeisteri*, *Aulodrilus pigueti*, *Rhyacodrilus simplex*.
Hirudinea: 1 species (indet.)
POLYZOA: Fredericellidae: *Fredericella ?australiensis*
MOLLUSCA: Gastropoda: Ferrissiidae: 1 species; Planorbidae: *Bulinus* sp., Planorbinae 2 spp.;
Physidae: *Physa* sp.; Hydrobiidae: *Potamopyrgus niger*.
Bivalvia: Hyriidae: *Hyridella depressa*; Sphaeriidae: *Sphaerium* sp.
CRUSTACEA: Amphipoda: Eusiridae: *Pseudomoera* ?sp.; Ceinidae: *Austrochiltonia australis*; Gammaridae: *Niphargus* ?sp., *Perthia* ?sp.
Decapoda: Atyidae: *Paratya australiensis*; Parastacidae: *Euastacus* sp.
ARACHNIDA: Hydracarina: 9 spp. + *Australiobates* sp.
INSECTA: Ephemeroptera: Siphlonuridae: *Coloburiscoides munionga haleuticus*, *C. giganteus*, *Tasmanophlebia lacus-coerulei*, *Ameletoides* sp.; Leptophlebiidae: *Atalophlebioides* 5 spp., *Atalonella* 4 spp., *A. australis*, *Atalophlebia* (nr) *longicaudata*, + *Atalophlebia* 5 spp., *Jappa* 2 spp., *Kirrara* sp.;
Caenidae: *Tasmanocoenis* sp.; Baetidae: *Baetis* 8 spp., *Cloeon* sp., *Centroptilum* sp.
Plecoptera: Eustheniidae: *Stenoperla australis*, *Eusthenia venosa*; Notonemouridae: *Austrocerca tasmanica*, *Austrocercella/Austrocercoides* spp., *Notonemoura lynchi*; Austroperlidae: *Acruroperla atra*, *Austropentura victoria*, *Austroheptura neboissi*, *A. picta*; Gripopterygidae: *Eunotoperla kershawi*, *Illiesoperla australis*, *Trinotoperla irrorata*, *Neboissoperla alpina*, *Neumanoperla thoreyi*, *Leptoperla neboissi*, *L. primitiva*, *L. kimminsi*, *Riekoperla tuberculata*, *R. rugosa*, *R. karki-reticulata* gp., *Dinotoperla christinae*, *D. arenaria*, *D. fontana*, *D. serricauda*.
Trichoptera: Leptoceridae: 11 species; Conoesucidae: 9 species; Helicophidae: 2 species;
Kokiridae: 1 species; Philorheithridae: 6 species; Helicopsychidae: 1 species; Calamoceratidae: 1 species; Limnophilidae: 2 species; Atriplectidae: 1 species; Hydroptilidae: 5 species;
Rhyacophilidae: *Taschorema rugulum*, + *Taschorema* 2 spp., *Austrochorema* 2 spp., *Apsilochorema obliquum*, *Psyllobettina cumberlandica*, *P. altunga*, *Ulmerochorema seona*, *U. rubiconum*, *U. onychion*;
Glossosomatidae: 2 species; Ecnomidae: *Ecnomus* sp., *E.* (nr.) *continentalis*, *Ecnomina* sp.;
Polycentropidae: 3 species; Philopotamidae: 1 species; Hydropsychidae: 4 species.

Diptera: Simulidae: *Austrosimulium (Austrosimulium) montanum, Austrosimulium (Novaustrosimulium) victoriae, A. (N.) furiosum, Simulium (Eusimulium) ornatipes, Cnephia aurantiacum*; Blephariceridae: *Edwardsina alticola, E. williamsi, E. affinis, E. polymorpha, Austrocnephia nicholsoni, Apistomyia tonnoiri*; Stratiomyidae: 2 species; Thaumaleidae: 3 species; Muscidae: 1 species; Tabanidae: 1 species; Culicidae: 1 species; Ephydridae: 1 species; Psychodidae: 1 species; Dixidae: 2 species; Rhagionidae: 1 species; Empididae: 6 species; Ceratopogonidae: *Nilobezzia sp.*, *Alludomyia sp., Dasheela sp., Atrichopogon sp., Bezzia sp.*; Tipulidae: 16 species; Chironomidae: 12 species + *Rheocricotopus sp., Sternochironomus sp., Cladotanytarsus sp., Stempellina 2 spp.*, *Ablabesmyia (?) notabilis*, nr *Borynoneura sp., Cricotopus sp., Polypedilum sp., Calopsectra sp.*, *Paraheptagyia sp., Macropelopia sp., Cricotopus sp., Ablabesmyia sp., Procladius 5 spp., Diamesa* 6 spp., *Symbiocladius sp.*

Coleoptera: (L = larvae; A = adults): Elmidae: *Kingolus 2 spp. (L), K. yarrensis (L), Simsonia sp. (L), S. nicholsoni (A), S. hopsoni (A), S. tasmanica (L), Notriolus maculatus (L)*, *N. quadraplegiatus (L), N. victoriae (L), N. allynensis (L), Notriolus sp. (L), Austrolimnius 3 spp. (L), Coxelmis novemnotata (L), Coxelmis sp. (L)*; Psephenidae: *Sclerocyphon maculatus, S. striatus*; Gyrinidae: *Macrogyrus oblongus apacior, M. australis, Aulonogyrus strigosus*; Haliplidae: 1 species; Ptilodactylidae: *Byrrocryptus sp.*; Hydraenidae: *Hydraena luridipennis (A), Ochthebius lividis (A)*, *O. (nr) clypeatus (A), hydraenid (L)*; Hydrochidae: *Hydrochus victoriae, Hydrochus 2 spp.*; Hydrophilidae: *Helochares australis, Paracymus pygmaeus, Nothydrus australis, Enochrus elongatus*, *Enochrus 1 sp., Chaetarthia australis, Limnoxenus zealandicus, Berosus involutus, B. (nr) flindersi*, *Hydrobius 1.sp.*; Dytiscidae: *Sternopriscus mundanus (A), S. multimaculatus (A), Australphilus saltus (A), Neterosoma penicillata (A), Liodessus amablis (A), Chostonectes gigas, Bidessus bistrigatus (A)*, *Megaporous hamatus (A), Platynectes decempunctatus (A), Rhantus suturalis (A), Antiporous blakei (A)*, *Carabhydrous sp. (A), Neterosoma sp. (L), Platynectes 2 spp. (L.), Rhantus 2 spp. (L), Lancestes (nr) lanceolatus (L), Laccophilus sp. (L)*; Helodidae: 3 species; *Cyphon sp. (A)*.

Odonata: Aeshnidae: *Austroaeschna 3 spp.*; Gomphidae: *Austrogomphus guerini*; Libellulidae: *Austrothemis nigrescens*; Coenagrionidae: *Ischnura heterosticta*; Chlorestidae: 1 species.

Megaloptera: Corydalidae: Archichauliodes 2 spp.

Neuroptera: Neurorthridae: *Austroneurorthus sp.*; Osmylidae: *Kempynus sp.*

Hemiptera: Velliidae: *Microvelia oceanica, M. distincta, M. peramoena, M. fluvialis fluvialis*; Mesoveliidae: *Mesovelia hungerfordi*; Naucoridae: *Naucoris (?) australicus*; Belostomatidae: *Diplonychus eques*; Gerridae: *Rheumatometra philarete, Tenagogerris euphrosyne, Gerris (Aquarius) antigone*; Notonectidae: *Anisops thienemanni, A. deanei, A. gratus, Anisops spp., Enithares bergrothi*; Corixidae: *Sigara (Tropocorixa) truncatipala, S. (T.) sublaevifrons, Sigara sp.*, *Agraptocorixa eurynome, A. parvipunctata, Micronecta robusta, M. annae annae, M. batilla.*

   Much has been written on the system, but with the exception of
three major groups of studies, rigorous biological investigations
designed to provide functional understanding and authoritative
descriptions are lacking. There is, however, no shortage of
material which stresses how interesting, distinctive, economically
valuable and polluted the system is, and what prophylactic
conservation measures are needed! Much of the biological literature
is repetitive, but useful background material on geomorphology,
hydrology, etc. is given by Frith and Sawer (1974) and Lawrence
and Smith (1975). These list many references. Special note
should be made of the proceedings of a Symposium on the Murray-
Darling system sponsored by the Royal Society of Victoria (Warren,
1978), and of several papers therein (especially by Smith and
Cadwallader).

   The three groups of studies which provide substantive
ecological information on the system are: that sponsored by the
cities Commission and the Albury–Wodonga Development Corporation
(Gutteridge et al., 1974; Croome et al., 1976; Walker and
Hillman, 1977; Walker et al., 1979; Walker, 1980); that by
R.J. Shiel (Koste, 1979; Shiel, 1978, 1979, 1981); and that by
Walker and students (Jones, 1978; Jones and Walker, 1979; Vickery,
1978; Millington, 1980; Walker, 1981). Walker (1979) has already
summarized the major results, so that remarks here need only be
brief.

   The first group of studies was primarily concerned with the
upper part of the River Murray and its tributaries in the
vicinity of the cities of Albury and Wodonga. Its basic objective
was to provide management advice to government agencies, and, as
such, emphasis was given to algal populations and nutrient
concentrations.

   The second group of studies was of potamoplankton (but mainly
the zooplankton) of the entire system and associated impoundments.
It demonstrated that the zooplankton of the Murray is essentially
a lacustrine one dominated by copepods and cladocerans,
reflecting its extensive impoundment and regulation, whereas the
zooplankton of the Darling, a river not yet significantly
impounded, is essentially a riverine one dominated by rotifers.
After the confluence of the Darling and the Murray, the zooplankton
becomes more complex with increased diversity and density.
Seasonal pulses occur beyond the confluence, and floods of the
Darling River produce rapid changes in the composition of the
zooplankton downstream; Figure 3 (from Shiel, 1981) indicates the
nature of such changes during the period July 1976 - August 1979.
Figure 4 illustrates a few of the more important zooplankters
recorded by Shiel (1978, 1979, 1981). Because of the size of the
river system and its relatively slow flow, Shiel (1981) argued
that much of its zooplankton can be regarded as true potamoplankton.

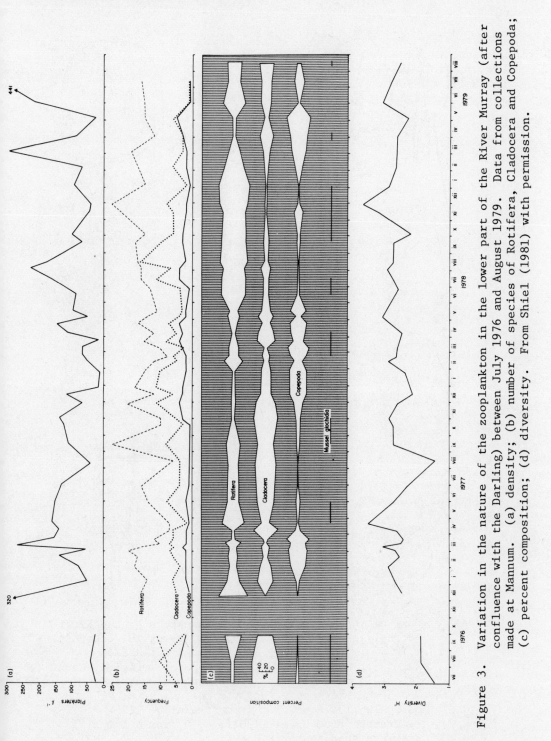

Figure 3. Variation in the nature of the zooplankton in the lower part of the River Murray (after confluence with the Darling) between July 1976 and August 1979. Data from collections made at Mannum. (a) density; (b) number of species of Rotifera, Cladocera and Copepoda; (c) percent composition; (d) diversity. From Shiel (1981) with permission.

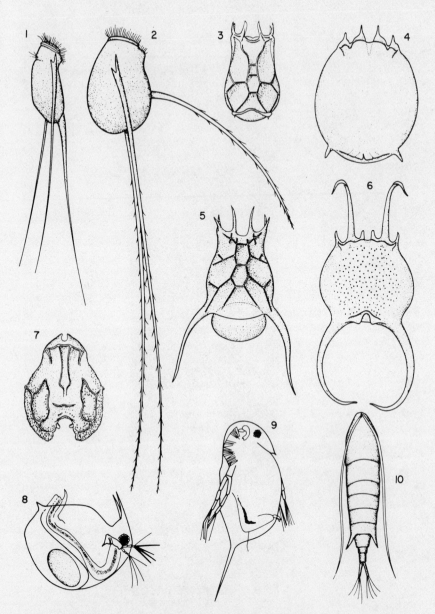

Figure 4.   Some of the more important zooplankters of the River
Murray.   1-7, Rotifera; 8,9, Cladocera; 10, Copepoda.
1, *Filinia pejleri*; 2, *Filinia australiensis*; 3,
Keratella shieli; 4, *Brachionus calcifloris*; 5,
Keratella australis; 6, *Brachionus falcatus*; 7,
Brachionus keikoa; .8, *Bosmina meridionalis*; 9, *Daphnia
carinata*; 10, *Boeckella triarticulata*.   1-7,
rearranged from Shiel (1981); 8-10 from original
drawings supplied by R.J. Shiel.

The third group of studies was of freshwater mussels. These studies concerned the potential of mussels as biomonitors (Jones, 1978; Millington, 1980), and the population ecology of River Murray species (Walker, 1981). The biomonitor studies identified several major problems yet to be overcome (e.g. unexplained intra-population variation in pollutant loads). Ecological studies showed that the two Murray species have distinctive attributes. *Alathyria jacksoni* is a true river species, favouring mainstream environments of moderate to strong flow, whereas *V. ambiguus* is a floodplain species, favouring slow or still waters. These differences are reflected in other ecological and physiological characteristics of the two species. It appears that impoundment of the Murray, largely accomplished between 1920-1940, displaced *A. jacksoni* from some areas and allowed the establishment of *V. ambiguus* in former river environments.

The extent to which the ecological character of the system reflects its distinctive hydrological character has yet to be determined fully. However, the low species diversity of the fish fauna, the strong development of opportunistic breeders, the occurrence of many zooplankters apparently endemic to the system, and the marked distinction between the communities of the river and floodplain environments appear to be some biological features involved.

ALLOCHTHONOUS ORGANIC INPUT

The nature of allochthonous organic input to forest streams in Australia is another distinctive characteristic of Australian running waters. In the northern hemisphere, allochthonous plant litter is an important source of energy for temperate forest streams, and, because north temperate forests are dominated by deciduous trees, is a precisely-timed phenomenon: Large amounts of organic energy arrive during a short autumnal period. A different situation prevails in Australian forests; trees are not autumnal shedders (nor indeed 'deciduous' in the northern sense of that word), but lose most material in summer and over a longer and less precisely-timed period.

The extent and effect of the difference so far as stream ecology is concerned have yet to be determined, but speculation has already begun. Williams and Wan (1972), for example, on the basis largely of work by H.B.N. Hynes on the life-cycles of Australian stoneflies (unpublished at that time), suggested that Australian stream communities may be less diverse and have less precisely-timed life-cycles than those of northern hemisphere streams. This suggestion was not invoked by Hynes and Hynes (1975) in their own explanation of why Australian stoneflies had more flexible life-cycles than their northern hemisphere counterparts: They believed that an important factor is the uncertain Australian climate. No

further insight is shed on this matter by a recent examination of
the life-cycles of two species of amphipods living in Australian
streams (Smith and Williams, 1981), nor by Suter's investigation of
the life-cycles of some South Australian mayflies (the latter, too,
have flexible life-cycles). The effects of the differences in the
timing of allochthonous energy inputs to Australian streams
remains a fruitful area for exploration.

Whatever the effects of timing differences, there are also
differences in the composition of the material (Barmuta, 1978;
Blackburn, 1976; Blackburn and Petr, 1979). The results of
Blackburn and Petr (1979), in particular, suggest that the
contribution of branch and bark litter is far more significant in
Australian temperate streams than in northern temperate streams,
where leaf litter is more significant. Their results suggest,
however, that the contribution of total biomass and energy of
Australian plant litter is comparable to that elsewhere.

THE IMPACT OF MAN

Man has had considerable impact upon running waters worldwide.
Such impact would require no discussion in the present context were
it not that it is especially obvious in Australia. Two reasons
can be advanced for this. Firstly, because of the large
variability in discharge of Australian rivers and streams, there
has been a need to impound flowing waters to a greater extent than
average discharge values would suggest. Already, some 340 large
dams and storages exist or are under construction. Secondly, many
aquatic plants and animals have been introduced and become
extremely successful over a relatively short period. The effects
of impoundments and introductions are of course supplementary to
that variety of other deleterious effects occasioned by man's use
of rivers. Thus, despite the large area of Australia and its
relatively small population, man's impact has been profound; few
permanent rivers survive in natural condition, and most of those
that do are under threat. As Lake (1980) trenchantly pointed out,
the conservation of rivers in Australia has been almost totally
ignored.

There has been no shortage of comment on the ecological effects
of stream impoundment in Australia, but until recently this has
mostly been speculative. That is not to say that early comment has
proven unreliable, but merely that it needed to rely more on work
from other continents than on the local product. That situation has
altered, and several studies now unequivocally demonstrate the
ecological impact of impoundment on Australian running waters.
Particular mention is made of the work of Cadwallader and Rogan
(1977), Davis (1977), Walker and Hillman (1977), Bishop and Bell
(1978), Shiel (1978, 1979, 1981), Yule (1978), Bishop (1979),

Walker et al. (1979), Blyth (1980) and Walker (1980). Some important, recent but more general comment has been made by Fraser (1972), Cadwallader (1978), and Lake (1980). All that need be said of this work is that it amply confirms early observations and predictions: Impoundment in Australia has indeed had profound ecological effects – an environment formerly characterised by wide variations in flow regime and the presence of ecological opportunists is now increasingly subject to flow regulation. Most of the documented adverse effects of impoundment in other continents (cf. Hynes, 1970; Ridley and Steel, 1975; Ward and Stanford, 1979) can now, unfortunately, be supported by Australian examples. It is fortunate that one example not provided is of an increase in the incidence of tropical water-borne human diseases. The potentiality for these to develop exists, however, as Stanley (1972) clearly spelt out. The Ord River impoundment (Lake Argyle) in northern Australia is considered vulnerable in this respect (Petr, 1980).

Compounding of the effects is now underway. *Lates calcarifer*, the giant perch of northern Australia, is a fish which lives in fresh waters for most of its life, but which breeds in estuaries. With increasing impoundment in northern Australia its populations are gradually declining since of course its breeding cycle is interrupted. This is of some importance since it is the only fresh-water fish in the area of both commercial and recreational significance. There are now persistent and powerfully supported proposals to introduce *Lates niloticus*, the nile perch, in order to offset the decline in the abundance of the giant perch. The nile perch breeds in fresh waters. There are strong ecological arguments against this (Williams, 1980d).

With regard to the impact of aquatic introductions, exotic fish and macrophytes are the main elements involved; few invertebrates other than snails and vertebrates other than fish have been introduced, and apart from the cane toad, *Bufo marinus*, none has yet had a significant impact. The cane toad has been discussed recently by Tyler (1980).

Until a few years ago, there was strong opinion that introduced fish had had little impact upon native fish (Weatherley and Lake, 1967; Lake, 1971). The argument suggested that any decline in the distribution and abundance of native fish was as attributable to physical environmental changes as to interaction with introduced fish. It was an argument founded largely on circumstantial evidence. There is now firmer evidence showing that in many cases a clear negative impact has been exerted upon some native fish by some introduced fish. Most of this evidence relates to the brown trout, *Salmo trutta*, introduced in 1864, and its effects on native galaxiids (Frankenberg, 1966; Tilzey, 1976; Cadwallader, 1979; Fletcher, 1979; and Jackson and Williams, 1980). Only Fletcher

(1979) has investigated the impact of brown trout on aquatic invertebrates. The carp, *Cyprinus carpio*, and the mosquito fish, *Gambusia affinis*, are two other introduced species for which there is mounting evidence of an impact in Australia (Myers, 1965; Wharton, 1971; Shearer and Mulley, 1978). Whatever the extent of the impact of introduced fish, many of them are now widespread and abundant (Tilzey, 1980; Mitchell, 1979). *Gambusia* has spread even into the Lake Eyre drainage basin (Glover and Sims, 1978). Proposals to introduce yet more (*Lates niloticus*) and the current importation of more than thirteen million aquarium fish per annum present additional threats (Senate Standing Committee on National Resources, 1979).

A variety of macrophytes has been introduced and many have become well-established in Australian running waters. Several have attained the status of pest species and are a considerable nuisance (Mitchell, 1978, 1980).

Quite why introduced species have prospered in Australian running waters remains a question to be answered fully. The changed nature of the environment is undoubtedly involved in the explanation in some cases (Weir, 1977), but invocations of 'competitive superiority' do no more than disguise inadequate knowledge of mechanisms. Only Tilzey (1977) has attempted to answer the question so far as trout are concerned. The sensitivity of Australian ecologists to the changing balance between introduced and indigenous aquatic species (Walker, 1979) has yet to be expressed in terms other than descriptive. Perhaps this again expresses the current conceptual divergence between most 'limnologists' and 'ecologists'. Increasingly the *leit-motiv* of many limnologists has become 'water quality', a subject less likely to interest ecologists than previous limnological themes.

ACKNOWLEDGMENTS

The criticism and comments of Dr. K.P. Walker, University of Adelaide, and Dr. Bryan Davies, Rhodes University, South Africa, are gratefully acknowledged; a great many improvements were suggested. Dr. Davies, a welcome visitor to the University of Adelaide at the time this chapter was prepared, is also acknowledged as having presented a target audience; kept in mind were the sorts of most likely questions a visiting stream ecologist would ask of a resident limnologist. No gross conflict exists between the nature of such questions and the original objectives of the chapter (though I acknowledge that not all of Dr. Davies' questions have been answered to his satisfaction!).

REFERENCES

Aldenhoven, J.A.   1975.   A comparative study of the stream fauna
    above and below a landslide lake in the Otway Ranges. B.Sc.
    Honours Thesis, Monash Univ., Melbourne.
Barmuta, L.A.   1978.   The decomposition of *Eucalyptus obliqua* leaf
    packs in small stream ecosystems. B.Sc. Honours Thesis, Univ.
    Adelaide, Adelaide.
Bayly, I.A.E. and Williams, W.D.   1973.   Inland Waters and their
    Ecology. Longman, Melbourne.
Benoit, R.J.   1969.   Geochemistry of eutrophication. *In*
    Eutrophication: Causes, Consequences, Correctives. Nat.
    Acad. Sci., Washington, D.C.
Bishop, K.A.   1979.   Fish and aquatic macroinvertebrate communities
    of a coastal river (Shoalhaven River, New South Wales) during
    the development of a water diversion scheme. M.Sc. Thesis,
    Macquarie Univ., Sydney.
Bishop, K.A. and Bell, J.D.   1978.   Observations on the fish fauna
    below Tallowa Dan (Shoalhaven River, New South Wales) during
    river flow stoppages. *Aust. J. Mar. Freshwat. Res.* 29:
    543-549.
Bishop, K.A. and Tilzey, R.D.J.   1978.   An investigation into the
    impact of the Welcome Reef impoundment upon the aquatic
    environment. *Snowy Mountains Hydro-Electric Authority, Cooma.*
Blackburn, W.M.   1976.   A study of forest litter in relation to a
    mountain stream ecosystem in Victoria. B.Sc. Honours Thesis,
    Monash Univ., Melbourne.
Blackburn, W.M. and Petr, T.   1979.   Forest litter decomposition
    and benthos in a mountain stream in Victoria, Australia.
    Arch. für Hydrobiol. 86: 453-498.
Blyth, J.D.   1980.   Environmental impact of reservoir construction:
    the Dartmouth Dam invertebrate survey: a case history.
    In An Ecological Basis for Water Resource Management.
    Williams, W.D., ed. Aust. Nat. Univ. Press, Canberra.
Buckney, R.T.   1974.   Chemical studies on inland waters in
    Tasmania. Ph.D. Thesis, Univ. Tasmania, Hobart.
Buckney, R.T.   1977.   Chemical dynamics in a Tasmanian river.
    Aust. J. Mar. Freshwat. Res. 28: 261-268.
Buckney, R.T.   1979.   Chemical loadings in a small river, with
    observations on the role of suspended matter in the nutrient
    flux. *Aust. Wat. Resources Tech. Pap.*, No. 40.
Buckney, R.T.   1980.   Chemistry of Australian waters: the basic
    pattern, with comments on some ecological implications. *In*
    An Ecological Basis for Water Resource Management. Williams,
    W.D., ed. Aust. Nat. Univ. Press, Canberra.
Buckney, R.T. and Tyler, P.A.   1973.   Chemistry of Tasmanian
    inland waters. *Internat. Revue ges. Hydrobiol.* 58: 61-78.
Cadwallader, P.L.   1978.   Some causes of the decline in range and
    abundance of native fish in the Murray-Darling River system.
    Proc. Roy. Soc. Victoria   90: 211-224.

Cadwallader, P.S. 1979. Native and introduced fishes of the Seven
    Creeks (Goulburn River System), Victoria. *Aust. J. Ecol.*
    4: 361-385.
Cadwallader, P.L. and Eden, A.K. 1977. Effect of a total solar
    eclipse on invertebrate drift in Snobs Creek, Victoria.
    Aust. J. Mar. Freshwat. Res. 28: 799-805.
Cadwallader, P.L. and Rogan, P.L. 1977. The Macquarie perch,
    Macquaria australasica (Pisces: Percichthyidae), of Lake Eildon,
    Victoria. *Aust. J. Ecol.* 2: 409-418.
Campbell, I.C. 1978. Inputs and putputs of water and phosphorus
    from Victorian catchments. *Aust. J. Mar. Freshwat. Res.* 29:
    577-784.
Campbell, I.C. 1978. A biological investigation of an organically
    polluted urban stream in Victoria. *Aust. J. Mar. Freshwat. Res.*
    29: 275-293.
Chessman, B.C. 1978. Ecological studies of freshwater turtles in
    Australia. Ph.D. Thesis, Monash Univ., Melbourne.
Croome, R.L., Tyler, P.A., Walker, K.F. and Williams, W.D. 1976.
    A limnological survey of the River Murray in the Albury-
    Wodonga area. *Search* 7 (1-2): 14-17.
Davis, T.L.O. 1977. Food habits of the freshwater catfish,
    Tandanus tandanus Mitchell, in the Gwydir River, Australia,
    and effects associated with impoundment of this river by the
    Copeton Dam. *Aust. J. Mar. Freshwat. Res.* 28: 455-465.
Department of National Resources, Australian Water Resources
    Council. 1976. Review of Australian Water Resources.
    Aust. Government Publ. Serv., Canberra.
Department of National Resources, Australian Water Resources Council
    1978a. Variability of Runoff in Australia. AWRC Hydrological
    Series, No. 11 Aust. Government Publ. Serv., Canberra.
Department of National Resources, Australian Water Resources Council
    1978b. Stream Gauging Information. Australia - Fourth Edition.
    Aust. Government Publ. Serv., Canberra.
Douglas, I. 1966. Denudation rates and water chemistry of selected
    catchments in eastern Australia and their significance for
    tropical geomorphology. Ph.D. Thesis, Aust. Nat. Univ., Canberra.
Douglas, I. 1968. The effects of precipitation chemistry and
    catchment area lithology on the quality of river water in
    selected catchments in eastern Australia. *Earth Sci. J.* 2:
    126-144.
Douglas, I. 1972. The geographical interpretation of river water
    quality data. *Prog. in Geography* 4: 1-81.
Douglas, I. 1973. Rates of denudation in selected small
    catchments in eastern Australia. *Uni. Hull, Occasional Pap.
    in Geography* No. 21: 1-127.
Fletcher, A.R. 1979. Effects of *Salmo trutta* on *Galaxias olidus*
    and macroinvertebrates in stream communities. M.Sc. Thesis,
    Monash Univ., Melbourne.
Frankerberg, R. 1966. Fishes of the family Galaxiidae. *Aust.
    Nat. Hist.* 15: 161-164.

Fraser, J.C. 1972. Regulated discharge and the stream environment. *In* River Ecology and Man. Ogelsby, R.T., Carlson, C.A. and McCann, J.A., eds. Academic Press, New York and London.

Frith, H.J. and Sawer, G. eds. 1974. Man, Nature and a River System. Angus and Robertson, Sydney.

Gibbs, R.J. 1970. Mechanisms controlling world water chemistry. *Science* 1970: 1088-1090.

Glover, C.J.M. and Sim, R.C. 1978. Studies on central Australian fishes: a progress report. *South Aust. Nat.* 52: 35-44.

Graham, W.A.E., Paton, G.J. and Waite, T.D. 1978. Report on water quality of the Mitta Mitta River catchment. *In* Dartmouth Dam Project: Report on Environmental Studies. State Rivers and Water Supply Commission, Melbourne.

Gutteridge, Haskins and Davey Pty. Ltd. 1974. River Murray in Relation to Albury-Wondonga. Cities Commission, Canberra.

Holmes, A.N., Williams, W.D. and Wood, G. 1980. Relationships between forms of nitrogen and hydrological characteristics in a small stream near Adelaide, South Australia. *Aust. J. Mar. Freshwat. Res.* 31: 297-317.

Hynes, H.B.N. 1970. The Ecology of Running Waters. Liverpool Univ. Press, Liverpool.

Hynes, H.B.N. 1974a. Comments on the taxonomy of Australian Austroperlidae and Gripopterygidae (Plecoptera). *Aust. J. Zool., Suppl.* 29: 1-36.

Hynes, H.B.N. 1974b. Observations on the adults and eggs of Australian Plecoptera. *Aust. J. Zool. Suppl.* 29: 37-52.

Hynes, H.B.N. 1976a. Tasmanian Antarctoperlaria (Plecoptera). *Aust. J. Zool.* 24: 115-143.

Hynes, H.B.N. 1976b. *Symbiocladius aurifodinae* sp. nov. (Diptera, Chironomidae), a parasite of nymphs of Australian Leptophlebiidae (Ephemeroptera). *Mem. Nat. Museum Victoria* 37: 47-52.

Hynes, H.B.N. 1978. An annotated key to the nymphs of the stone-flies (Plecoptera) of the State of Victoria. *Aust. Soc. Limnol. Spec. Publ.* No. 2.

Hynes, H.B.N. and Hynes, M.E. 1975. The life histories of many of the stoneflies (Plecoptera) of south-eastern mainland Australia. *Aust. J. Mar. Freshwat. Res.* 26: 113-153.

Hynes, H.B.N. and Hynes, M.E. 1980. The endemism of Tasmanian stoneflies. *Aquatic Insects* 2(2): 81-89.

Jackson, P.D. 1978. Benthic invertebrate fauna and feeding relationships of brown trout, *Salmo trutta* Linnaeus, and river blackfish, *Gadopsis marmoratus* Richardson, in the Aberfeldy River, Victoria. *Aust. J. Mar. Freshwat. Res.* 29: 725-742.

Jackson, P.D. and Williams, W.D. 1980. Effects of brown trout, *Salmo trutta* L., on the distribution of some native fishes in three areas of southern Victoria. *Aust. J. Mar. Freshwat. Res.* 31: 61-67.

Johnson, W.D. and Muir, G.L. 1977. Chemistry of the Castlereagh River, New South Wales. *Aust. J. Mar. Freshwat. Res.* 28: 683-692.

Jolly, V.H. and Chapman, M.A. 1966. A preliminary biological study of the effects of pollution on Farmer's Creek and Cox's River. New South Wales. *Hydrobiologia* 27: 160-192.

Jones, W.G. 1978. The freshwater mussel, *Velesunio ambiguus* (Philippi), as a biological monitor of heavy metals. M.Env.St. Thesis, Univ. Adelaide, Adelaide.

Jones, W.G. and Walker, K.F. 1979. Accumulation of iron, manganese, zinc and cadmium by the Australian freshwater mussel *Velesunio ambiguus* (Philippi) and its potential as a biological monitor. *Aust. J. Mar. Freshwat. Res.* 30: 74-751.

Kendrick, G.W. 1976. The Avon: faunal and other notes on a dying river in south-western Australia. *Western Aust. Nat.* 13(5): 97-114.

Koste, W. 1979. New Rotifera from the River Murray, south-eastern Australia, with a review of the Australian species of *Brachionus* and *Keratella*. *Aust. J. Mar. Freshwat. Res.* 30: 237-253.

Lake, J.S. 1967. Principal fishes of the Murray-Darling River system. *In* Australian Inland Waters and Their Fauna: Eleven Studies. Weatherley, A.H., ed. Aust. Nat. Univ. Press, Canberra.

Lake, J.S. 1971. Freshwater Fishes and Rivers of Australia. Nelson, Sydney.

Lake, P.S. 1980. Conservation. *In* An Ecological Basis for Water Resource Management Williams, W.D., ed. Aust. Nat. Univ. Press, Canberra.

Lawrence, G.V. and Smith, G.K. eds 1975. The Book of the Murray. Rigby, Adelaide.

Lewis, A.D. and Ellway, C.P. 1971. Fishes of Tallebudgera Creek, South Queensland. *Operculum* 1(3): 60-63.

Livingstone, D.A. 1963. Chemical composition of rivers and lakes. Chapter 9. Data of Geochemistry. 6th Edition. *Professional Pap., U.S. Geol. Surv.* 440-449.

Macmillan, L.A. 1975. Longitudinal zonation of benthic invertebrates in the Acheron River, Victoria. B.Sc. Honours Thesis, Monash Univ., Melbourne.

Marchant, R. 1980. The macroinvertebrates of Magela Creek, Northern Territory. *Report for the Office of the Supervising Scientist, August 1980, Adelaide.*

McIvor, C.C. 1976. The effects of organic and nutrient enrichment on the benthic macroinvertebrate community of Moggill Creek, Queensland. *Water* 3(4): 16-21.

Metzeling, L.H. 1977. An investigation of the distribution of aquatic macroinvertebrates found in sections of streams flowing through areas with differing amounts of vegetation cover. B.Sc. Honours Thesis, Monash Univ., Melbourne.

Millington, P.J. 1980. The Australian freshwater mussel *Velesunio ambiguus* (Philippi) as a biological monitor, with regard for zinc. M.Env.St. Thesis, Univ. Adelaide, Adelaide.

Mitchell, B.D. 1979. Aspects of growth and feeding in golden
    carp, *Carassius auratus*, from South Australia. *Trans. Roy.
    Soc. South Aust.* 103: 137-144.
Mitchell, D.S. 1978. Aquatic Weeds in Australian Inland Waters.
    Aust. Governmental Publ. Serv., Canberra.
Mitchell, D.S. 1980. Aquatic weeds. *In* An Ecological Basis for
    Water Resource Management Williams, W.D. ed. Aust. Nat.
    Univ. Press, Canberra.
Morrissy, N.M. 1967. The ecology of trout in South Australia.
    Ph.D. Thesis, Univ. Adelaide, Adelaide.
Morrissy, N.M. 1974. Reversed longitudinal salinity profile of a
    major river in the south-west of Western Australia. *Aust.
    J. Mar. Freshwat. Res.* 25: 327-335.
Morrissy, N.M. 1979. Inland (non-estuarine) halocline formation
    in a Western Australian river. *Aust. J. Mar. Freshwat. Res.*
    30: 343-353.
Myers, G.C. 1965. *Gambusia*, the fish destroyer. *Aust. Zool.*
    13: 102.
Norris, R.H. 1978. The ecological effects of mine effluent on
    the South Esk River (north east Tasmania). Ph.D. Thesis,
    Univ. Tasmania, Hobart.
Norris, R.H. 1979. Response of the benthic macroinvertebrates of
    the Lower Latrobe River to heated water discharge from
    Yallourn power station. Interim Report to the State
    Electricity Commission of Victoria, Melbourne.
Norris, R.H., Lake, P.S. and Swain, R. 1980. Ecological effects
    of mine effluents on the South Esk River, north-eastern
    Tasmania. I. Study area and basic water characteristics.
    Aust. J. Mar. Freshwat. Res. 31: 817-827.
Petr, T. 1980. Medically important diseases with aquatic vectors
    and hosts. *In* An Ecological Basis for Water Resource
    Management Williams, W.D., ed. Aust. Nat. Univ. Press,
    Canberra.
Pollard, D.A. 1974. The freshwater fishes of the Alligator Rivers
    'Uranium Province' area (Top End, Northern Territory), with
    particular reference to the Magela Creek Catchment (East
    Alligator River System). *Aust. Atomic Energy Comm. Rep. E305.*
Pollard, D.A., Llewellyn. L.C. and Tilzey. R.D.J. 1980. Management
    of freshwater fish and fisheries. *In* An Ecological Basis for
    Water Resource Management Williams, W.D., Ed. Aust. Nat.
    Univ. Press, Canberra.
Ridley, J.E. and Steel, J.A. 1975. Ecological aspects of river
    impoundments. *In* River Ecology Whitton, B.A., ed. Blackwell
    Scientific Publications, Oxford.
Rippingale, R.J. 1981. The ecology of plankton fauna in saline
    river pools. *In* Salt Lakes: Proceedings of an International
    Symposium on Athalassic (Inland) Salt Lakes Williams, W.D.
    ed. Junk, The Hague.
Rixon, A.J. ed. (in press) Proceedings of the Salinity and Water
    Quality Symposium. Darling Downs Institute of Advanced
    Education, Toowoomba, Queensland.

Robinson, D.P.   1977.   The study of the invertebrate fauna of
    three mountain streams in relation to the type of land use in
    each of the three stream catchments. B.Sc. Honours Thesis,
    Monash Univ., Melbourne.

Ross, C.W. and Connell, D.W.   1979.   Some spatial and temporal
    factors affecting water quality in the Condamine River.
    Search   10(3): 88-90.

Senate Select Committee. 1970. Water Pollution in Australia.
    Commonwealth Government Printing Office, Canberra.

Senate Standing Committee on National Resources. 1979. Report
    on the Adequacy of Quarantine. Aust. Government Publ. Serv.
    Canberra.

Shearer, K.D. and Mulley, J.C.   1978.   The introduction and
    distribution of the carp, *Cyprinus carpio* Linnaeus, in
    Australia. *Aust. J. Mar. Freshwat. Res.*   29: 551-563.

Shiel, R.J.   1978.   Zooplankton communities of the Murray-Darling
    system. *Proc. Roy. Soc. Victoria*   90: 193-202.

Shiel, R.J.   1979.   Synecology of the Rotifera of the River Murray,
    South Australia. *Aust. J. Mar. Freshwat. Res.*   30: 255-263.

Shiel, R.J.   1981.   Plankton of the Murray-Darling River system,
    with particular reference to the zooplankton. Ph.D. Thesis,
    Univ. Adelaide, Adelaide.

Smith, B.J.   1978.   Molluscs of the Murray-Darling River system.
    Proc. Roy. Soc. Victoria. 90: 211-224.

Smith, B.J., Malcolm, H.E. and Morison, P.B.   1977.   Aquatic
    invertebrate fauna of the Mitta Mitta Valley, Victoria.
    Victorian Naturalist   94: 228-238.

Smith, B.J., Malcolm, H.E. and Morison, P.B.   1978.   Report on the
    survey of the invertebrate fauna of the Mitta Mitta Valley.
    In Dartmouth Dam Project: Report on Environmental Studies for
    River Murray Commission. State Rivers and Water Supply
    Commission, Melbourne.

Smith, M.J. and Williams, W.D.   1981.   Reproductive strategies in
    some freshwater amphipods in southern Australia. *Aust. Museum
    Records.* (in press).

Stanley, N.F. 1972. Ord River ecology. *Search*   3:(1-2):
    7-12.

Suter, P.J.   1979.   Interim report to the State Electricity
    Commission of Victoria on the qualitative benthic survey of
    the La Trobe River catchment. *Nat. Museum Victoria, Melbourne.*

Suter, P.J.   1980.   The taxonomy and ecology of the Ephemeroptera
    (mayflies) of South Australia. Ph.D. Thesis, Univ.
    Adelaide, Adelaide.

Suter, P.J. and Williams, W.D.   1977.   Effect of a total eclipse on
    stream drift. *Aust. J. Mar. Freshwat. Res.*   28: 793-799.

Talsma, T. and Philip, J.R.   1971.   Salinity and Water Use. Aust.
    Acad. Sci., Canberra.

Thorp, V.J. and Lake, P.S.   1973.   Pollution of a Tasmanian river
    by mine effluents. II. Distribution of macroinvertebrates.
    Internat. Revue ges. Hydrobiol.   58: 885-892.

Tilzey, R.D.J.   1976.   Observations on interactions between indigenous Galaxiidae and introduced Salmonidae in the Lake Eucumbene catchment, New South Wales. *Aust. J. Mar. Freshwat. Res.*   27: 551-564.

Tilzey, R.D.J.   1977.   The key factors in the establishment and success of trout in Australia. *Proc. Ecol. Soc. Aust.* 10: 97-105.

Tilzey, R.D.J.   1980.   Introduced fish. *In* An Ecological Basis for Water Resource Management   Williams, W.D., ed.   Aust. Nat. Univ. Press, Canberra.

Tyler, M.J.   1980.   Introduced amphibians. *In* An Ecological Basis for Water Resource Management   Williams, W.D., ed., Aust. Nat. Univ. Press, Canberra.

Tyler, M.J. and Crook, G.A.   1980.   Frogs of the Magela Creek system, Alligator Rivers Region, Northern Territory, Australia. *Report to the Office of the Supervising Scientist, Alligator Rivers Region., Adelaide.*

Vickery, A.H.   1978.   The response of two freshwater mussel species to changing environmental salinity. B.Sc. Honours Thesis, Univ. Adelaide, Adelaide.

Walker, K.F.   1979.   Regulated streams in Australia: the Murray-Darling River system. *In* Ecology of Regulated Streams   Ward, J.V. and Stanford, J.A. ed.   Plenum Press, New York and London.

Walker, K.F.   1980.   The downstream influence of Lake Hume on the River Murray. *In* An Ecological Basis for Water Resource Management   Williams, W.D., ed.   Aust. Nat. Univ. Press, Canberra.

Walker, K.F.   1981.   Ecology of freshwater mussels in the River Murray. *Aust. Wat. Resources Council Tech. Pap.*   (in press).

Walker, K.F. and Hillman, T.H.   1977.   Limnological survey of the River Murray in relation to Albury-Wodonga, 1973-1976. Albury-Wodonga Development Corporation and Gutteridge, Haskins and Davey, Melbourne.

Walker, K.F., Hillman, T.J. and Williams, W.D.   1979.   The effects of impoundment on rivers: an Australian case study. *Verh. int. Verein. theor. angew. Limnol.*   20: 1695-1701.

Ward, J.V. and Stanford, J.A. eds.   1979.   The Ecology of Regulated Streams.   Plenum Press, New York and London.

Warren, J. ed.   1978.   The Murray-Darling River System Symposium. *Proc. Roy. Soc. Victoria*   90(1): 1-224.

Weatherley, A.H. and Lake, J.S.   1967.   Introduced fish species in Australian inland waters. *In* Australian Inland Waters and Their Fauna: Eleven Studies   Weatherley, A.H. ed.   Aust. Nat. Univ. Press, Canberra.

Weir, J.S.   1977.   Exotics: past, present and future. *Proc. Ecol. Soc. Aust.*   10: 4-14.

Wetzel, R.G.   1975.   Limnology.   Saunders, Philadelphia.

Wharton, J.C.F.   1971.   European carp in Victoria. *Fur. Feathers and Fins*, 130: 3-11.   (Fisheries and Wildlife Division, Victoria.

Whelan, B.R. 1977. Nutrient levels in the Walsh and Barron Rivers
    draining agricultural catchments in north Queensland.
    *Proc. Hydrol. Symp., The Institution of Engineers, Australia,
    Brisbane, June 1977:* 192-193.
Williams, W.D. 1980a. Distinctive features of Australian water
    resources. *In* An Ecological Basis for Water Resource
    Management Williams, W.D., ed. Aust. Nat. Univ. Press,
    Canberra.
Williams, W.D. 1980b. Catchment management. *In* An Ecological
    Basis for Water Resource Management Williams, W.D., ed.
    Aust. Nat. Univ. Press, Canberra.
Williams, W.D. 1980c. Australian Freshwater Life. The
    Invertebrates of Australian Inland Waters. Second edition.
    Macmillan, Melbourne.
Williams, W.D. 1980d. On the proposed introduction of *Lates
    niloticus* (L.) to Australia: some further comments.
    Proc. 15th Assembly Aust. Fresh Water Fishermen, 9-18.
Williams, W.D. and Buckney, R.T. 1976. Chemical composition of
    some inland surface waters in South, Western and northern
    Australia. *Aust. J. Mar. Freshwat. Res.* 27: 379-397.
Williams, W.D. and Wan, H.F. 1972. Some distinctive features of
    Australian inland waters. *Wat. Res.* 6: 829-836.
Wood, G. 1975. An assessment of eutrophication in Australian
    inland waters. *Aust. Wat. Resources Council Tech. Pap.*
    No. 15.
Woodyer, K.D. 1978. Sediment regime of the Darling River.
    Proc. Roy. Soc. Victoria 90: 139-147.
Yule, C.M. 1978. Fauna-Substrate relationships in a Victorian
    coastal stream with consideration of the impact of dam
    construction. B.Sc. Honours Thesis, Monash Univ., Melbourne.

CONTRIBUTORS

J.B.E. Awachie
Hydrobiology/Fishery Research Unit
Department of Zoology
University of Nigeria
Nsukka, Nigeria
West Africa

David R. Barton
Department of Biology
University of Waterloo
Waterloo, Ontario
Canada N2L 3G1

Glen A. Bird
Department of Environmental Biology
University of Guelph
Guelph, Ontario
Canada N1G 2W1

Kenneth W. Dance
Ecologistics Limited
309 Lancaster Street West
Kitchener, Ontario
Canada N2H 4V4

P.P. Harper
Départment de Sciences Biologiques
Université de Montréal
Case Post ale 6128
Montréal 101, Québec
Canada

Narinder K. Kaushik
Department of Environmental Biology
University of Guelph
Guelph, Ontario
Canada N1G 2W1

Maurice A. Lock
School of Animal Biology
University College of North Wales
Bangor, Gwynedd
United Kingdom LL57 2UW

Donald L. Lush
Beak Consultants Limited
6870 Goreway Drive
Mississauga, Ontario
Canada L4V 1L9

Richard Marchant
Survey Department
National Museum of Victoria
71 Victoria Crescent
Abbotsford, Victoria
South Australia 3067

Letitia E. Obeng
United Nations Environment Programme
Regional Office for Africa
P.O. Box 30552
Nairobi, Kenya
East Africa

J.B. Robinson
Department of Environmental Biology
University of Guelph
Guelph, Ontario
Canada N1G 2W1

W.N. Stammers
School of Engineering
University of Guelph
Guelph, Ontario
Canada N1G 2W1

William D. Taylor
Department of Biology
University of Waterloo
Waterloo, Ontario
Canada N2L 3G1

Ronald R. Wallace
Canstar Oil Sands Limited
#805 - 605 Fifth Avenue S.W.
Calgary, Alberta
Canada T2P 3H5

Peter M. Wallis
Kananaskis Centre for Environmental Research
University of Calgary
Calgary, Alberta
Canada T2N 1N4

H.R. Whiteley
School of Engineering
University of Guelph
Guelph, Ontario
Canada N1G 2W1

D. Dudley Williams
Division of Life Sciences
Scarborough College
University of Toronto
West Hill, Ontario
Canada M1C 1A4

Nancy E. Williams
Division of Life Sciences
Scarborough College
University of Toronto
West Hill, Ontario
Canada M1C 1A4

W.D. Williams
Department of Zoology
University of Adelaide
P.O. Box 498
Adelaide
South Australia 5001

LIST OF GRADUATE STUDENTS AND POSTDOCTORAL FELLOWS

SUPERVISED BY DR. H.B.N. HYNES

Awachie, J.B.E.

Barton, D.R.

Bhajan, W.R.

Bilyj, B.

Bird, G.A.

Bishop, J.E.

Coleman, M.J.

Cressa, C.

Crowther, R.A.

Dance, K.W.

Deacon, K.

Dunn, D.R.

Fuller, R.L.

Godbout, L.

Harper, P.P.

Holland, P.D.G.

Jefferies, D.J.

Kakonge, S.A.K.

Kaushik, N.K.

Khoo, S.G.

Lee, D.R.

Lock, M.A.

Lush, D.L.

Marchant, R.

Obeng, L.E.

Prestt, I.

Pugsley, C.W.

Rowe, V.L.

Semple, D.

Spence, J.A.

Stocker, Z.S.J.

Taylor, W.D.

Thomas, M.P.

Wallace, R.R.

Wallis, P.M.

Williams, D.D.

Williams, N.E.

Williams, T.R.

Williams, W.D.

PUBLICATIONS OF DR. H.B.N. HYNES

Hynes, H.B.N. 1940. A key to the British species of Plecoptera
(Stoneflies) with notes on their ecology. *Sci. Pub. Freshw.
Biol. Ass.* 2: 39 p.

Hynes, H.B.N. 1941. The taxonomy and ecology of the nymphs of
British Plecoptera with notes on the adults and eggs. *Trans.
R. Ent. Soc. Lond.* 91: 459-557.

Hynes, H.B.N. 1942. A study of the feeding of adult stoneflies
(Plecoptera). *Proc. R. Ent. Soc. Lond. A* 17: 81-82.

Hynes, H.B.N. 1942. Lepidopterous pests of maize in Trinidad.
Trop. Agric. 19: 194-202.

Hynes, H.B.N. 1946. Control of the Desert Locust in Somalia.
Government Press, Magadishu 15 p.   (This was also published
in Italian)

Hynes, H.B.N. 1947. Use of "Gammexane" dust against domestic
cockroaches. *Nature, Lond.* 159: 200-201.

Hynes, H.B.N. 1947. Observations on *Systoechus somali* (Diptera:
Bombyliidae) attacking the eggs of the desert locust
(*Schistocerca gregaria*) (Forskal) in Somalia. *Proc. R. Ent.
Soc. Lond. A* 22: 79-85.

Hynes, H.B.N. 1948. Notes on the Aquatic Hemiptera – Heteroptera
of Trinidad and Tobago, B.W.I., with a description of a new
species of *Martarega* B. White (Notonectidae). *Trans. R. Ent.
Soc. Lond.* 99: 341-360.

Hynes, H.B.N. 1948. The nymph of *Anacroneuria aroucana* Kimmins
(Plecoptera, Perlidae). *Proc. R. Ent. Soc. Lond. A* 23:
105-110.

Hynes, H.B.N. 1950. The food of fresh-water sticklebacks
(*Gasterosteus aculeatus* and *Pygosteus pungitius*), with a
review of methods used in studies of the food of fishes.
J. Anim. Ecol. 19: 35-58.

Jones, J.W. and Hynes, H.B.N. 1950. The age and growth of
Gasterosteus aculeatus, Pygosteus pungitius and *Spinachia
vulgaris*, as shown by their otoliths. *J. Anim. Ecol.* 19:
59-73.

Hynes, H.B.N. 1950. Preliminary report on freshwater species of
Gammarus in the Isle of Man. *Ann. Rep. Biol. Sta. Port Erin*
62: 24-25.

Hynes, H.B.N.   1951.   Distribution of British freshwater Amphipoda. *Nature, Lond.*   167: 152-153.

Hynes, H.B.N.   1951.   Distribution écologique de *Gammarus* d'eau douce en Grande Bretagne. *Verh. Int. Ver. Limnol.*   11: 210-212.

Hynes, H.B.N.   1952.   The Neoperlinae of the Ethiopian region. (Plecoptera, Perlidae). *Trans. R. Ent. Soc. Lond.*   103: 85-108.

Hynes, H.B.N.   1952.   Perlidae (Plecoptera). *Mission G.F. de Witte. Parc National de l'Upemba. Brussels*   8: 3-12.

Hynes, H.B.N.   1952.   The Plecoptera of the Isle of Man. *Proc. R. Ent. Soc. Lond. A*   27: 71-76.

Hynes, H.B.N.   1953.   The Plecoptera of some small streams near Silkeborg, Jutland. *Ent. Medd.*   26: 289-294.

Hynes, H.B.N.   1953.   The nymph of *Neoperla spio* (Newman) (Plecoptera: Perlidae). *Proc. R. Ent. Soc. Lond. A*   28: 93-99.

Hynes, H.B.N. and Jones, J.W.   1953.   Zoology. A scientific survey of Merseyside. *British Association for the Advancement of Science.*   p. 86-89.

Stephensen, K. and Hynes, H.B.N.   1953.   Notes on some Belgian freshwater and brackish water *Gammarus*. *Vid. Medd. Dansk Naturhist. Foren.*   115: 290-304.

Hynes, H.B.N.   1953.   A comparative study of anti-locust baits with special reference to base materials. *Bull. ent. Res.*   44: 693-702.

Hynes, H.B.N.   1954.   The ecology of *Gammarus duebeni* Lilljeborg and its occurrence in fresh water in Western Britain. *J. Anim. Ecol.*   23: 38-84.

Hynes, H.B.N.   1954.   Identity of *Gammarus tigrinus* Sexton 1939. *Nature Lond.*   174: 563.

Hynes, H.B.N.   1955.   Distribution of some freshwater Amphipoda in Britain. *Verh. Int. Ver. Limnol.*   12: 620-628.

Hynes, H.B.N.   1955.   Biological notes on some East African aquatic Heteroptera. *Proc. R. Ent. Soc. Lond. A*   30: 43-54.

Hynes, H.B.N.   1955.   The nymphs of the British species of *Capnia* (Plecoptera). *Proc. R. Ent. Soc. Lond. A*   30: 91-96.

Hynes, H.B.N.   1955.   The reproductive cycle of some British freshwater Gammaridae. *J. Anim. Ecol.*   24: 352-387.

Hynes, H.B.N.   1955.   A note on the stoneflies of Iceland. *Proc. R. Ent. Soc. Lond. A*   30: 164-166.

Hynes, H.B.N.   1956.   British freshwater shrimps. *New Biology.*   21: 25-42.

Hynes, H.B.N.   1957.   Note sur les *Gammarus* de Suisse. *Rev. Suisse Zool.*   64: 215-217.

Hynes, H.B.N.   1957.   The British Taeniopterygidae (Plecoptera). *Trans. R. Ent. Soc. Lond.*   109: 233-243.

Hynes, H.B.N.   1957.   River pollution - biological effects. *In*: Isaac P.C.F. (ed). The treatment of trade waste waters and the prevention of river pollution. University of Durham. p. 27-36.

Hynes, H.B.N. and Nicholas, W.L. 1957. The development of
    Polymorphus minutus (Goeze, 1782) (Acanthocephala) in the
    intermediate host. *Ann. Trop. Med. Parasitol.* 51: 380-391.
Hynes, H.B.N. 1958. Notes on the freshwater fauna of Bardsey
    Island. *Rep. Bardsey Field Observ.* 1957: 41-42.
Hynes, H.B.N. 1958. A key to the adults and nymphs of British
    Stoneflies (Plecoptera) with notes on their ecology and
    distribution. *Sci. Pub. Freshw. Biol. Ass.* 17: 86 p.
Hynes, H.B.N. 1958. The effect of drought on the fauna of a
    small mountain stream in Wales. *Verh. Int. Ver. Limnol.*
    13: 826-833.
Nicholas, W.L. and Hynes, H.B.N. 1958. Studies on *Polymorphus
    minutus* (Goeze, 1782) (Acanthocephala) as a parasite of the
    domestic duck. *Ann. Trop. Med. Parasitol.* 52: 36-47.
Hynes, H.B.N. and Nicholas, W.L. 1958. The resistance of
    Gammarus spp. to infection by *Polymorphus minutus* (Goeze, 1782)
    (Acanthocephala). *Ann. Trop. Med. Parasitol.* 52: 376-383.
Hynes, H.B.N. 1959. On the occurrence of *Gammarus duebeni*
    Lilljeborg in fresh water and of *Asellus meridianus*
    Racovitza in Western France. *Hydrobiologia* 13: 152-155.
Hynes, H.B.N. 1959. The use of invertebrates as indicators of
    river pollution. *Proc. Linn. Soc. Lond.* 170: 165-170.
Hynes, H.B.N. 1959. The biological effects of water pollution.
    In: Yapp W.B. (ed). The effects of pollution on living
    material. *Symp. Inst. Biol.* 8: 11-24.
Hynes, H.B.N., Macan, T.T. and Williams, W.D. 1960. A key to the
    British species of Crustacea: Malacostraca occurring in fresh
    water. *Sci. Pub. Freshw. Biol. Ass.* 19: 1-35.
Hynes, H.B.N. 1960. A plea for caution in the use of DDT in the
    control of aquatic insects in Africa. *Ann. Trop. Med.
    Parasitol.* 54: 331-332.
Hynes, H.B.N. 1960. The biology of polluted waters. Liverpool
    University Press. 202 p. (reprinted 1962, 1967, 1971, 1976).
Hynes, H.B.N. 1961. Perlidae (Plecoptera). *Mission H. de Saeger.
    Parc National de la Garamba. Brussels* 97: 3-5.
Hynes, H.B.N. 1961. The biological assessment of river pollution.
    Year Book River Boards' Ass. 9: 83-93. (This was also
    published in Italian: Il biologo de fronte al fenomeno di
    polluzione dei corsi d' acqua. *Acqua Industriale.* 1961.
    1-4).
Williams, T.R., Hynes, H.B.N. and Kershaw, W.E. 1961. Maintenance
    and diet of Africa freshwater crabs associated with
    Simulium naevei. *Ann. Soc. Belge. Méd. trop.* 4: 291-292.
Williams, T.R., Hynes, H.B.N. and Kershaw, W.E. 1961. The fauna
    in a Welsh hillstream containing *Simulium*. *Ann. Soc. Belge
    Méd. trop.* 4: 293-298.
Williams, T.R., Connolly, R., Hynes, H.B.N. and Kershaw, W.E. 1961.
    The size of particulate material ingested by *Simulium* larvae.
    Ann. Soc. Belge. Méd. trop. 4: 299-302.

Hynes, H.B.N.   1961.   The effect of water-level fluctuations on
     littoral fauna.   *Verh. Int. Ver. Limnol.*   14: 652-656.
Hynes, H.B.N.   1961.   The effect of sheep-dip containing the
     insecticide BHC on the fauna of a small stream, including
     Simulium and its predators.   *Ann. trop. Med. Parasitol.*
     55: 192-196.
Hynes, H.B.N., Williams, T.R. and Kershaw, W.E.   1961.   Freshwater
     crabs and *Simulium neavei* in East Africa.   I. Preliminary
     observations made on the slopes of Mount Elgon in December,
     1960, and January 1961.   *Ann. trop. Med. Parasitol.*   55:
     197-201.
Williams, T.R., Connolly, R., Hynes, H.B.N. and Kershaw, W.E.   1961.
     Size of particles ingested by *Simulium* larvae.   *Nature, Lond.*
     189: 78.
Hynes, H.B.N.   1961.   The invertebrate fauna of a Welsh mountain
     stream.   *Arch. Hydrobiol.*   57: 344-388.
Hynes, H.B.N. and Williams, T.R.   1961.   The effect of DDT on the
     fauna of a Central African stream.   *Ann. trop. Med. Parasitol.*
     56: 78-91.
Hynes, H.B.N.   1962.   The hatching and growth of the nymphs of
     several species of Plecoptera.   *Int. Congr. Ent.*   11:
     3. 269-273.
Hynes, H.B.N. and Roberts, F.W.   1962.   The biological effects of
     synthetic detergents in the River Lee, Hertfordshire.
     Ann. appl. Biol.   50: 779-790.
Nicholas, W.L. and Hynes, H.B.N.   1963.   Embryology, post
     embryonic development, and phylogeny of the Acanthocephala.
     In: E.C. Dougherty (ed).   The Lower Metazoa.   University of
     California Press.   p. 385-402.
Hynes, H.B.N.   1963.   *Isogenus nubecula* Newman in Britain (Plecoptera:
     Perlodidae).   *Proc. R. Ent. Soc. Lond. A*   38: 12-14.
Hynes, H.B.N.   1963.   The gill-less Nemourid nymphs of Britain
     (Plecoptera).   *Proc. R. Ent. Soc. Lond. A*   38: 70-76.
Hynes, H.B.N. and Nicholas, W.L.   1963.   The importance of the
     Acanthocephalan *Polymorphus minutus* as a parasite of domestic
     ducks in the United Kingdom.   *J. Helminth.*   37: 185-198.
Hynes, H.B.N.   1963.   Imported organic matter and secondary
     productivity in streams.   *Int. Congr. Zool.*   16: 324-329.
Nicholas, W.L. and Hynes, H.B.N.   1963.   The embryology of
     Polymorphus minutus (Acanthocephala).   *Proc. Zool. Soc. Lond.*
     141: 791-801.
Hynes, H.B.N.   1964.   The use of biology in the study of water
     pollution.   *Chemistry and Industry.*   1964:   435-436.
Hynes, H.B.N.   1964.   Some Australian Plecopteran nymphs.   *Gewässer
     Abwässer* 34/35: 17-22.
Hynes, H.B.N.   1964.   The interpretation of biological data with
     reference to water quality.   Environmental Measurements.
     U.S. Pub. Hlth. Serv. Publ.   999-AP-15: 289-298.

Hynes, H.B.N.   1965.   A survey of water pollution problems.
Ecology and the Industrial Society. *Symp. Brit. Ecol. Soc.*
5: 49-63.

Hynes, H.B.N. and Williams, W.D.   1965.   Experiments on competition
between two *Asellus* species (Isopoda Crustacea). *Hydrobiologia*
26: 203-210.

Kershaw, W.E., Williams, T.R., Frost, S. and Hynes, H.B.N.   1965.
Selective effect of particulate insecticides on *Simulium*
among stream fauna. *Nature. Lond.* 208: 199.

Hynes, H.B.N.   1965.   The significance of macroinvertebrates in the
study of mild river pollution. Biological problems in water
pollution. *U.S. Pub. Hlth. Serv.* 999-WP-25: 235-243.

Hynes, H.B.N.   1966.   Excess nutrients as a pollution problem in
streams. *Ann. Tech. Seminar. Prof. Engnrs Ont. Wat. Res.
Comm.*   3: 1-13 (mimeo).

Hynes, H.B.N. and Greib, B.J.   1967.   Experiments on control of
release of phosphate from lake muds. *Proc. 10th Conf. Gt.
Lakes Res.*   357-362.

Hynes, H.B.N.   1967.   A key to the adults and nymphs of British
Stoneflies (Plecoptera) with notes on their ecology and
distribution. *Sci. Pub. Fresh. Biol. Ass.*   17: 1-91.

Kaushik, N.K. and Hynes, H.B.N.   1968.   Experimental study on the
role of autumn-shed leaves in aquatic environments. *J.
Ecol.*   56: 229-243.

Hynes, H.B.N.   1968.   The scientific results of the Hungarian Soil
Zoological Expedition to the Brazzaville-Congo.   36.   The
Plecoptera species *Neoperla spio* (Newman). *Opusc. Zool.
Budapest.*   8: 353-356.

Hynes, H.B.N.   1968.   Further studies on the invertebrate fauna
of a Welsh mountain stream. *Arch. Hydrobiol.*   65: 360-379.

Hynes, H.B.N. and Coleman, M.J.   1968.   A simple method of
assessing the annual production of stream benthos. *Limnol.
Oceanogr.*   13: 569-573.

Bishop, J.E. and Hynes, H.B.N.   1969.   Upstream movements of the
benthic invertebrates in the Speed River, Ontario. *J. Fish.
Res. Bd. Canada.*   26: 279-298.

Bishop, J.E. and Hynes, H.B.N.   1969.   Downstream drift of the
invertebrate fauna in a stream ecosystem. *Arch. Hydrobiol.*
66: 56-90.

Hynes, H.B.N. and Kaushik, N.K.   1969.   The relationship between
dissolved nutrient salts and protein production in submerged
autumnal leaves. *Verh. Int. Ver. Limnol.*   17: 95-103.

Hynes, H.B.N.   1969.   The enrichment of streams. *In*:
Eutrophication: Causes, Consequence, Corrections. National
Academy of Science, Washington p. 188-196.

Hynes, H.B.N.   1969.   Life in freshwater communities in Man-made
lakes. The Accra Symposium. Ghana Academy of Science. L.E.
Obeng, (ed). p. 25-31.

Hynes, H.B.N.   1970.   The ecology of stream insects.   *Ann. Rev. Ent.*
    15: 25–42.

Hynes, H.B.N.   1970.   The ecology of flowing waters in relation to
    management.   *J. Wat. Pollut. Contr. Fed.*   42: 418–424.

Coleman, M.J. and Hynes, H.B.N.   1970.   The vertical distribution of
    the invertebrate fauna in the bed of a stream.   *Limnol.
    Oceanog.*   15: 31–40.

Hynes, H.B.N. and Greib, B.J.   1970.   Movement of phosphate and other
    ions from and through lake muds.   *J. Fish. Res. Bd. Canada.*
    27: 653–668.

Coleman, M.J. and Hynes, H.B.N.   1970.   The life histories of some
    Plecoptera and Ephemeroptera in a Southern Ontario stream.
    Can. J. Zool.   48: 1333–1339.

Harper, P.P. and Hynes, H.B.N.   1970.   Diapause in the nymphs of
    Canadian winter Stoneflies.   *Ecology.*   51: 925–927.

Hynes, H.B.N.   1970.   Why lakes become green and slimy.   *Globe
    and Mail.*   6 June p. 7.

Hynes, H.B.N.   1970.   The ecology of running waters.   University of
    Toronto Press.   555 p. (reprinted 1972, 1975).

Hynes, H.B.N. and Macphie, F.M.   1970.   The maxillae of the British
    species of *Isoperla* (Plecoptera: Perlodidae).   *Proc. R. Ent.
    Soc. Lond. A* 45: 123–124.

Spence, J.A. and Hynes, H.B.N.   1971.   Differences in benthos up-
    stream and downstream of an impoundment.   *J. Fish. Res. Bd.
    Canada.*   28: 35–43.

Spence, J.A. and Hynes, H.B.N.   1971.   Differences in fish populations
    upstream and downstream of a mainstream impoundment.   *J. Fish.
    Res. Bd. Canada.*   28: 45–46.

Kaushik, N.K. and Hynes, H.B.N.   1971.   The fate of dead leaves that
    fall into streams.   *Arch. Hydrobiol.*   68: 465–515.

Harper, P.P. and Hynes, H.B.N.   1971.   The Leuctridae of Eastern
    Canada (Insecta: Plecoptera).   *Can. J. Zool.*   49: 915–920.

Harper, P.P. and Hynes, H.B.N.   1971.   The Capniidae of Eastern
    Canada (Insecta: Plecoptera).   *Can. J. Zool.*   49: 921–940.

Harper, P.P. and Hynes, H.B.N.   1971.   The nymphs of the
    Taeniopterygidae of Eastern Canada (Insecta: Plecoptera).
    Can. J. Zool.   49: 941–947.

Harper, P.P. and Hynes, H.B.N.   1971.   The nymphs of the Nemouridae
    of Eastern Canada (Insecta: Plecoptera).   *Can. J. Zool.*
    49: 1129–1142.

Hynes, H.B.N.   1971.   Zonation of the invertebrate fauna in a West
    Indian stream.   *Hydrobiologia.*   38: 1–8.

Williams, T.R. and Hynes, H.B.N.   1971.   A survey of the fauna of
    streams on Mount Elgon, East Africa, with special reference to
    the Simuliidae (Diptera).   *Freshwat. Biol.*   1: 227–248.

Hynes, H.B.N. and Harper, F.   1972.   The life histories of *Gammarus
    lacustris* and *Gammarus pseudolimnaeus* in Southern Ontario.
    Crustaceana Suppl.   3: 329–341.

Bhajan, W.R. and Hynes, H.B.N. 1972. Experimental study on the ecology of *Bosmina longirostris* (O.F. Muller) (Cladocera). *Crustaceana*. 23: 133-140.

Harper, P.P. and Hynes, H.B.N. 1972. Life-histories of Capniidae and Taeniopterygidae (Plecoptera). *Arch. Hydrobiol. Suppl.* 40: 274-314.

Williams, N.E. and Hynes, H.B.N. 1973. Microdistribution and feeding of the net-spinning caddisflies (Trichoptera) of a Canadian stream. *Oikos*. 24: 73-84.

Wallace, R.R., West, A.S., Downe, A.E.R. and Hynes, H.B.N. 1973. The effects of experimental blackfly (Diptera: Simuliidae) larviciding with abate, dursban, and Methoxychlor on stream invertebrates. *Can. Ent.* 105: 817-831.

Lush, D.L. and Hynes, H.B.N. 1973. The formation of particles in freshwater leachates of dead leaves. *Limnol. Oceanogr.* 18: 968-977.

Hynes, H.B.N. 1973. The effects of sediment on the biota in running water. *Proc. Ninth Canad. Hydrol. Symp. Edmonton.* pp. 652-663.

Hynes, H.B.N. 1974. Further studies on the distribution of stream animals within the substratum. *Limnol. Oceanogr.* 19: 92-99.

Hynes, H.B.N., Kaushik, N.K., Lock, M.A., Lush, D.L., Stocker, Z.S.J., Wallace, R.R. 1974. Benthos and allochthonous matter in streams. *J. Fish Res. Bd Canada.* 31: 545-553.

Williams, D.D., Williams, N.E. and Hynes, H.B.N. 1974. Observations on the life history and burrow construction of the crayfish *Cambarus fodiens* (Cottle) in a temporary stream in southern Ontario. *Can. J. Zool.* 52: 365-370.

Williams, D.D. and Hynes, H.B.N. 1974. The occurrence of benthos deep in the substratum of a stream. *Freshwat. Biol.* 4: 233-256.

Hynes, H.B.N. 1974. Comments on the taxonomy of Australian Austroperlidae and Gripopterygidae (Plecoptera). *Aust. J. Zool. Suppl.* 29: 1-36.

Hynes, H.B.N. 1974. Observations on the adults and eggs of Australian Plecoptera. *Aust. J. Zool. Suppl.* 29: 37-52.

Rowe, V.L., Hynes, H.B.N. and Tyler, A.V. 1975. Reproductive timing by the polychaetes *Clymenella torquata* and *Ptaxillela praetermissa* in Passamaquoddy Bay, New Brunswick. *Can. J. Zool.* 53: 292-296.

Hynes, H.B.N. and Hynes, M.E. 1975. The life histories of many of the stoneflies (Plecoptera) of southeastern mainland Australia. *Aust. J. Mar. Freshwat. Res.* 26: 113-154.

Wallace, R.R. and Hynes, H.B.N. 1975. The catastrophic drift of stream insects after treatment with methoxychlor (1, 1, 1-trichloro-2-2 bis p-methoxyphenyl ethane). *Environ. Pollut.* 8: 255-268.

Hynes, H.B.N. 1975. Edgardo Baldi Memorial Lecture. The stream and its valley. *Verh. Int. Verein. theor. angew. Limnol.* 19: 1-15.

Lock, M.A. and Hynes, H.B.N. 1975. The disappearance of four leaf leachates in a hard and soft water stream in southwestern Ontario. *Int. Rev. ges. Hydrobiol.* 60: 847-855.

Wallace, R.R., Hynes, H.B.N. and Kaushik, N.K. 1975. Laboratory experiments affecting activity of *Gammarus pseudolimnaeus*. Bousfield. *Freshwat. Biol.* 5: 533-546.

Hynes, H.B.N. 1-76. Biology of Plecoptera. *Ann. Rev. Entomol.* 21: 135-153.

Hynes, H.B.N. 1976. Tasmanian Antarctoperlaria (Plecoptera). *Aust. J. Zool.* 24: 115-143.

Stocker, Z.S.J. and Hynes, H.B.N. 1976. Studies on the tributaries of Char Lake, Cornwallis Island, Canada. *Hydrobiologia.* 49: 97-102.

Williams, D.D. and Hynes, H.B.N. 1976. Stream habitat selection by aerially colonizing invertebrates. *Can. J. Zool.* 54: 685-693.

Williams, D.D. and Hynes, H.B.N. 1976. The recolonization mechanisms of stream benthos. *Oikos.* 27: 265-272.

Hynes, H.B.N., Williams, D.D. and Williams, N.E. 1976. Distribution of the benthos within the substratum of a Welsh mountain stream. *Oikos* 27: 307-310.

Lock, M.A. and Hynes, H.B.N. 1976. The fate of "dissolved" organic carbon derived from autumn-shed maple leaves (*Acer saccharum*) in a temperate hard-water stream. *Limnol. Oceanogr.* 21: 436-443.

Wallace, R.R., Hynes, H.B.N. and Merritt, W.F. 1976. Laboratory and field experiments with methoxychlor as a larvicide for Simuliidae (Diptera). *Environ. Pollut.* 10: 251-269.

Hynes, H.B.N. 1976. *Symbiocladius aurifodinae* sp. nov. (Diptera, Chironomidae), a parasite of nymphs of Australian Leptophlebiidae (Ephemeroptera). *Mem. nat. Mus. Victoria.* 37: 47-52.

Williams, D.D. and Hynes, H.B.N. 1976. The ecology of temporary streams I. The fauna of two Canadian streams. *Int. Revue ges. Hydrobiol.* 61: 761-788.

Barton, D.R. and Hynes, H.B.N. 1976. The distribution of Amphipoda and Isopoda on the exposed shores of the Great Lakes. *J. Gt. Lakes Res.* 2: 207-214.

Williams, D.D. and Hynes, H.B.N. 1977. The ecology of temporary streams II. General remarks on temporary streams. *Int. Revue ges. Hydrobiol.* 62: 53-61.

Williams, D.D. and Hynes, H.B.N. 1977. Benthic community development in a new stream. *Can. J. Zool.* 5: 1071-1076.

Hynes, H.B.N. 1977. A key to the adults and nymphs of British stoneflies (Plecoptera) *Sci. Pub. Freshwater Biol. Ass.* 17: 91 p. (3rd edn).

Lock, M.A., Wallis, P.M. and Hynes, H.B.N. 1977. Colloidal organic carbon in running waters. *Oikos.* 29: 1-4.

Crowther, R.A. and Hynes, H.B.N. 1977. The effect of road deicing salt on the drift of stream benthos. *Environ. Pollut.* 14: 113-126.

Barton, D.R. and Hynes, H.B.N. 1978.   Seasonal study of the fauna
    of bedrock substrates in the wave-zones of lakes Huron and
    Erie. *Can. J. Zool.* 56: 48-54.
Barton, D.R. and Hynes, H.B.N. 1978. Wave-zone macrobenthos of
    the exposed Canadian shores of the St. Lawrence Great Lakes.
    J. Gt. Lakes Res. 4: 27-45.
Barton, D.R. and Hynes, H.B.N. 1978. Seasonal variations in
    densities of macrobenthic populations in the wave-zone of
    north-central Lake Erie. *J. Gt. Lakes Res.* 4: 50-56.
Lee, D.R. and Hynes, H.B.N. 1978. Identification of groundwater
    discharge zones in a reach of Hillman Creek in Southern
    Ontario. *Water Poll. Res. Canada.* 13: 121-133.
Hynes, H.B.N. 1978. Annotated key to the stonefly nymphs (Plecoptera)
    of Victoria. *Spec. Pub. Australian Soc. Limnol.* 2: 63 pp.
Lush, D.L. and Hynes, H.B.N. 1978. Particulate and dissolved
    organic matter in a small partly forested Ontario stream.
    Hydrobiologia. 60: 177-185.
Lush, D.L. and Hynes, H.B.N. 1978. The uptake of dissolved
    organic matter by a small spring stream. *Hydrobiologia.*
    60: 271-275.
Williams, D.D. and Hynes, H.B.N. 1979. Reply to comments by
    Exner and Davies on the use of a standpipe corer. *Freshwat.*
    Biol. 9: 79-80.
Dance, K.W., Hynes, H.B.N. and Kaushik, N.K. 1979. Seasonal
    drift of solid organic matter in two adjacent streams.
    Arch. Hydrobiol. 87: 139-151.
Dance, K.W. and Hynes, H.B.N. 1979. A continuous study of the
    drift in adjacent intermittent and permanent streams. *Arch.*
    Hydrobiol. 87: 253-261.
Dance, K.W. and Hynes, H.B.N. 1980. Some effects of agricultural
    land use on stream insect communities. *Environ. Pollut.*
    (ser. A) 22: 19-28.
Hynes, H.B.N. and Hynes, M.E. 1980. The endemism of Tasmanian
    stoneflies (Plecoptera). *Aquatic Insects.* 2: 81-89.
Hynes, H.B.N. 1980. A name change in the secondary production
    business. *Limnol. Oceanogr.* 25: 778.
Bird, G.A. and Hynes, H.B.N. 1981. Movements of adult aquatic
    insects near streams in southern Ontario. *Hydrobiologia.*
    77: 65-69.
Bird, G.A. and Hynes, H.B.N. 1981. Movement of immature aquatic
    insects in a lotic habitat. *Hydrobiologia.* 77: 103-112.
Marchant, R. and Hynes, H.B.N. 1981. Field estimates of feeding
    rate for *Gammarus pseudolimnaeus* (Crustacea: Amphipoda) in
    the Credit River, Ontario. *Freshwat. Biol.* 11: 27-36.

Acari, 323
Achromobacter sp., 122
Acroneuria abnormis, 184
Acroneuria lycorias, 185
Activated sludge, 131
Adsorption, 27, 31
Adsorption of dissolved and colloidal organic matter, 27
Adsorption processes, 30, 116
Aeromonas sp., 358
Africa - characteristics of major dammed lakes, 271
         - major dammed rivers, 268
Africa running waters - Annual/Biannual flow regimes, 345
         - benthos, 350-353
         - benthos, altitudinal zonation of *Simulium* spp., 352
         - conductivity/salinity, 348-349
         - distribution, 340
         - dry season/flood plains, 344
         - fish, composition and distribution, 354
         - fish disease and management, 358-359
         - fish, ectoparasites/endoparasites, 358
         - fish, feeding, growth and reproduction, 355
         - fish migration, 355
         - fish mortality, 356
         - fish, response to flooding/dry conditions, 355
         - geomorphology, 341-344
         - human disease vectors, 359
         - hydrological characteristics of flood plains, 343
         - hydrological features, 345-347
         - oxygen, 347
         - pH, 348
         - "reservoir"/"flood" rivers, 346
         - sources of flood, 345
         - temperature, 347
         - zooplankton, 349-350
         - zooplankton, "reservoir populations", 350
         - zooplankton, seasonal dynamics, 350
African man-made lakes - physico-chemical features, 268-272
Afronurus sp., 352